# Mammalian Egg Transfer

Editor

**Cyril E. Adams**
Agricultural Research Council
Institute of Animal Physiology
Animal Research Station
Cambridge, England

CRC Press
Taylor & Francis Group
Boca Raton London New York

CRC Press is an imprint of the
Taylor & Francis Group, an **informa** business

First published 1982 by CRC Press
Taylor & Francis Group
6000 Broken Sound Parkway NW, Suite 300
Boca Raton, FL 33487-2742

Reissued 2018 by CRC Press



**Library of Congress Cataloging in Publication Data**
Main entry under title:

Mammalian egg transfer.

Bibliography: p.
Includes index.
1. Ovum implantation. 2. Mammals—
Reproduction. I. Adams, Cyril Edwin.
[DNLM: 1. Mammals 2. Ovum—Transplantation.
SF 105.7 M265]
QL739.2.M348 599.03'2 82-4212
ISBN 0-8493-6140-0 AACR2

A Library of Congress record exists under LC control number: 82004212

Publisher's Note
The publisher has gone to great lengths to ensure the quality of this reprint but points out that some imperfections in the original copies may be apparent.

Disclaimer
The publisher has made every effort to trace copyright holders and welcomes correspondence from those they have been unable to contact.

ISBN 13: 978-1-315-89513-0 (hbk)
ISBN 13: 978-1-351-07423-0 (ebk)

Visit the Taylor & Francis Web site at http://www.taylorandfrancis.com and the
CRC Press Web site at http://www.crcpress.com

# PREFACE

Egg transfer was first performed in 1890, but for half a century it received scant attention. However, since 1950 the technique has become increasingly widely used — in the laboratory for fundamental studies and more recently in practice, both veterinary and medical, to boost the reproductive potential of genetically superior cattle and to overcome sterility due to impaired tubal function in women.

As a result, a considerable body of literature has accumulated, totaling well in excess of a thousand references. But till now there has not been a single comprehensive text devoted solely to this subject. The present work was designed to meet that need at a time when the field is fast expanding with new techniques and approaches constantly being evolved. One need only cite the tremendous rate of progress in human egg transfer in the last three years.

The work embraces laboratory and farm animals and primates, including man, altogether representing a total of 16 species. In general, each one has been treated on an individual basis by authors with an expert knowledge of particular species. Additional chapters are concerned with the uses and applications of egg transfer in relation to basic research and practice. It is regretted that chapters on egg transfer in the pig and on the short- and long-term storage of eggs failed to materialize in time for inclusion.

As Editor, I wish to thank the many individuals, editors, and publishers for granting their permission to reproduce figures; Jennifer Constable, Barbara Hall, and Maureen Mansfield for secretarial assistance, and Judith Banks, Peter Philby, and Wendy Reynolds for help in obtaining library material. I also wish to express my appreciation to the Publishers, CRC Press, Inc., not least for their patience and understanding as deadlines came and went.

C.E. Adams

# THE EDITOR

Cyril E. Adams, Ph.D., is a member of staff of the Agricultural Research Council Institute of Animal Physiology at the Animal Research Station, Cambridge, England, having received his first appointment there in 1952. He has also worked in the United States and Sweden and acted as a consultant in West Germany, Switzerland, Brazil, and Mexico.

Dr. Adams graduated in 1948 from the University College of North Wales, Bangor, with a B.Sc. (Hons.) degree in Agriculture and obtained his Ph.D. degree in 1953 from the University of Cambridge, where he was supervised by Dr. John Hammond.

Dr. Adams is a member of the Society for Endocrinology, Society for the Study of Fertility, Society for the Study of Reproduction, Society for Developmental Biology, Association for the Study of Animal Behaviour, Laboratory Animal Science Association, honorary member of the Fur Breeders' Association, and a corresponding member of the Societa Italiana per il Progresso della Zootecnica, Milan.

Dr. Adams has traveled widely attending international meetings and giving lectures in Europe, the Americas, and recently Japan as a Fellow of the Japanese Society for the Promotion of Science. He has served on several editorial boards, including that of the *Journal of Reproduction and Fertility,* of which company he is an Executive Director.

Dr. Adams has published well over 100 research papers, including a considerable number on various aspects of egg transfer. His current research interests are concerned with factors and mechanisms affecting the initiation of pregnancy and their influence on fertility and fecundity.

## CONTRIBUTORS

**W. R. Allen, BVSc., Ph.D., MRCVS**
Agricultural Research Council
Institute of Animal Physiology
Animal Research Station
Cambridge, England

**Zeev Dickmann, MA, Ph.D.**
Professor, Department of
Gynecology and Obstetrics
University of Kansas Medical Center
Kansas City, Kansas

**W. Richard Dukelow, Ph.D.**
Professor and Director
Endocrine Research Unit
Michigan State University
East Lansing, Michigan

**Ian Gordon, BSc., MA, Ph.D.**
Professor and Head
Department of Agriculture
University College
Dublin, Ireland

**N. W. Moore, MAgr. Sc., Ph.D.**
Associate Professor in Animal Husbandry
University of Sydney
Camden, N.S.W., Australia

**Raymond Newcomb, BVSc., MSc., FRCVS**
Veterinary Advisor
Milk Marketing Board (England and Wales)
Thames Ditton
Surrey, England

**S.M. Willadsen, DVMd., VLd.**
Principal Scientific Officer
Agricultural Research Council
Institute of Animal Physiology
Animal Research Station
Cambridge, England

**Ian Wilmut, BSc., Ph.D.**
Principal Scientific Officer
ABRO Field Laboratory
Midlothian, Scotland

Dedicated to the Memory

of

Sir John Hammond

TABLE OF CONTENTS

Chapter I

# EGG TRANSFER: HISTORICAL ASPECTS

C. E. Adams

## TABLE OF CONTENTS

# I. INTRODUCTION

Although egg transfer was first successfully performed in the last century,[1] the technique attracted relatively little attention prior to 1950, and, as a result, probably in excess of 95% of the published work in this field has appeared in the last 30 years. Inevitably, therefore, any survey will be concerned with very recent times. Nevertheless, such has been the pace of events that even quite recent findings have already assumed a degree of historical perspective.

In the last decade, in particular, progress in egg transfer has become increasingly dependent upon technical advances in a variety of ancillary fields. Some of these will be considered in the present review.

Tabulated and annotated data on egg transfer both within and between species, spanning the years 1890 to 1960, have been presented by Austin,[2] whose book, *The Mammalian Egg*, contains a section on "Manipulation of Eggs".

Reviews on various aspects of egg transfer include those of Willett,[3] Rowson,[4,5] Chang and Pickworth,[6] Dziuk,[7] Foote and Onuma,[8] McLaren and Kerr,[9] Dickmann,[10] David et al.,[11] Newcomb,[12] Trounson and Rowson,[13] Betteridge,[14,15] Murray,[16] Polge,[17] Whittingham,[18] and Seidel.[19]

Several conferences devoted specifically to egg transfer have taken place; the first was held on April 14, 1949 at the Foundation of Applied Research, San Antonio, Texas.[20] Others have been held at Hållsta, Sweden[21] (September 24 and 25, 1974); Cambridge, England[22] (December 9 to 12, 1975); Krakow, Poland[23] (July 13, 1976); Galway, Eire[24] (September 27 to 30, 1977); and in Kiel, West Germany[25] (September 24 to 27, 1980). The International Embryo Transfer Society has been meeting annually since 1975 in Denver, Colorado; its proceedings are now published in *Theriogenology*. This society, formed on May 26, 1974, had its origins in a course held in Cambridge in 1972.[26,27]

# II. EGG TRANSFER

## A. Historical Development

*"In this preliminary note I wish merely to record an experiment by which it is shown that it is possible to make use of the uterus of one variety of rabbit as a medium for the growth and complete foetal development of fertilized ova of another variety of rabbit."*

With these now celebrated lines, Walter Heape[1] introduced the first paper on egg transfer, which he read before members of the Royal Society on November 27, 1890. The actual transfer had been performed on April 27, and the young were born on May 29, 1890. Heape undertook the experiment "to determine in the first place what effect, if any, a uterine foster mother would have upon her foster children, and whether or not the presence and development of foreign ova in the uterus of a mother would affect the offspring of that mother born at the same time." He concluded "so far as this single case goes, the evidence is negative."

In a second paper published in 1897, Heape[28] described his attempts to extend the work in which he was, however, largely frustrated by deaths resulting from the anesthesia — chloroform was used—and infertile stock. A total of eight litters was produced but only one of them possibly contained young from transferred eggs.

Heape's experiments were first repeated by Biedl in 1911 to 1913, but due to the war and his moving from Vienna to Prague the work was not submitted for publication until May 1920.[29] Apparently, only one litter was born — during the night of May 29, 1912 — but even that was destroyed by the mother.

Pincus[30] described the transfer of eggs cultured in vitro to the oviducts of pseudopregnant does, but no offspring were obtained "for a number of reasons, chiefly technical." However, one of three control transfers was successful. These experiments were carried out in 1929 while Pincus was working in Cambridge, England, where the transfers were performed by Hammond and Walton. At about the same time Asdell and Hammond[31] obtained three litters from a series of 21 transfers.

Nicholas[32] transferred tubal rat eggs (two- and four-cells) to the uterus, repeating Heape's experiments "as a test for egg viability and its capacity for continued development." In so doing Nicholas also varied the synchrony between the donors and recipients and observed that "definite differences in result were obtained". Thus, in his synchronous transfers 11 out of 18 eggs developed into normal embryos; 16 out of 20 eggs developed when the donor was in heat after the recipient, but in the reverse combination only 2 out of 20 eggs developed. Though this must rank as the first observation on synchronization, it was not until 1950 that Chang[33] investigated this aspect in a systematic manner and thereby highlighted its significance. In 1936, Pincus[34] described techniques for transferring mouse and rabbit eggs to the oviducts and, incidentally, demonstrated that one- and two-cell rabbit eggs could develop to the blastocyst stage in the estrous rabbit. Moreover, he recognized "it is necessary that a progestational uterus be available for ensuring differentiation of uterine stages." Pincus[34] also transferred single blastomeres of two-cell eggs to the oviducts and obtained "normally differentiating but small-sized blastodermic vesicles," noting "the physiological processes occurring in such embryos are of extraordinary interest and certainly deserve further investigation." In collaboration with E. V. Enzmann, Pincus used egg transfer to determine whether rabbit eggs "fertilized" and cultured in vitro might give rise to normal offspring.[35] The 1930s also witnessed the first ectopic transplantation of mammalian eggs when Nicholas[36] placed isolated blastomeres of two-cell rat eggs under the kidney capsule.

The transfer of an egg of a larger species — that of a goat — was first successfully performed on December 8, 1932[37,38] (Figure 1). Warwick and Berry's[38] description of the event is worthy of recall: "This animal (a 4-celled egg) seemed so beautiful to us that we could not bear to let it die of neglect so we hastily decided to put it back in its original home for five months." They admitted that they had never seen a living mammalian egg until 5 days previously and their apparent lack of confidence in egg recovery had deterred them from preparing a recipient, thus explaining why the egg was returned to its donor. The birth of a lamb from a transferred egg soon followed.[37]

In 1933 Warwick et al.[37] also transferred eggs reciprocally between sheep and goats — the first recorded cases of interspecies egg transfer, — though apparently without success. In two further series of experiments in 1941 and 1943, they discovered that some of the recipients were becoming pregnant only to undergo resorption between 30 and 50 days of gestation.[38]

The reader will have noted that the use of egg transfer was still confined to very few laboratories and its application no more than sporadic. Another decade would pass before the technique began to attract more widespread interest. After the end of the 1939 to 1945 war came a growing realization of the technique's potential not only in laboratory investigations, but also for practical application to livestock improvement and production. The prospects for egg transfer were being boosted by developments in the use of gonadotrophins for superovulation and, paradoxically, by the failure to augment in a controlled manner the fecundity of treated animals.

Following the discovery of the gonadotrophic hormones, the ability of the ovary to respond with greatly increased numbers of ovulations (superovulation) was soon recognized in laboratory animals. Observations were later extended to sheep and cattle[39-41] where potential practical applications were envisaged. In the rat and mouse the ovu-

FIGURE 1. (A) Lamb from auto-transfer. The lamb was examined as an eight-celled egg and injected back into its own mother. (B) Kid from auto-transfer. The mother goat was operated on the second day after breeding, the oviduct was removed, and the four-celled egg was washed out, examined, and injected into the opposite horn of the uterus of its mother. (C) Lamb with its "uterine foster mother". The lamb was removed as an eight-celled egg from the ovdiuct of its mother and developed to birth in this foster mother. (D) Kid with uterine foster mother. (Reproduced from Warwick, B. L. and Berry, R. O., *J. Hered.*, 40, 297, 1949. With permission of American Genetic Association.)

lation of 20 to 48 follicles was recorded,[42] and in the rabbit 54 to 80 eggs were obtained from single individuals.[43,44] Furthermore, it was established that the eggs were fertilizable and capable of developing normally, at least to implantation. Thus, in the mouse 19 and 29 implantations were found on the 9th to 10th day of pregnancy,[45,46] in rats up to 34 implantations,[47] and in rabbits up to 36 normal embryos[43] or 38 implantations.[44]

However, it soon became apparent that increased numbers of implantations did not generally result in larger litters than normal; indeed the resulting litters were more often well below normal in size.[47,44] Evans and Simpson[47] concluded "the birth of few or no young in pregnancies which are inaugurated with an abnormally high number of successfully implanted and apparently wholly normal embryos is almost surely attributable to defects in uterine function." Later work confirmed this point. For example, Chang[48] obtained 46 normal young from 53 two-cell eggs recovered from one superovulated doe and transferred to 4 others, and Adams[49] demonstrated differential embryonic mortality between horns within rabbits, depending upon the number of eggs transferred.

Therefore, in order to exploit superovulation for practical purposes it would be necessary to recover the eggs and distribute them between several recipients. For livestock improvement it was logical to think in terms of superovulating superior females in order to maximize the yield of their offspring with less valuable stock acting as recipients. By this means Hammond, who together with Walton had pioneered the use of artificial insemination (A.I.) in cattle in England, envisaged egg transfer "doing for the female what A.I. had done for the male." For contemporary reviews and an assessment of the problems, see Hammond.[50,51]

In 1949 Umbaugh,[52] at the Foundation of Applied Research, San Antonio, Texas, reported the use of 750 cows over a period of 8 years in the investigation of superovulation and egg transfer. He described three pregnancies resulting from egg transfer but each failed, ending in abortion at 65, 104, and 112 days. Umbaugh also attempted to fertilize cow eggs in the oviducts of rabbits. The first calf to be produced as a result of egg transfer was born in Wisconsin in 1951.[53]

An event which was to have a considerable influence on the future development of egg transfer was the establishment in 1949 of the Unit of Animal Reproduction in Cambridge by the Agricultural Research Council. The council's primary objective in setting up the unit was to investigate the possibility of transferring fertilized cow eggs from one cow to another. According to contemporary sources it was realized from the start that this problem presented formidable difficulties and that the objective might not be attained. Thirty years on, in the light of actual achievements and future portents, these words seem unduly pessimistic, but they do serve to put the problem into perspective.

The Unit of Animal Reproduction was disbanded in 1954 when Hammond, its honorary director, reached retiring age (then 65), and in its place the Unit of Reproductive Physiology and Biochemistry came into being under the direction of Dr. Thaddeus Mann, who retired in 1976. During those 22 years a considerable portion of the Unit's budget and effort went into work on egg transfer, especially in farm animals, under the direction of L.E.A. Rowson. Eight of the present authors were attached to one or both of these units, including five as Ph.D. students.

The growth of interest in egg transfer coincided with and gained impetus from studies involving mammalian eggs, which attracted an increasing number of students from 1950 onwards. The initial trickle of papers from just a few laboratories grew at first slowly and then apace so that by 1977, nearly 1000 articles on egg transfer had been published,[54-56] with an ever increasing number on cattle.

The first records of successful egg transfers in different species are listed in Table 1,[57-70] which shows that of the total of 19 species, no fewer than 9 have been added since 1970. Apart from achieving successful transfers in three primate species, including man, the most notable development during the past decade has been the establishment of commercial embryo transfer organizations, particularly in North America. There the number of pregnancies resulting from egg transfer has grown from 20 in 1972 to 10,000 in 1978, with a projected 50,000 by 1985.[71]

Recently an English embryo transfer organization paid a United Kingdom record price of 32,000 guineas (U.S. $67,000) for a 7-year-old dairy cow, Ullswater Beataxus 8th, for use as a donor. At the same sale, even a 16½-year-old cow, which had not been in calf for 2 years, fetched 3000 guineas, (U.S. $6300).[72]

The fertilization in vitro of human eggs,[73] an essential step in the development of human egg transfer, culminated in the sensational birth of Louise Joy Brown on July 25, 1978.[67] By virtue of having been conceived outside her mother's body, on November 11, 1977 she became the world's first test-tube baby. Her birth, filmed by the Central Office of Information for the purposes of assisting in the evaluation of the baby at birth and to provide an historical and medical record, was given unprecedented coverage by all the media. However, the formal account of the birth took the form of a letter to the editor of the *Lancet*, appearing on August 12, 1978.[67] It began as follows: "We wish to report that one of our patients, a 30-year-old nulliparous married woman, was safely delivered by caesarian section on July 25, 1978 of a normal healthy infant girl, weighing 2700 g."

Table 1
FIRST RECORDS OF SUCCESSFUL EGG TRANSFERS

| Date | Species | Ref. |
|---|---|---|
| 1891 | Rabbit | 1 |
| 1932 | Goat[a] | 37,38 |
| 1933 | Rat | 32 |
| 1933 | Sheep | 37 |
| 1942 | Mouse | 57 |
| 1949 | Goat | 38 |
| 1951 | Cow | 53 |
| 1951 | Pig | 58 |
| 1963 | Quokka | 59 |
| 1964 | Golden hamster | 60 |
| 1968 | Ferret | 61 |
| 1970 | Tammar wallaby | 62 |
| 1974 | Horse | 63 |
| 1975 | Mink | 64 |
| 1976 | Baboon | 65 |
| 1977 | Rhesus monkey[a] | 66 |
| 1978 | Man[a] | 67 |
| 1978 | Cat | 68 |
| 1979 | Dog | 69 |
| 1981 | Mongolian gerbil | 70 |

[a] Auto-transfers.
Adapted from Betteridge, K. J., *J. Reprod. Fert.*, 62, 1, 1981. With permission.

Further accounts of the work are now available.[74,75] The possibility that women with blocked fallopian tubes might be able to have an "artificial pregnancy", i.e., by egg transfer, was suggested by Hammond in 1950.[51]

The ability to fertilize the human oocyte in vitro and transfer it successfully has inevitably brought to the fore questions of medical ethics, moral and legal aspects, as anticipated at a Ciba Symposium on "legal and other aspects of artificial insemination by donor (AID) and embryo transfer," held on December 1, 1972.[76] At that time, Edwards and Steptoe[77] noted that the "only potential cause for concern for a long time to come appears to be the use of host mothers for the convenience of an embryo donor." But in 1979 at the meeting of the British Association for the Advancement of Science in Edinburgh, Steptoe himself suggested that an egg might be transferred to a "stand-in" mother if the real mother was unable to have the baby herself.[78] Within hours the president of the British Medical Association, Dame Josephine Barnes, was warning of the dangers that might attend such transfers,[79] and elsewhere[80] the need for new legislation on host mothers had already been expressed. At the Ciba Symposium there were two contributions on "English law in relation to AID and embryo transfer"[81] and "legal aspects of artificial insemination and embryo transfer in French domestic law and private international law."[82] Legal approaches to "test-tube reproduction" have also been discussed.[83]

### B. Terminology

Several terms, including "egg-" or "ovum-transfer", "egg-" or "ovum transplantation", "implantation",[84] or "reimplantation" of ova[67] have been or are currently being used to describe the transfer of eggs from one individual to another. Consideration of the title of articles shows that terminology has tended to vary with the epoch and species. For example, "transplantation of ova", first used by Heape,[1] remained in use until the early 1950s, when "egg transfer" was adopted, particularly by those working with laboratory animals. On the other hand, those working with sheep and cattle continued to use "ovum" or "ova" but substituted "transfer" for "transplantation" which, as a technique, was coming to be more associated with organs. Recently, the term "embryo transfer" has gained favor, especially with reference to cattle under commercial conditions. Betteridge[14] uses the term "embryo" synonymously with "fertilized egg", "fertilized ovum", "morula", "blastocyst", or "expanded blastodermic vesicle". Such variations in terminology will be found in the present work reflecting the preferences of individual authors.

Beatty[85] suggested that the transfer of eggs directly through the cervical opening into a host uterus, i.e., without surgery, could be called "inovulation" by analogy with "insemination". However, the term never caught on, possibly because of confusion with the ovulation process. Similarly, the terms "implantation" and "reimplantation of ova", as used to describe the transfer of human eggs to the uterus via the cervix, are capable of confusion with the implantation process and, therefore, hardly to be recommended.

The transfer, 5 days after fertilization, of a human embryo from a fertile donor to the uterus of an infertile recipient has been called "artificial embryonation"[86] or "embryo adoption" if donor semen was used instead of the semen of the recipient's husband.[87] Whether these terms will gain acceptance remains to be seen.

### C. Major Contributors

Four individuals (Figure 2) who have made outstanding contributions to the development of egg transfer are Walter Heape (1855 to 1929), John Hammond (1889 to 1964), Min Chueh Chang, and Lionel ("Tim") Rowson. In the case of Heape and Hammond, biographical data are to be found elsewhere,[88-90] together with accounts of their work.

#### *1. Min Chueh Chang, B.Sc., Dip. Agric. Sci., PhD., Sc.D., D.Sc.*

Chang was born in Taiyuan, Shansi, China on October 10, 1908. He graduated from Tsing Hua University, Peking, in 1933, remaining there as a research assistant and instructor for the next 5 years. In 1938 he came to study at Edinburgh University, and in 1939 moved to Cambridge, where he was attached to the School of Agriculture, gaining his Ph.D. in 1941. In 1945 Chang left for the U.S. to join the Worcester Foundation for Experimental Biology, Shrewsbury, Massachusetts, where he is still engaged in research, having held a Career Award of the National Institutes of Health since 1962 and a Professorship of Boston University since 1961.

Chang has published 335 papers including several concerned with highly original and fundamental aspects of egg transfer. A truly personal and entertaining account of his work can be found elsewhere.[91]

He was awarded the Sc.D. degree by the University of Cambridge (1969) and the D.Sc. degree of the Worcester Foundation (1974). His achievements have also been recognized by a number of awards from professional societies, including the Hartmann Award by the Society for the Study of Reproduction in 1970[92] and the Marshall Medal by the Society for the Study of Fertility in 1971.

#### *2. Lionel Edward Aston Rowson, O.B.E., F.R.S., F.R.C.V.S., M.A.*

Tim Rowson was born on May 28, 1914 at Stafford, England. He qualified at the Royal College of Veterinary Surgeons, London, in 1938, following which he was appointed to the Institute of Animal Pathology, Cambridge. In 1942, he became director of the newly established Cambridge and District Artificial Insemination Centre, a position he still occupies today. His work on egg transfer in cattle and sheep, which he pursued until retiring in 1979, commenced in 1950 when he joined Hammond. He was

FIGURE 2. (A) John Hammond; (B) Walter Heape; (C) Min-Chueh-Chang; (D) Lionel Rowson.
The photograph of John Hammond is reproduced by permission of Godfrey Argent Ltd., and that of Water Heape by permission of the Royal Society.

appointed deputy director of the Unit of Reproductive Physiology and Biochemistry at its inception in 1955 and, when it was disbanded in 1976, became officer-in-charge of the Animal Research Station. Rowson, who was awarded the O.B.E. in 1955, was made F.R.C.V.S. in 1972, elected F.R.S. in 1973, and a Fellow of Wolfson College the same year. In 1974 he was named lifetime honorary president of the International Egg Transfer Society.

FIGURE 3. Two 5-month-old Border Leicester ram lambs born in South Africa following their importation from Great Britain as fertilized eggs. The eggs were transported from Cambridge to Pietermaritzburg in the ligated oviducts of live rabbits and transferred on arrival to the uteri of Dorper ewes, shown here with the lambs. (Reproduced from Hunter, G. L., Bishop, G. P., Adams, C. E., and Rowson, L. E. A., *J. Reprod. Fert.*, 3, 33, 1962. With permission.)

### D. Long Distance Transport of Eggs

The first successful long distance shipment of mammalian eggs, from Shrewsbury, Massachusetts to Cambridge, England, was accomplished in 1952 by Marden and Chang,[93] who sent fertilized rabbit eggs "on ice" in a vacuum flask. Three earlier attempts — from Uppsala, Sweden to Cambridge (February 26, 1950), Bangor, North Wales to Cambridge (August 9, 1950), and from Cambridge to Uppsala (September 23, 1950) — had ended in failure, with none of the transferred eggs developing to term.[94]

On February 16, 1953 three four- to eight-cell sheep eggs were transferred to one oviduct of a rabbit, which had been given an ovulating injection 24 hr earlier. Five days later when the eggs were recovered from the rabbit's uterus, it was found that they had developed to the late morula stage.[95] Further work was carried out during the next breeding season, resulting in the demonstration that such eggs were capable of developing into normal embryos if transferred to a suitable recipient ewe.[96] Subsequently, this finding was put to practical use for exporting "sheep" from England to South Africa, where the first lambs were born in March 1961[97,98] (Figure 3). This exercise had been first attempted in November 1957 (Figure 4), but it proved unsuccessful largely due to a series of mishaps.

The development of methods for the long-term preservation of embryos[99,100] led to the first international shipment of deep frozen eggs.[101]

## III. EGG RECOVERY

Egg recovery is an integral part of the egg transfer procedure. Broadly speaking, there are two main approaches to egg recovery: (1) in vivo, either by flushing the tract at laparotomy or nonsurgically, and (2) by flushing the appropriate region of the excised tract at autopsy or after surgery. In recent years the technique of laparoscopy[102,103] has been adopted for the recovery of human oocytes.[104]

The recovery of mammalian eggs — those of a rabbit killed 72 hr p.c. — was first

UNIE VAN SUID-AFRIKA. UNION OF SOUTH AFRICA.

Gelieve in u antwoord u verwys na
In reply please quote
No. V.A.21/4

**KANTOOR VAN DIE—OFFICE OF THE**

DIRECTOR OF VETERINARY SERVICES.
PRIVATE BAG 138.
PRETORIA.

The Director,
Natal Agricultural Region,
PIETERMARITZBURG.
Natal.

With reference to your minute No. N.32/1 of the 8th May, 1957, you are hereby authorised to introduce into the Union of South Africa, six Rabbits from the Agricultural Research Council's Animal Research Station, Huntingdon Road, Cambridge, England, by air from London to Louis Botha Airport, Natal, for research purposes at the Agricultural Research Institute, Pietermaritzburg.

This authority is granted subject to the following conditions:-

1. That the rabbits are adequately secured during the journey to the Union;

2. That the rodents are strictly isolated at the Agricultural Research Institute, Pietermaritzburg, until slaughtered, and the carcases disposed of to the satisfaction of the Sub-Director of Veterinary Services, Natal.

This permit is valid until the 30th September, 1957.

PRINCIPAL VETERINARY OFFICER.
DIRECTOR OF VETERINARY SERVICES.

FIGURE 4. Copy of letter authorizing the importation into South Africa of rabbits carrying fertilized sheep eggs.

described in 1672 by de Graaf, who found "one egg in the middle of the right oviduct and at the end of the right horn two very small ones" plus one egg from the left side "at the very beginning of the horn."[105] De Graaf observed that when pricked with a needle the eggs released a clear liquid and his translators, therefore, noted "these were probably blastocysts."[105] But at 72 hr p.c., the rabbit egg is normally only just beginning to form a blastocoel,[106] and very little fluid is present. Elsewhere, Catchpole[107] commented "it is difficult to decide the status of de Graaf's very earliest ova."

An interesting account of the early attempts at egg recovery has been provided by Squier,[108] who cites a technique described by Barry,[109] based on the earlier report of Cruikshank,[110] as follows: "Search in vain for the ova on the right side; at last, by drawing a probe gently over the Fallopian tube on the left side, before it was opened, more than an inch on the side next to the uterus, I pressed out several ova, which seemed to come from about its middle, as I began the pressure there and the ova did not appear till the very last." According to Barry, Cruikshank used this method only once, otherwise he opened the rabbit oviducts with scissors, as much later also did Bischoff[111] (1852), Reichert[112] (1862) working with guinea pigs, and possibly also

Heape.[1] Squier[108] noted that the guinea pig oviduct was too small to be flushed with fluid in spite of his use of very fine glass cannulae; he postulated that the flushing fluid failed to pass due to "mucosal folds and tubal contracture." Squier, therefore, adopted a different approach; he placed the oviduct in a shallow dish filled with isotonic saline solution, cut it into quarters, and then gently stroked each piece longitudinally with round, curved needles. Hansson[113] employed a similar technique to recover mink eggs by "massaging the uterine tubes." Nowadays, sufficiently fine hypodermic needles are available to make possible the flushing of even the finest tubes, but a considerable degree of skill and patience are needed.

Hensen[114] in 1876 was the first to use flushing fluid, albeit a fixative, to recover eggs from the uterus. "In order to obtain eggs from the uterus, I inserted a fairly wide, thin walled tube into the cervix, and injected Müller's solution into the uterine cavity from the tubal end." In 1883, Spee used 0.75% sodium chloride solution as the flushing medium.

Allen et al.[115] were the first to describe a technique for the recovery of eggs in vivo by flushing the oviduct; they developed the technique in the dog, using animals 1 to 2 days after mating, and flushing the tubes with isotonic saline in the following way: "A hypodermic needle was passed through the walls of the tube close to the uterus, the injections being made toward the ovary and the washings recovered in watch glasses held under the fimbriated ends of the tubes." After 1 to 2 weeks the tubes were removed and found to be patent and serial sections of the isthmus revealed "very little if any scar."

Subsequently, the technique, suitably modified, was used for the recovery of human ova. The cervix was clamped with a special clamp and "isotonic saline was injected directly into the uterine cavity from above the hypodermic syringe while first one and then the other oviduct was gently pinched by the assistant," the fluid being collected into a watch glass via the fimbriae.

Techniques for the recovery of tubal rabbit eggs from the living animal were described by Avis and Sawin,[116] Chang,[117] who together with Pincus had shown the procedure on film in 1949,[118] and Adams.[119] Shortly afterwards, two of these techniques,[117,119] which resembled that of Allen et al.,[115] were modified for use in sheep[120] and later applied to other farm animals.

Recently other techniques, particularly nonsurgical ones, have been developed for egg recovery in cattle. An apparatus for the nonsurgical recovery of cow eggs was first described in 1949.[121]

## IV. SYNCHRONIZATION OF ESTRUS

Chang's[33] unequivocal demonstration of the necessity for a close correspondence between the time of ovulation in the donor and recipient animals was later extended to and found to apply equally in other species, including the mouse,[122,123] rat,[124] ferret,[125] pig,[126] sheep,[127,128] and cow.[129,130] In fact, synchronization has long been accepted as a *sine qua non* for successful transfer, though whether it is preferable for the donor to be slightly in advance of the recipient or vice versa may still cause debate.

In species like the rabbit in which ovulation is nonspontaneous and easy to arrange, either by mating or the injection of hCG or LHRH, synchronization presents no problem. However, in those having spontaneous ovulation, particularly farm animals whose numbers are often insufficient to permit natural synchronization, artificial methods have had to be developed to achieve synchronization. The evolution of methods for synchronizing estrus, through "squeezing" the corpus luteum,[131] treatment with synthetic progesterone or progestagens,[132] or prostaglandins,[16,133] presents a fascinating story as it reflects achievements in gaining control over basic reproductive mechanisms.

## V. CULTURE OF EGGS IN VITRO AND LONG-TERM STORAGE

In 1929 Lewis and Gregory[134] succeeded in culturing rabbit eggs from initial cleavage through to the blastocyst stage and, remarkably, used time-lapse photography to record the events, including rupture and "healing" of the zona pellucida. Comparatively, the rabbit egg is easy to culture; those of several other species still defy attempts at culture, and thereby illustrate the wide variation in the behavior of mammalian eggs.

Pincus and Enzmann[35] early recognized that "the certain tests of the viability and normal development of mammalian ova subject to experimental manipulation in vitro may be made by transplanting these ova into a properly prepared female and obtaining young which have developed from the transplanted ova". They were the first to obtain normal-term young from eggs cultured in vitro; five one-cell rabbit eggs, which had been cultured in Carrel flasks for 20 hr, were transferred to a pseudopregnant doe which bore two young. No details of the medium used or the stage of development of the eggs at transfer were given. The next reports on combined culture and egg transfer did not appear until 1956 (rabbit)[135] and 1958 (mouse),[136] but then defined media were used, coinciding with the successful use of Krebs' Ringer bicarbonate and bovine plasma albumin, fraction V for culturing mouse eggs.[137]

Tabulated and annotated data on in vitro culture, covering the period 1912 to 1958, have been presented by Austin.[2] Among several recent reviews are those of Foote and Onuma,[8] Seidel,[138] and Maurer.[139]

In 1947 Chang[140] reported the birth of rabbits following the transfer of eggs which had been stored at 0, 5, or 10°C for 4 to 6 days and, 25 years on, the long-term preservation of mouse embryos at −196°C was accomplished.[99,100] For recent reviews of the low temperature preservation of embryos see Whittingham.[18,141]

## VI. FERTILIZATION IN VITRO

The early attempts to fertilize mammalian eggs in vitro have been critically reviewed by Smith,[142] whose account embraces a comprehensive history of the subject as also does that of Chang.[143] Smith concluded "At the present time, therefore, there is no conclusive evidence that the complete process of fertilization of the mammalian egg has been accomplished. Nevertheless, one difficulty after another is being surmounted, and there is little doubt that an effective technique is in sight." This assessment proved well founded for within a few years Chang[144] succeeded in fertilizing rabbit eggs in vitro and within a decade the human egg followed.[73]

In vitro fertilization has proved a popular subject with reviewers;[145-148] the more recent advances have been reviewed by Chang[149] and Blandau.[150]

## VII. MICROMANIPULATION OF EGGS

In recent years rapid progress has been made in the micromanipulation of mammalian eggs, leading to the production of chimaeras[151,152] and identical twins.[153] At the same time our knowledge and understanding of embryonic differentiation have been greatly increased.

## VIII. CONCLUSION

Within the span of 90 years — indeed mostly within a mere 3 decades — egg transfer has truly come a long way, greatly assisted and catalyzed by achievements in ancillary fields. The impact of micromanipulation, genetic engineering, and cloning have yet to be felt.

## REFERENCES


1. **Heape, W.**, Preliminary note on the transplantation and growth of mammalian ova within a uterine foster mother, *Proc. R. Soc.*, London 48, 457, 1891.
2. **Austin, C. R.**, *The Mammalian Egg*, Blackwell Scientific, Oxford, 1961.
3. **Willett, E. L.**, Egg transfer and superovulation in farm animals, *Iowa State Coll. J. Sci.*, 28, 83, 1953.
4. **Rowson, L. E. A.**, Egg transfer, *Discovery*, 25, 46, 1964.
5. **Rowson, L. E. A.**, Recent developments in ovum transplantation in the cow, *Atti 7 Simp. Int. Zootec.*, Milan, p.924, 1972.
6. **Chang, M. C. and Pickworth, S.** Egg transfer in the laboratory animal, in *The Mammalian Oviduct Comparative Biology and Methodology*, Hafez, E. S. E. and Blandau, R. J., Eds., University Chicago Press, Chicago, 1969, 389.
7. **Dziuk, P. J.**, Egg transfer in cattle, sheep, and pigs, in *The Mammalian Oviduct Comparative Biology and Methodology*, Hafez, E. S. E. and Blandau, R. J., Eds., University Chicago Press, Chicago, 1969, 407.
8. **Foote, R. H. and Onuma, H.**, Superovulation, ovum collection, culture and transfer. A review, *J. Dairy Sci.*, 53, 1681, 1970.
9. **McLaren, A. and Kerr, M. G.**, Egg transplantation, *Sci. J.*, 6, 51, 1970.
10. **Dickmann, Z.**, Egg transfer, in *Methods in Mammalian Embryology*, Daniel, J. C., Jr., Ed., W. H. Freeman, San Francisco, 1971, 133.
11. **David, J. S. E., Jones, W. A., Newcomb, R., Smith, G. F., and Wishart, D. F.**, Embryo Transfer with Particular Reference to Cattle, Br. Vet. Assoc., London, 1976.
12. **Newcomb, R.**, Fundamental aspects of ovum transfer in cattle, *Vet. Rec.*, 99, 40, 1976.
13. **Trounson, A. O. and Rowson, L. E. A.**, Research on embryo transfer in domestic animals, *J. R. Agric. Soc. (England)*, 137, 77, 1976.
14. **Betteridge, K. J., Ed.**, Embryo Transfer in Farm Animals — A Review of Techniques and Applications, Monograph 16, Canada Department of Agriculture, Ottawa, 1977.
15. **Betteridge, K. J.**, An historical look at egg transfer, *J. Reprod. Fert.*, 62, 1, 1981.
16. **Murray, F. A.**, Embryo transfer in large domestic animals, in *Methods in Mammalian Reproduction*, Daniel, J. C., Jr., Ed., Academic Press, New York, 1978, 285.
17. **Polge, C.**, Embryo transfer and embryo preservation, in *Artificial Breeding of Non-Domestic Animals*, Watson, P. F., Ed., Academic Press, London, 1978, 303.
18. **Whittingham, D. G.**, In vitro fertilization, embryo transfer and storage, *Br. Med. Bull.*, 35, 105, 1979.
19. **Seidel, G. E., Jr.**, Applications of embryo preservation and transfer, in *Beltsville Symposium in Agricultural Research. III. Animal Reproduction*, Hawk, H., Ed., Allanheld, Osmun & Co., Montclair, N.J., 1979, 195.
20. Foundation of Applied Research, Ed., Proc. Natl. Egg Transfer Breeding Conf., April 14, 1949, San Antonio, Tex., 1951.
21. **Swensson, T. and Hewett, C., Eds.**, Symposium on Egg Transplanation, *Med. Svensk. Husdjurskōtseln*, No. 75, 1975.
22. **Rowson, L. E. A., Ed.**, *Egg Transfer in Cattle*, Commission of the European Communities, Luxembourg, 1976.
23. **Betteridge, K. J., Ed.**, Embryo transfer in farm animals, *Proc. 8th Int. Congr. Anim. Reprod. & A. I., Krakow*, 5, 1218, 1976.
24. **Sreenan, J. R., Ed.**, *Control of Reproduction in the Cow*, Martinius Nijhoff, The Hague, 1978.
25. **Hafez, E. S. E. and Semm, K.**, Instrumental insemination, in vitro fertilization and embryo transfer, World Conf., Kiel, W. Germany, Sept. 24 to 27, 1980; *Arch. Androl.*, 5, 1, 1980.
26. **Carmichael, R. A.**, History of the International Embryo Transfer Society. I, *Theriogenology*, 13, 3, 1980.
27. **Schultz, R. H.**, History of the International Embryo Transfer Society. II, *Theriogenology*, 13, 7, 1980.
28. **Heape, W.**, Further note on the transplantation and growth of mammalian ova within a uterine foster mother, *Proc. R. Soc.*, 62, 178, 1897-98.
29. **Biedl, A., Peters, H., and Höfstatter, R.**, Experimentelle Studien über die Einnistung und Weiterentwicklung des Eies im Uterus, *Ztschr. Geburtshilfe Gynäk.*, 84, 59, 1921.
30. **Pincus, G.**, Observations on the living eggs of the rabbit, *Proc. R. Soc. B*, 107, 132, 1930.
31. **Asdell, S. A. and Hammond, J.**, cited by Pincus, G., Observations on the living eggs of the rabbit, *Proc. R. Soc. B*, 107, 132, 1930.
32. **Nicholas, J. S.**, Development of transplanted rat eggs, *Proc. Soc. Exp. Biol. Med.*, 30, 1111, 1933.
33. **Chang, M. C.**, Development and fate of transferred rabbit ova or blastocyst in relation to the ovulation time of recipients, *J. Exp. Zool.*, 114, 197, 1950.

34. **Pincus, G.**, *The Eggs of Mammals*, Macmillan, New York, 1936.
35. **Pincus, G. and Enzmann, E. V.**, Can mammalian eggs undergo normal development in vitro?, *Proc. Natl. Acad. Sci.*, 20, 121, 1934.
36. **Nicholas, J. S.**, The development of rat embryonic tissues after transplantation of the egg to the kidney, *Anat. Rec.*, 55 (Abstr. 31), 1933.
37. **Warwick, B. L., Berry, R. O., and Horlacher, W. R.**, Results of mating rams to angora female goats, *Rec. Proc. Am. Soc. Anim. Prod. 27th Ann. Meet.*, p. 225, 1935.
38. **Warwick, B. L. and Berry, R. O.**, Inter-generic and intraspecific embryo transfers, *J. Hered.*, 40, 297, 1949.
39. **Loginova, N. V. and Lopyrin, A. I.**, Increasing fertility in ewes by the use of mare pregnancy serum, *Prob. Anim. Husb. Moscow*, 10, 114, 1938; cited by Hammond, J. in *Marshall's Physiology of Reproduction*, Vol. 2, Parkes, A. S., Ed., Longmans, Green & Co., London, 1952, 648.
40. **Lopyrin, A. I., Loginova, N. V., and Babicev, I. G.**, Field experiments on raising the proportion of multiple births in Karakul sheep, *Sovetsk. Zooteh.*, 1, 82, 1940; cited in *Anim. Breed. Abstr.*, 9, 325, 1941.
41. **Folley, S. J. and Malpress, F. M.**, The response of the bovine ovary to pregnant mares' serum and horse pituitary extract, *Proc. R. Soc. B*, 132, 164, 1944.
42. **Smith, P. E. and Engle, E. T.**, Experimental evidence reporting the role of the anterior pituitary in the development and regulation of the genital system, *Am. J. Anat.*, 40, 159, 1927.
43. **Pincus, G.**, Superovulation in rabbits, *Anat. Rec.*, 77, 1, 1940.
44. **Parkes, A. S.**, Induction of superovulation and superfecundation in rabbits, *J. Endocrinol.*, 3, 268, 1943.
45. **Engle, E. T.**, Pregnancy following superovulation in the mouse, *Proc. Soc. Exp. Biol. Med.*, 25, 85, 1927.
46. **Engle, E. T.**, Pituitary-gonadal relationship and problem of precocious sexual maturity, *Endocrinology*, 15, 405, 1931.
47. **Evans, H. M. and Simpson, M. E.**, Experimental superfecundity with pituitary gonadotropins, *Endocrinology*, 27, 305, 1940.
48. **Chang, M. C.**, Transplantation of fertilized ova: the effect on viability of age, in vitro storage period, and storage temperature, *Nature (London)*, 161, 978, 1948.
49. **Adams, C. E.**, Studies on prenatal mortality in the rabbit, Oryctolagus cuniculus: the effect of transferring varying numbers of eggs, *J. Endocrinol.*, 26, 470, 1962.
50. **Hammond, J.**, The possibility of artificial pregnancy in cattle, *Agriculture (London)*, 51, 67, 1950.
51. **Hammond, J.**, Problems concerning the transplantation of fertilized ova or "artificial pregnancy", *Anal. Fac. Med. (Montevideo)*, 35, 810, 1950.
52. **Umbaugh, R. E.**, Superovulation and ovum transfer in cattle, *Am. J. Vet. Res.*, 10, 295, 1949.
53. **Willett, E. L., Black, W. G., Casida, L. E., Stone, W. H., and Buckner, P. J.**, Successful transplantation of a fertilized bovine ovum, *Science*, 113, 247, 1951.
54. **Adams, C. E. and Abbott, M.**, Bibliography on recovery and transfer of mammalian eggs and ovarian transplantation. Literature 1887 to 1971, *Bibliography*, No. 45, Reproduction Research Information Service, Cambridge, 1971.
55. **Adams, C. E.**, Bibliography on recovery and transfer of mammalian eggs. Literature 1971 to 1974, *Bibliography*, No. 89, Reproduction Research Information Service, Cambridge, 1975.
56. **Adams, C. E.**, Bibliography on recovery and transfer of mammalian eggs. Literature 1975 to 1977, *Bibliography*, No. 101, Reproduction Research Information Service, Cambridge, 1977.
57. **Fekete, E. and Little, C. C.**, Observations on the mammary tumour incidence of mice born from transferred ova, *Cancer Res.*, 2, 525, 1942.
58. **Kvasnickii, A. V.**, Interbreed ova transplantation, *Sovetsk. Zooteh.*, p. 36, 1951; *Anim. Breed. Abstr.*, 19, 224, 1951.
59. **Tyndale-Biscoe, C. H.**, Blastocyst transfer in the marsupial Setonix brachyurus, *J. Reprod. Fert.*, 6, 41, 1963.
60. **Blaha, G. C.**, Effect of age of the donor and recipient on the development of transferred golden hamster ova, *Anat. Rec.*, 150, 413, 1964.
61. **Chang, M. C.**, Reciprocal insemination and egg transfer between ferrets and mink, *J. Exp. Zool.*, 168, 49, 1968.
62. **Tyndale-Biscoe, C. H.**, Resumption of development by quiescent blastocysts transferred to primed ovariectomized recipients in the marsupial, Macropus eugenii, *J. Reprod. Fert.*, 23, 25, 1970.
63. **Oguri, N. and Tsutsumi, Y.**, Non-surgical egg transfer in mares, *J. Reprod. Fert.*, 41, 313, 1974.
64. **Adams, C. E.**, unpublished data, 1975.
65. **Kraemer, D. C., Moore, G. T., and Kramen, M. A.**, Baboon infant produced by embryo transfer, *Science (N.Y.)*, 192, 1246, 1976.
66. **Marston, J. H., Penn R., and Sivelle, P. C.**, Successful autotransfer of tubal eggs in the rhesus monkey (Macaca mulatta), *J. Reprod. Fert.*, 49, 175, 1977.

67. **Steptoe, P. C. and Edwards, R. G.**, Birth after the reimplantation of a human embryo, *Lancet*, 2, 366, 1978.
68. **Schriver, M. D. and Kraemer, D. C.**, Embryo transfer in the domestic feline, *Am. Assoc. Lab. Anim. Sci.*, 78-4, 12, 1978.
69. **Kinney, G. M., Pennycook, J. W., Schriver, M. D., Templeton, J. W., and Kraemer, D. C.**, Surgical collection and transfer of canine embryos, *Biol. Reprod.*, 20(Suppl. 1), 96A, 1979.
70. **Norris, M. L.**, personal communication, 1981.
71. **Anon.**, Minister backs embryo transfer, *Vet. Rec.*, 107, 409, 1980.
72. **Anon.**, *Farmers Weekly (London)*, p. 51, May 15, 1981.
73. **Edwards, R. G., Steptoe, P. C., and Purdy, J. M.**, Fertilization and cleavage in vitro of preovulatory human oocytes, *Nature (London)*, 227, 1307, 1970.
74. **Edwards, R. G., Steptoe, P. C., and Purdy, J. M.**, Establishing full-term human pregnancies using cleaving embryos grown in vitro, *Br. J. Obstet. Gynaecol.*, 87, 737, 1980.
75. **Steptoe, P. C., Edwards, R. G., and Purdy, J. M.**, Clinical aspects of pregnancies established with cleaving embryos grown in vitro, *Br. J. Obstet. Gynaecol.*, 87, 757, 1980.
76. **Wolstenholme, G. E. W. and Fitzsimons, D. W., Eds.**, *Law and Ethics of A.I.D. and Embryo Transfer*, Ciba Foundation Symposia 17 (new series), Elsevier, Amsterdam, 1973.
77. **Edwards, R. G. and Steptoe, P. C.**, Biological aspects of embryo transfer, in *Law and Ethics of A.I.D. and Embryo Transfer*, Ciba Foundation Symposia 17 (new series), Wolstenholme, G. E. W. and Fitzsimons, D. W., Eds., Elsevier, Amsterdam, 1973, 11.
78. **Langley, C.**, *The Daily Star*, London, p. 1, September 7, 1979.
79. **Phillips, M.**, Dame Josephine Barnes warns of dangers. BMA leader uneasy over test tube babies, *Guardian (London)*, p. 4, September 8, 1979.
80. **Silcock, B.**, Test tube baby experiments may lead to new laws, *Sunday Times*, London, p. 5, July 30, 1978.
81. **Stone, O. M.**, English law in relation to A.I.D. and embryo transfer, in *Law and Ethics of A.I.D. and Embryo Transfer*, Wolstenholme, G. E. W. and Fitzsimons, D. W., Eds., Elsevier, Amsterdam, 1973, 69.
82. **Revillard, M.**, Legal aspects of artificial insemination and embryo transfer in French domestic law and private international law, in *Law and Ethics of A.I.D. and Embryo Transfer*, Wolstenholme, G. E. W. and Fitzsimons, D. W., Eds., Elsevier, Amsterdam, 1973, 77.
83. **Dickens, B. M.**, Legal approaches to new biology, in *Population and the New Biology*, Benjamin, B., Cox, P. R., and Peel, J., Eds., Academic Press, London, 1974, 159.
84. **Lopata, A., Johnson, I. W. H., Hoult, I. J., and Speirs, A. I.**, Pregnancy following intra-uterine implantation of an embryo obtained by in vitro fertilization of a preovulatory egg, *Fertil. Steril.*, 33, 117, 1980.
85. **Beatty, R. A.**, Transplantation of mouse egg, *Nature (London)*, 168, 994, 1951.
86. **Seed, R. G. and Seed, R. W.**, Artificial embryonation — human embryo transplant, *Arch. Androl.*, 5, 80, 1980.
87. **Seed, R. G and Weiss, R.**, Embryo adoption — technical, ethical and legal aspects, *Arch. Androl.*, 5, 119, 1980.
88. **Marshall, F. H. A.**, Obituary notice of Walter Heape (with portrait) 1855—1929, *Proc. R. Soc. B*, 106, xv, 1930.
89. **Pomeroy, R. W.**, Sir John Hammond, CBE, FRS, (obituary), *Nature (London)*, 206, 135, 1965.
90. **Slater, W. K. and Edwards, J.**, John Hammond 1889—1964, *Biogr. Mem. Fellows R. Soc. R. Soc. (London)*, 11, 101, 1965.
91. **Chang, M. C.**, Experimental studies of mammalian spermatozoa and eggs, *Biol. Reprod.*, 4, 3, 1971.
92. **VanDemark, N. L.**, Introduction of the second Carl G. Hartmann lecture, *Biol. Reprod.*, 4, 1, 1971.
93. **Marden, W. G. R. and Chang, M. C.**, The aerial transport of mammalian ova for transplantation, *Science (N.Y.)*, 115, 705, 1952.
94. **Adams, C. E.**, unpublished data, 1950.
95. **Adams, C. E.**, unpublished data, 1953.
96. **Averill, R. L. W., Adams, C. E., and Rowson, L. E. A.**, Transfer of mammalian ova between species, *Nature (London)*, 176, 167, 1955.
97. **Adams, C. E., Rowson, L. E. A, Hunter, G. L., and Bishop, G. P.**, Long distance transport of sheep ova, *Proc. 4th Int. Congr. Anim. Reprod. A. I. (The Hague)*, 2, 381, 1961.
98. **Hunter, G. L., Bishop, G. P., Adams, C. E., and Rowson, L. E. A.**, Successful long distance aerial transport of fertilized sheep ova, *J. Reprod. Fert.*, 3, 33, 1962.
99. **Whittingham, D. G., Leibo, S. P., and Mazur, P.**, Survival of mouse embryos frozen to −196°C and −269°C, *Science (N.Y.)*, 178, 411, 1972.
100. **Wilmut, I.**, Effect of cooling rate, warming rate, cryoprotective agent and stage of development on survival of mouse embryos during freezing and thawing, *Life Sci.*, 11, 1071, 1972.
101. **Willadsen, S. M. and Tischner, M.**, Successful long distance transport by car of sheep embryos stored in liquid nitrogen, *Bull. Acad. Sci. Ser. Biol.*, 26, 725, 1978.

102. **Steptoe, P. C.,** *Laparoscopy in Gynaecology,* E. & S. Livingstone, Edinburgh, 1967.
103. **Dukelow, W. R.,** Laparoscopic research techniques in mammalian embryology, in *Methods in Mammalian Reproduction,* Daniel, J. C., Jr., Ed., Academic Press, New York, 1978, 437.
104. **Steptoe, P. C. and Edwards, R. G.,** Laparoscopic recovery of pre-ovulatory human oocytes after priming of ovaries with gonadotrophins, *Lancet,* 1, 683, 1970.
105. **Jocelyn, H. D. and Setchell, B. P.,** Regnier de Graaf on the human reproductive organs. An annotated translation of Tractatus De Virororum Organis Generationi Inservientibus (1668) and De Mulierum Organis Generationi Inservientibus Tractatus Novus (1672), in *J. Reprod. Fert.,* Suppl., No. 17, Blackwell Scientific, Oxford, 1972.
106. **Adams, C. E.,** Egg development in the rabbit: the influence of postcoital ligation of the uterine tube and of ovariectomy, *J. Endocrinol.,* 16, 283, 1958.
107. **Catchpole, H. R.,** Regnier de Graaf 1641—1673, *Bull. Hist. Med.,* 8, 1261, 1940.
108. **Squier, R. R.,** The living egg and early stages of its development in the guinea pig, Contrib. Embryol. No. 137, Carnegie Inst., Washington, D.C., 32, 323, 1932.
109. **Barry, M.,** Researches in embryology, 2nd series, *Phil. Trans. R. Soc. (London),* 129, 307, 1839.
110. **Cruikshank, W.,** Experiments in which, on the third day after impregnation, the ova of rabbits were found in the fallopian tubes, and on the fourth day after impregnation in the uterus itself; with the first appearance of the foetus, *Phil. Trans. R. Soc. (London),* 87, 197, 1797.
111. **Bischoff, T. L. W.,** *Entwickelungsgeschichte des Meerschweinchens,* J. Rickersche Buchhandlung, Giessen, 1852.
112. **Reichert, C. B.,** Beiträge zur Entwickelungsgeschichte des Meerschweinchens, *Abh. Königl. Akad. Wissensch. Z. (Berlin),* p. 97, 1862.
113. **Hansson, A.,** The physiology of reproduction in the mink (Mustela vison, Schreb) with particular reference to delayed implantation, *Acta Zool.,* 28, 1, 1947.
114. **Hensen, V.,** Beobachtungen über die Befruchtung und Entwicklung des Kaninchens und Meerschweincheins, *Zeitschr. Anat. Entwick.,* 1, 213, 1876.
115. **Allen, E., Pratt, J. P., Newell, Q. J., and Bland, L. J.,** Human tubal ova: related early corpora lutea and uterine tubes, Contrib. Embryol. No. 127, Carnegie Inst., Washington, D.C., 45, 1930.
116. **Avis, F. R. and Sawin, P. B.,** A surgical technique for the reciprocal transplantation of fertilized rabbit eggs, *J. Hered.,* 62, 259, 1951.
117. **Chang, M. C.,** Fertilizability of rabbit ova and the effects of temperature in vitro on their subsequent fertilization and activation in vivo, *J. Exp. Zool.,* 121, 351, 1952.
118. **Chang, M. C. and Pincus, G.,** Artificial insemination of rabbits and transplantation of rabbit eggs (motion picture), *Fed. Proc.,* 8, March, 1949.
119. **Adams, C. E.,** Some aspects of ovulation, recovery and transplantation of ova in the immature rabbit, in *Mammalian Germ Cells,* Wolstenholme, G. E. W., Ed., J. & A. Churchill, London, 1953, 198.
120. **Hunter, G. L., Adams, C. E., and Rowson, L. E. A.,** Inter-breed ovum transfer in sheep, *J. Agric. Sci.,* 46, 143, 1955.
121. **Rowson, L. E. A. and Dowling, D. F.,** An apparatus for the extraction of fertilized eggs from the living cow, *Vet. Rec.,* 61, 171, 1949.
122. **McLaren, A. and Michie, D.,** Studies on the transfer of fertilized mouse eggs to uterine foster-mothers. I. Factors affecting the implantation and survival of native and transferred eggs, *J. Exp. Biol.,* 33, 394, 1956.
123. **Doyle, L. L., Gates, A. H., and Noyes, R. W.,** Asynchronous transfer of mouse ova, *Fertil. Steril.,* 14, 215, 1963.
124. **Noyes, R. W. and Dickmann, Z.,** Relationship of ovular age to endometrial development, *J. Reprod. Fert.,* 1, 186, 1960.
125. **Chang, M. C.,** Development of transferred ferret eggs in relation to age of corpora lutea, *J. Exp. Zool.,* 171, 459, 1969.
126. **Webel, S. K., Peters, J. B., and Anderson, L. L.,** Synchronous and asynchronous transfer of embryos in the pig, *J. Anim. Sci.,* 30, 565, 1970.
127. **Moore, N. W. and Shelton, J. N.,** Egg transfer in sheep. Effect of degree of synchronization between donor and recipient, age of egg, and site of transfer on the survival of transferred eggs, *J. Reprod. Fert.,* 7, 145, 1964.
128. **Rowson, L. E. A. and Moor, R. M.,** Embryo transfer in the sheep: the significance of synchronizing oestrus in the donor and recipient animal, *J. Reprod. Fert.,* 11, 207, 1966.
129. **Rowson, L. E. A., Moor, R. M., and Lawson, R. A. S.,** Fertility following egg transfer in the cow; effect of method, medium and synchronization of oestrus, *J. Reprod. Fert.,* 18, 517, 1969.
130. **Newcomb, R. and Rowson, L. E. A.,** Conception rate after transfer of cow eggs in relation to synchronization of oestrus and age of eggs, *J. Reprod. Fert.,* 43, 539, 1975.
131. **Hammond, J.,** *The Physiology of Reproduction in the Cow,* Cambridge University Press, New York, 1927.

132. **Hansel, W.**, Control of the ovarian cycle in cattle, in *Reproduction in the Female Mammal,* Lamming, G. E. and Amoroso, E. C., Eds., Butterworths, London, 1967, 419.
133. **Philippo, M. and Rowson, L. E. A.**, Prostaglandins and superovulation in the bovine, *Ann. Biol. Anim. Biochem. Biophys.,* 15, 233, 1975.
134. **Lewis, W. H. and Gregory, P. W.**, Cinematography of living developing rabbit eggs, *Science (N.Y.),* 69, 226, 1929.
135. **Adams, C. E.**, Egg transfer and fertility in the rabbit, *Proc. 3rd Int. Congr. Anim. Reprod. A. I.,* Section 2, Cambridge, England, 1956, 5.
136. **McLaren, A. and Biggers, J. D.**, Successful development and birth of mice cultivated in vitro as early embryos, *Nature (London),* 182, 877, 1958.
137. **Whitten, W. K.**, Culture of tubal mouse ova, *Nature (London),* 177, 96, 1956.
138. **Seidel, G. E., Jr.**, Short-term maintenance and culture of embryos, in *Embryo Transfer in Farm Animals,* Betteridge, K. J., Ed., Canada Dept. Agric., Monograph 16, 1977, 20.
139. **Maurer, R. R.**, Advances in embryo culture, in *Methods in Mammalian Reproduction,* Daniel, J. C., Jr., Ed., Academic Press, New York, 1978, 259.
140. **Chang, M. C.**, Normal development of fertilized rabbit ova stored at low temperature for several days, *Nature (London),* 159, 602, 1947.
141. **Whittingham, D. G.**, Low temperature preservation of embryos, in *Embryo Transfer in Farm Animals,* Betteridge, K. J., Ed., Canada Dept. Agric., 1977, 50.
142. **Smith, A. V.**, Fertilization in vitro of the mammalian egg, in *The Biochemistry of Fertilization and the Gamete,* Williams, R. T., Ed., Biochem. Soc. Symp. No. 7, Cambridge University Press, New York, 1951.
143. **Chang, M. C.**, In vitro fertilization of mammalian eggs, *J. Anim. Sci.,* 27 (Suppl. 1), 15, 1968.
144. **Chang, M. C.**, Fertilization of rabbit eggs in vitro, *Nature (London),* 184, 466, 1959.
145. **Austin, C. R.**, Fertilization of mammalian eggs in vitro, *Int. Rev. Cytol.,* 12, 337, 1961.
146. **Thibault, C.**, In vitro fertilization of the mammalian egg, in *Fertilization, Comparative Morphology, Biochemistry, Immunology,* Metz, C. B. and Monroy, A., Eds., Academic Press, New York, 1969, 405.
147. **Tyler, E. T., Tejada, R., and Marik, J.**, Current status of in vitro fertilization and embryo transplantation, *J. Reprod. Med.,* 11, 5, 1973.
148. **Brackett, B. J.**, In vitro fertilization and its assessment with embryo culture, in *Animal Reproduction,* Hawk, H. W., Ed., Allanheld, Osmun & Co., Montclair, N. J., 1979, 171.
149. **Chang, M. C.**, Recent advances in fertilization of mammalian eggs in vitro, in *The Functional Anatomy of the Spermatozoon,* Afzelius, B. A., Ed., Pergamon Press, Oxford, 1974, 97.
150. **Blandau, R. J.**, In vitro fertilization and embryo transfer, *Fertil. Steril.,* 33, 3, 1980.
151. **Tarkowski, A. K.**, Mouse chimaeras developed from fused eggs, *Nature (London),* 190, 557, 1961.
152. **Gardner, R. L.**, Production of chimaeras by injecting cells into the blastocyst, in *Methods in Mammalian Embryology,* Daniel, J. C., Jr., Ed., Academic Press, New York, 1978, 137.
153. **Willadsen, S. M.**, A method for culture of micromanipulated sheep embryos and its use to produce identical twins, *Nature (London),* 277, 298, 1979.

Chapter 2

# EGG TRANSFER IN THE MOUSE AND RAT*

**Zeev Dickmann**

## TABLE OF CONTENTS



* I dedicate this article to Dr. Robert W. Noyes, who introduced me to the egg transfer technique.

# I. INTRODUCTION

The purpose of this chapter is to update and complement a previous review article entitled "Egg Transfer".[1] In that article, I discussed problems of egg transfer in general, and I outlined in detail the steps involved in the recovery of eggs from donors, their maintenance in vitro, and their transfer to recipients, in the rat. The egg transfer technique is essentially the same for rats and mice, with minor differences that will be pointed out later. The above-mentioned article was written and submitted in 1969; however, the book containing it was published in 1971. Although we are living in an era of rapid advances in science and technology, with the exception of nonsurgical transfers in mice, there have been practically no improvements in the technique of egg transfer in rats and mice. However, progress has been made in utilizing the egg transfer method. The present chapter is divided into two main parts: (1) technique and (2) applications.

# II. TECHNIQUE

## A. Comments on Surgical Transfers to the Rat Uterus

The method I previously described[1] is used by most workers. It consists of two basic steps: (1) the uterine wall is pierced with a needle so as to create a passage into the uterine lumen, and (2) the micropipette containing the eggs is inserted through this passage, and the eggs are then gently released into the uterine lumen. Instead of using two instruments, i.e., a needle and a pipette, Wu[2] accomplishes transfers by employing a single instrument — a hypodermic needle attached to a 5-$\mu\ell$ Hamilton syringe. He stabs through the uterine wall with the hypodermic needle, which contains the eggs to be transferred, and then operates the plunger to expel the eggs. The advantage of this method is that the number of steps is reduced from two to one. The disadvantage is that one has less control, because, unlike in the standard method in which a glass pipette is used, the operator cannot observe the movement of the medium containing the eggs when it is expelled from the hypodermic needle. This shortcoming could be overcome by substituting the hypodermic needle with a glass needle. However, the glass needle has to be sufficiently sturdy so that it does not break during the stabbing of the uterine wall.

## B. Nonsurgical Transfers to the Rat Uterus

The only paper on this subject that I have been able to find in the literature is by Vickery et al.[3] The procedure they used is as follows. The recipient is anesthetized by an intraperitoneal injection of sodium pentobarbitone. The cervices are visualized with the aid of an otoscope (an instrument used in human medicine for inspecting the ear, especially the drum membrane). If cellular debris are present around the cervices, they are cleaned by using a cotton-tipped swab moistened with saline. The pipette containing the eggs is then passed through one of the cervices and into the uterine horn, where the eggs are expelled.

In theory, this transfer method seems highly attractive, because it is simple and it avoids surgical trauma. In practice, surgical transfers are still preferred because their yield (number of term fetuses divided by number of eggs transferred) is much higher — 70 to 80% vs. 30% for the nonsurgical transfers. The reason for the relatively low yield is not known. It is possible that passing the pipette through the cervix and part of the uterus stimulates uterine contractions which, in turn, expel a portion of the eggs from the uterus into the vagina.

It seems obvious that, if possible, one would want to avoid anesthesia, surgery, and stress when carrying out any experimental procedure on an animal. However, rats are highly resistant to the procedures necessary to carry out surgical egg transfers. In my extensive experience of doing surgical transfers,[1] I have not noticed any adverse effects resulting from the surgery. Nevertheless, the nonsurgical method would undoubtedly gain popularity if the technique was improved so that the yield matches that of surgical transfers.

### C. Comments on Transferring Eggs to the Mouse Oviduct

The somewhat elaborate procedure I have described for transferring eggs into the rat oviduct[1] is not necessary for the mouse. In the rat, at least in the strains with which I worked, the bursal membrane is very vascular. In order to get to the fimbriated end of the oviduct, a small hole has to be made in the bursal membrane. In most cases, making the hole with a pair of iridectomy scissors results in bleeding — a condition under which the transfer cannot be carried out. To avoid bleeding, the hole in the bursa is made with an electro-scalpel. In the mouse, an area free of blood vessels in the bursa can readily be found. Thus, unlike in the rat, in the mouse the procedure does not require the use of an electro-scalpel; iridectomy scissors will do.

### D. Nonsurgical Transfers into the Mouse Uterus

Beatty[4] and Tarkowski[5] reported that term fetuses resulted from their nonsurgical transfers to the uteri of mice. However, because the yields in these studies were low, investigators kept on using the surgical approach. More recently, a number of studies have been carried out demonstrating that yields resulting from nonsurgical transfers can be improved and brought to the same level as surgical transfers.

Before describing the transfer techniques, I wish to point out an anatomical difference between rat and mouse. The rat has two cervices which can be visually identified. The mouse has one external os and two internal ora (plural for os). Hence, after the transfer pipette is inserted beyond the external os in the mouse, it is then directed either to the right or the left internal os.

The method used by Marsk and Larsson[6] is as follows. A micropipette with an outer diameter (o.d.) of 0.6 mm is mounted in a polyethylene tube, which in turn is connected to a micrometer syringe. The recipient is anesthetized with an intraperitoneal injection of 0.2 mℓ of 2% tribromoethanol. A 10-mm plastic tube (4 mm o.d.) is inserted into the vagina and serves as a speculum. By gentle probing with the transfer pipette, the external os is blindly searched for, but is usually rapidly found. The pipette is then inserted about 10 mm into the uterus, and the eggs are expelled. The authors state that most of the eggs landed in the right uterine horn, and their explanation for this was that the operator was right-handed. From their writing, it is not clear whether they were aware of the fact that the mouse has one external os and two internal ora. Nevertheless, their results were good. In the group which gave the highest yield, they transferred 345 blastocysts into 40 recipients resulting in 185 fetuses — a 54% yield, which is similar to optimal results obtained with surgical transfers. They had other groups, all of which were interstrain transfers. The authors point out that the yield can vary depending on the strain combination. Thus, in their case, CBA mice were better recipients than A/Jax mice, and A/Jaxes were better donors than CBAs. They did not comment on intrastrain transfers.

The study by Aragon and Markert[7] aimed to determine whether anesthetizing the recipient is beneficial for nonsurgical transfers. They taped recipients in a vertical position directly below a Zeiss techniscope equipped with vertical illumination. A 20× magnification was used to visualize the cervix. A quantity of 477 blastocysts were transferred to the uteri of 30 anesthetized recipients, and 436 blastocysts were transferred into 30 unanesthetized recipients. Of the blastocysts, 30% implanted in the anesthetized mice, whereas only 8% implanted in the unanesthetized mice. After implantation, about 85% of the implanted embryos developed normally in both groups. Apparently, in the hands of these authors the anesthesia was beneficial. However, their best yield of 25% is not a good result. Because of the brevity of the description given by the authors, the reason for the low yield cannot be assessed.

Using a remarkably simple technique, Moler et al.[8] achieved a 60% yield, which compares favorably with surgical transfers. Two different mouse lines were used, but eggs from a given line were always transferred to recipients of the same line. Micropipettes were made by drawing the tip of 3-$\mu \ell$ capillary tubes (Wire-trol. Drummond Scientific Co., Broomal, Pennsylvania). A stainless steel plunger is included with each set of capillary tubes, permitting their use as microsyringes. A glass speculum was made by breaking and fire-polishing a 10-mm portion at the center junction area of a Pasteur pipette. Recipients are not anesthetized. In preparation for a transfer, the operator holds the recipient by the tail and lifts the hindquarters, while the front legs are resting on a surface which can be easily grasped by the paws. (This is a position frequently used for taking vaginal smears). In this position, the mouse is reasonably still enabling the operator to insert the speculum into the vagina. The speculum fits snugly around the cervix, thereby preventing the operator from running the tip of the microsyringe into the fornix. The external os of the cervix is located by gentle probing with the microsyringe. Once past the external os, the operator directs the pipette either to the right or left to enter one of the uteri. On reaching the uterus, the plunger is depressed to expel the eggs. The pocedure is then repeated in order to transfer eggs to the contralateral uterine horn. To develop the skill of transferring eggs to a predetermined uterine horn, it is recommended to transfer dye, kill the animal, open her up, and see whether the dye reached the desired horn.

Because of its simplicity, the transfer method described is likely to supersede the surgical method, providing other workers can obtain as good a yield as did Moler et al.[8]

## E. Logistics for Harvesting Eggs

Many investigators in the "egg business" inject donor females with gonadotrophins for the purpose of superovulating them. It was already known more than 50 years ago that a substance(s) contained in the anterior pituitary gland can induce superovulation. Smith and Engle[9] showed that superovulation occurred when anterior pituitary fragments were injected into the leg muscle of immature and mature female rats and mice. They recorded as many as 48 eggs in a single oviduct of an immature mouse. The foundation for the modern method of inducing superovulation was established by Rowlands,[10] who treated immature rats with sequential injections of PMSG and hCG. Following Rowlands' lead, others induced superovulation in immature mice. For some years, researchers seemed unaware that adult females can also be superovulated as had been shown by Smith and Engle.[9] Studies such as those by Fowler and Edwards[11] and by Husain and Saucier[12] reestablished that mature mice and rats can be superovulated. In gonadotrophin-treated rats and mice, the average number of eggs ovulated is determined by the following factors: the animal's strain, whether it is mature or immature, the dose of PMSG and of hCG, the time interval between PMSG and hCG injections, and the time interval between hCG injection and the recovery of eggs. If a maximal number of eggs per animal is required, immature animals should be used. Thus, in one of their treatment groups, Wilson and Zarrow[13] obtained an average of $76.2 \pm 5.4$ eggs in a group of six immature mice, all of which had ovulated. Gates[14] claimed that under optimal conditions he could regularly obtain an average crop of 80 eggs per treated immature mouse. By contrast, the highest harvest from adult mice was only $29.8 \pm 3.3$ eggs.[15]

Most investigators begin the hormonal treatment of adult mice on a random day, that is, without regard to the day of the cycle. It is convenient to do that, because most mice have irregular cycles. In contrast to mice, rats have regular cycles. Reviewing two studies in which adult rats were used, I could not ascertain whether starting PMSG treatment on a given day of the cycle was an advantage. Welschen and Rutte[16] injected PMSG on the day of estrus. Among their experimental groups, they varied the dose of PMSG, hCG, and the time interval between the two injections. Their highest crop of 43.4 ± 5.8 eggs was obtained with 50 IU PMSG followed 3 days later with 15 IU hCG. In a study by Husain and Saucier,[12] the highest average crop obtained was only 24 eggs. They gave the PMSG injection without regard to the day of the cycle, and hCG was injected 2 days later. Thus, there were two basic differences in procedure between this study and that of Welschen and Rutte:[16] the day on which the hormonal treatment was started, and the time interval between PMSG and hCG injections.

The incidence of mating is another consideration in harvesting fertilized eggs. Usually no more than 65% of gonadotrophin-treated animals mate.[15] Can the mating efficiency be increased? Watson et al.[17] addressed this problem and solved it by substituting mating with artificial insemination. In their hands, the crop of embryos was doubled through the use of artificial insemination.

### F. General Comments

As regards all types of transfers — be they transfers to the oviduct, uterus, surgical, or nonsurgical — some outstanding questions concerning the yield of fetuses remain to be resolved. Kirby[18] was of the opinion that immunological dissimilarity between the blastocyst and the mother is an advantage for implantation. In other words, blastocysts genetically dissimilar to the mother result in a higher yield than blastocysts genetically similar. Additional studies are needed to either confirm or refute this allegation. The yield can also be affected by other genetic factors. Thus, some strains of mice seem to make better recipients and/or donors than other strains. This point was illustrated in the experiments of Marsk and Larsson[6] in which A/Jax blastocysts transferred to CBA recipients resulted in a significantly higher yield than CBA blastocysts transferred to A/Jax recipients. The specific factor(s) contributing to the making of a good donor or recipient is not known. We may further postulate that certain individuals, within the same strain, are superior as donors and/or recipients. If such factors operate, they will have to be taken into consideration when analyzing transfer data in both experimental and applied work, e.g., in farm animals and in man.

## III. APPLICATIONS

To perform egg transfers in rats and mice requires a fair degree of dexterity. Considerably less dexterity is needed when using larger animals such as rabbits. The use of the egg transfer method is indispensable for solving a wide array of problems in reproductive biology, developmental biology, and teratology. The following are examples of studies in which the egg transfer technique had to be used in order to investigate particular problems.

### A. Investigations of Ectopic Pregnancies

The incidence of human ectopic pregnancies varies from 1 in 40 to 1 in 300, depending on the location and the type of population surveyed. Even the lower figure constitutes a medical problem of considerable magnitude, because an ectopic pregnancy endangers the life of the mother, and embryos rarely survive. Since there is much to be learned about this problem, it would be worthwhile to determine to what extent laboratory rodents could be used as research models. In rats and mice, ectopic pregnancies almost never occur naturally, but they can be induced experimentally. Kirby[19] made major contributions to the exploration of artificially induced ectopic pregnancies in rats and mice. His studies involved the transfer of blastocysts and cleavage eggs to the kidney, spleen, testis, brain, and anterior chamber of the eye. Kirby was interested in implantation and postimplantation development at these sites. Blastocysts recovered from the uterus and transferred to ectopic sites showed fairly normal embryonic development for a few days after implantation. However, when embryos younger than blastocysts were transferred, only trophoblast and extraembryonic membranes developed, but no embryos.

Normal looking blastocysts are formed in the oviduct, when fertilized eggs are prevented from entering the uterus by placing a ligature around the uterotubal junction. When such oviductal blastocysts are transferred to ectopic sites, they give rise to trophoblast and extraembryonic membranes, but not to embryos.[19] Apparently, postblastocyst embryogenesis is dependent on a uterine factor which either promotes development or neutralizes an inhibitory factor in the embryo.

Let us now consider the mouse oviduct as a possible research model for human tubal pregnancies which constitute the vast majority of extrauterine pregnancies — 97.7% in a large survey conducted in New Jersey.[20] Can mouse uterine blastocysts implant in the oviduct? Bronson and Cunnane[21] investigated this problem. A quantity of 109 blastocysts were transferred to the oviducts of 24 day-1 pseudopregnant recipients. Following transfers, the uterotubal junctions were ligated. One week later, the mice were killed, the oviducts were fixed, sectioned, stained, and examined to determine whether implantation had occurred. There were no implantations. Although these experiments gave negative results, one cannot rule out the possibility that different experimental approaches might succeed in making the mouse oviduct a useful model for the problem at hand.

### B. Teratological Studies

If a pregnant mother was treated with a teratogen during the preimplantation period and adverse effects resulted, we could determine whether the teratogen acted directly, or via the mother, on the embryo. The determination would be done through the use of the egg transfer technique. Austin[22] stated that " . . . no teratogenic agent, to my knowledge, has ever been conclusively shown to affect the cleavage embryo in such a way as to lead to the ultimate appearance of a congenital defect. Either the embryo dies because too many of its cells have been injured or destroyed, or else it survives and undergoes the necessary adjustment to give rise to an apparently normal individual at birth." Indeed, the prevalent opinion is that chemical or physical insults during the preimplantation period do not result in birth defects. However, Spielmann[23] argued that " . . . most of the statements on the absence of malformations after treatment during the preimplantation period are confined to gross morphologic examination at term and do not take into account embryonic death during organogenesis or abnormalities which manifest themselves postnatally." Furthermore, there are studies showing that treatment during the cleavage stages does result in birth defects.[24] It is possible that in such cases the preimplantation embryo is not damaged, but it absorbs and retains the toxic substance which manifests its deleterious effects at a later time, e.g., during organogenesis. Because human egg transfer work has been started,[25] it is of prime importance to resolve the question whether an insult to the preimplantation embryo can result in birth defects.

### C. Investigations to Discern Whether Morphological Features of the Newborn Are Determined by the Uterine Environment of the Mother or by the Genotype of the Embryo

Mice of the C57BL/How strain generally have six lumbar vertebrae, whereas C3H/

Bi mice generally have five lumbar vertebrae. Through the use of the appropriate reciprocal egg transfers, McLaren and Michie[26] demonstrated that the uterine environment is the factor determining the number of vertebrae in the newborn.

Treatment of pregnant A/Jax mice with cortisone induces cleft palate in 100% of the embryos. In contrast, when C57BL or CBA mice are treated in the same manner, only 20% of their embryos are affected. Does the uterine environment determine the frequency of cleft palate? To answer this question, Marsk et al.[27] carried out reciprocal blastocyst transfers among the three strains. Their results show that the genetic constitution of the embryo is responsible, and that the uterus has no influence on the incidence of cleft palate.

### D. Relationship of Embryonic Age to Endometrial Development

In mice[28] and in rats,[29,30] egg transfer studies established the importance of synchronization between the development of the uterus and that of the preimplantation embryo. As is well known, uterine development is controlled by estrogen and progesterone. Additional investigations on this subject were carried out by Wu,[2] who studied the fate of two-cell embryos transferred into the uteri of female rats which had been ovariectomized, and then received either no treatment, progesterone, or estrogen. It was shown that most of the two-cell embryos did not develop into viable blastocysts. However, the incidence of survival to the blastocyst stage was somewhat higher in recipients receiving daily injections of progesterone.

### E. Fertility Decline in Aging Female Mice

Talbert and Krohn[31] carried out a study whose purpose was to determine whether the cause for fertility decline in aging females was the uterus or the egg. To achieve their objective, they transferred preimplantation embryos recovered from young mice into the uteri of either young or old recipients, and preimplantation embryos recovered from old mice into the uteri of young mice. The data of the resulting term fetuses showed that the fertility decline was due to an unfavorable uterine environment in the aging mice, and not due to defective preimplantation embryos. Gosden[32] repeated the above study with the modification of having an internal control; preimplantation embryos from both young and old donors were transferred to the opposite uterine horns of the same recipient. His results confirmed those of Talbert and Krohn.[31]

### F. Interspecific Transfer of Eggs

There are many techniques for studying intrinsic embryonic vs. maternal factors that control early embryonic development and implantation. One of these techniques is interspecific transfer of embryos, which was pioneered in rodents by Briones and Beatty.[33] Other authors have subsequently used various species combinations for interspecific transfers, but our concern here will be limited to rat → mouse and mouse → rat transfers. Tarkowski[34] demonstrated that development from the two-cell stage to blastocyst can occur in the foreign reproductive tract. Moreover, foreign blastocysts induce a decidual response, which is followed by the initial phases of implantation. However, postimplantation embryonic development is short, the upper limit being the 7th day in the mouse and the 8th day in the rat.

Beyer and Zeilmaker[35] transferred one-cell embryos into the oviducts of the foreign species. The recipients were prepuberal, cyclic, and pregnant mice, and prepuberal rats. The embryos of both species were recovered as morulae 72 hr after transfer. The normality of these morulae was demonstrated by transferring them to recipients of their own species and letting them develop to term fetuses.

In order to create uterine conditions of delayed implantation, recipient mice were bilaterally ovariectomized and subsequently given daily injections of progesterone.[36]

One day after ovariectomy, rat blastocysts were transferred into one uterine horn and mouse blastocysts into the contralateral horn. On the 5th or 6th day after ovariectomy, estradiol was injected to induce implantation. The mice were killed on the 9th day after ovariectomy. The mouse blastocysts which developed were normal egg-cylinders. The rat blastocysts induced decidualization as frequently as did the mouse blastocysts, but rat postimplantation development did not occur; only retarded egg-cylinders were observed.

Although rat and mouse blastocysts can implant in the foreign uterus, the implantation is not normal according to Tarkowski.[34] An anomalous implantation could be a logical cause for the impediment of postimplantation development, but there could be other causes. It would be of great interest to determine whether immunological factors are involved in the implantation process.

### G. Concluding Statement

The examples of the six categories of experiments cited above serve to illustrate the indispensability of the egg transfer technique in the study of various problems in reproductive biology, developmental biology, and teratology.

## REFERENCES


1. **Dickmann, Z.**, Egg transfer, in *Methods in Mammalian Embryology*, Daniel, J. C., Jr., Ed., W. H. Freeman, San Francisco, 1971, chap. 8.
2. **Wu, J. T.**, Development of two-cell eggs transferred into the uteri of ovariectomized rats, *Biol. Reprod.*, 13, 211, 1975.
3. **Vickery, B. H., Erickson, G. I., and Bennett, J. P.**, Nonsurgical transfer of eggs through the cervix in rats, *Endocrinology*, 85, 1202, 1969.
4. **Beatty, R. A.**, Transplantation of mouse eggs, *Nature*, 168, 995, 1951.
5. **Tarkowski, A. K.**, Experiments on the transplantation of ova in mice, *Acta Theriol.*, 2, 251, 1959.
6. **Marsk, L. and Larsson, K. S.**, A simple method for nonsurgical blastocyst transfer in mice, *J. Reprod. Fert.*, 37, 393, 1974.
7. **Aragon, C. M. G. and Markert, C. L.**, Trans-cervical transfer of mouse blastocysts, in Proc. 9th Annu. Meet. Soc. Study Reprod., Philadelphia, Abstr. No. 129, 1976.
8. **Moler, T. L., Donahue, S. E., and Anderson, G. B.**, A simple technique for nonsurgical embryo transfer in mice, *Lab. Anim. Sci.*, 29, 353, 1979.
9. **Smith, P. E. and Engle, E. T.**, Experimental evidence regarding the role of the anterior pituitary in the development and regulation of the genital system, *Am. J. Anat.*, 40, 159, 1927.
10. **Rowlands, I. W.**, The production of ovulation in the immature rat, *J. Endocrinol.*, 3, 384, 1944.
11. **Fowler, R. E. and Edwards, R. G.**, Induction of superovulation and pregnancy in mature mice by gonadotrophins, *J. Endocrinol.*, 15, 374, 1957.
12. **Husain, S. M. and Saucier, R.**, Induction of superovulation in mature rats with gonadotrophins, *Can. J. Physiol. Pharmacol.*, 48, 196, 1970.
13. **Wilson, E. D. and Zarrow, M. X.**, Comparison of superovulation in the immature mouse and rat, *J. Reprod. Fert.*, 3, 148, 1962.
14. **Gates, A. H.** Maximizing yield and developmental uniformity of eggs, in *Methods in Mammalian Embryology*, Daniel, J. C., Jr., Ed., W. H. Freeman, San Francisco, 1971, chap. 4.
15. **Spindle, A. I. and Goldstein, L. S.**, Induced ovulation in mature mice and developmental capacity of the embryos in vitro, *J. Reprod. Fert.*, 44, 113, 1975.
16. **Welschen, R. and Rutte, M.**, Ovulation in adult rats after treatment with pregnant mare serum gonadotrophin during oestrus, *Acta Endocrinol.*, 68, 41, 1971.
17. **Watson, J. G., Wright, R. W., Jr., and Chaykin, S.**, Collection and transfer of preimplantation mouse embryos, *Biol. Reprod.*, 17, 453, 1977.
18. **Kirby, D. R. S.**, The egg and immunology, *Proc. R. Soc. Med.*, 63, 59, 1970.
19. **Kirby, D. R. S.**, The role of the uterus in the early stages of mouse development, in *Preimplantation Stages of Pregnancy*, Wolstenholme, G. E. W. and O'Conner, M., Eds., Churchill Livingstone, London, 1965, 325.

20. **Breen, J. L.**, A 21-year survey of 654 ectopic pregnancies, *Am. J. Obstet. Gynecol.*, 106, 1004, 1970.
21. **Bronson, R. and Cunnane, M.**, Transfer of uterine implantation blastocysts to the oviduct in mice, *Fert. Steril.*, 26, 455, 1975.
22. **Austin, C. R.**, Developmental anomalies arising from errors of fertilization and cleavage, *Curr. Top. Pathol.*, 62, 3, 1976.
23. **Spielmann, H.**, Embryo transfer technique and action of drugs on the preimplantation embryo, *Curr. Top. Pathol.*, 62, 87, 1976.
24. **Gottschewski, G. H. M.**, Diencephalic malformations due to treatment of rabbit and hamster females during preimplantation stages of pregnancy, *Teratology*, 10, 310, 1974.
25. **Steptoe, P. C. and Edwards, R. G.**, Birth after the reimplantation of a human embryo, *Lancet*, 2, 366, 1978.
26. **McLaren, A. and Michie, D.**, Factors affecting vertebral variations in mice. IV. Experimental proof of the uterine basis of a maternal effect, *J. Embryol. Exp. Morphol.*, 6, 645, 1958.
27. **Marsk, L., Theorell, M., and Larsson, K. S.**, Transfer of blastocysts as applied in experimental teratology, *Nature*, 234, 358, 1971.
28. **McLaren, A. and Michie, D.**, Studies on the transfer of fertilized mouse eggs to uterine foster-mothers. I. Factors affecting the implantation and survival of native and transferred eggs, *J. Exp. Biol.*, 33, 394, 1956.
29. **Noyes, R. W. and Dickmann, Z.**, Relationship of ovular age to endometrial development, *J. Reprod. Fert.*, 1, 186, 1960.
30. **Dickmann, Z. and Noyes, R. W.**, The fate of ova transferred into the uterus of the rat, *J. Reprod. Fert.*, 1, 197, 1960.
31. **Talbert, G. B. and Krohn, P. L.**, Effect of maternal age on viability of ova and uterine support of pregnancy in mice, *J. Reprod. Fert.*, 11, 399, 1966.
32. **Gosden, R. G.**, Survival of transferred C57BL mouse embryos: effects of age of donor and recipient, *Fert. Steril.*, 25, 348, 1974.
33. **Briones, H. and Beatty, R. A.**, Inter-specific transfers of rodent eggs, *J. Exp. Zool.*, 125, 99, 1954.
34. **Tarkowski, A. K.**, Inter-specific transfers of eggs between rat and mouse, *J. Embryol. Exp. Morphol.*, 10, 476, 1962.
35. **Beyer, G. and Zeilmaker, G. H.**, Development of mouse and rat zygotes following transfer to non-synchronized rat and mouse oviducts, *J. Reprod. Fert.*, 33, 141, 1973.
36. **Copp, A. J. and Rossant, J.**, Effect of implantation delay on transfer of rat embryos to mice, *J. Reprod. Fert.*, 52, 119, 1978.

Chapter 3

# EGG TRANSFER IN THE RABBIT

C. E. Adams

## TABLE OF CONTENTS

# I. INTRODUCTION

To the rabbit belongs the distinction of having been the first species in which egg transfer was performed.[1] Moreover, it has remained a popular subject for fundamental studies on egg transfer, contributing very significantly over the years to the development and application of the technique.

The rabbit is eminently suitable for work involving egg recovery and transfer, and the experimental manipulation of pregnancy. It is freely available and easily managed under laboratory conditions; it breeds throughout the year and is easy to mate or artificially inseminate. Ovulation, which is nonspontaneous and normally induced by mating, can also be induced by treatment with LHRH or human chorionic gonadotrophin (hCG), which it follows after 10 to 12 hr.[2,3] Superovulation can be induced and very high levels of fertilization are the rule, irrespective of the number of ovulations. Egg recovery is quite simple to perform, either at autopsy or in vivo. The gestation period is short, 31 to 32 days; litter size averages eight to ten young in the large breeds. Several genetic markers are available. Above all there is a voluminous literature on the rabbit, especially its reproduction.

The present review is divided into two main sections; the first is concerned with the methodology of superovulation, egg recovery and transfer, including results, and the second with factors affecting the success of transfer.

# II. SELECTION AND MANAGEMENT OF DONORS AND RECIPIENTS

Only fertile, healthy animals should be selected for use as donors or recipients. They should be kept in separate cages for at least 3 weeks prior to use. If virgin does are to be used as recipients it may be advantageous to put them through a pseudopregnancy, i.e., by inducing ovulation 20 days before use. There is always a risk of bought-in stock being sexually immature unless age is specified.

## A. The Donor

### *1. Superovulation*

#### *a. Use of Pituitary Gonadotrophins*

Donors can be induced to superovulate by treatment with anterior pituitary gland preparations, as described by Pincus[4] (two injections s.c. daily for 3 days) or Parkes[5] (one injection s.c. daily for 5 days). The rabbit is mated with one or two fertile males 12 to 24 hr after the final priming injection. To reduce the risk of ovulation failure 30 IU hCG (Gonadotrophon, Paines & Byrne Ltd., Greenford, Middlesex) may be injected intravenously near the time of mating. Instead of mating donors with fertile males, artificial insemination may be used in conjunction with hCG or LHRH treatment to induce ovulation. The collection of semen and the technique of insemination have been described elsewhere.[6,7]

The ovarian response of rabbits to horse anterior pituitary gland preparations (HAP) tends to be reliable and very satisfactory,[8] as the following examples show. In 12 does, weighing 3.9 to 4.4 kg, treated with HAP (2 mg × 5) at 9 months of age, a total of 554 ovulations was counted at laparotomy 24 hr p.c.[9] In another series of 42 donors, 36.9 ± 2.7 (8 to 94) eggs were recovered either at 30 or 60 hr p.c.; 78 (5%) of the eggs were unfertilized, 3 appeared abnormal, and 4 others showed retarded cleavage. Of 1386 eggs recovered 60 hr p.c., all but 52 (3.75%) had developed normally into morulae.[10]

In a further series of 114 superovulated does 96.1% of the 3659 eggs recovered were fertilized compared with 98.6% of 221 eggs in 22 contemporary control does. Fertilization failed completely in 8 (6%) out of 122 does.[11] In yet another series of 21 supero-

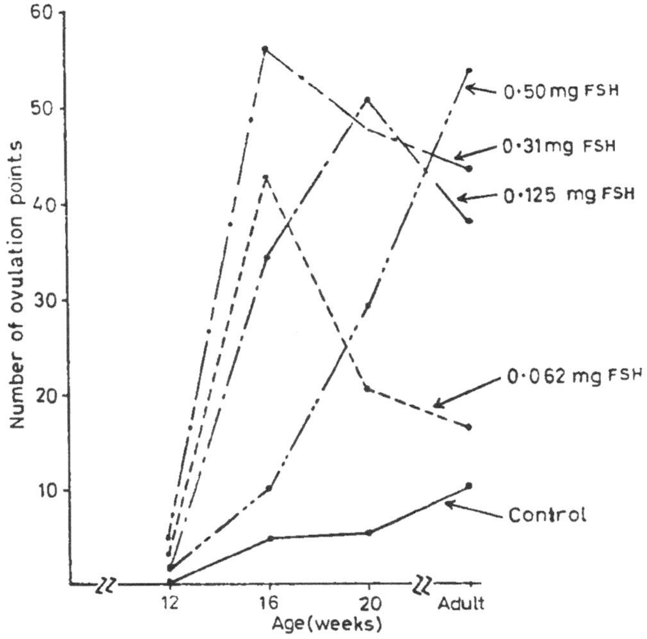


FIGURE 1. Relationship between the level of FSH administered to rabbits at different ages and the number of ovulations. (From Kennelly, J. J. and Foote, R. H., *J. Reprod. Fert.*, 9, 177, 1965. With permission.)

vulated does (mean 40.0 ± 2.6 eggs recovered), 95.2% of the eggs were fertilized; excluding two does in which few eggs were fertilized, the overall fertilization rate was 97.7% (772/790).[12]

The response of sexually immature does to HAP treatment has been extensively studied.[5,13,14] Kennelly and Foote[15] examined the superovulatory response of Dutch belted does treated with several standard gonadotrophin preparations at 12, 16, or 20 weeks of age, and in adult animals; the results are shown in Figure 1. The effects of age and the treatment × age interaction on the number of ovulations were highly significant ($p < 0.01$). In ageing does there is a decline in ovarian response.[8]

### *b. Use of Pregnant Mares Serum Gonadotrophin (PMSG)*

Experience has shown that treating rabbits with PMSG to induce superovulation is often unsatisfactory due to luteinization and blood follicle formation. However, with certain preparations satisfactory responses have been reported, e.g., with a single injection i.m. of 200 IU PMSG and mating 80 hr later; Tsutsumi et al.[16] recorded an average of 61 corpora lutea (range 35 to 96) in 20 Japanese white rabbits compared with 11 (range 8 to 16) in control animals. Hafez[17] investigated various schedules and sources of PMSG in an attempt to maximize superovulatory response; his best results came from three subcutaneous injections (day 1, 1600 hr; day 2, 1600 hr; day 3, 0800 hr), each containing 50 IU PMSG, followed on day 4 (1600 hr) by an ovulating injection. In 77% of the donors, the mean number of ovulations, recorded 16 to 24 hr p.c., was 38.5 (6 to 135); however, 23% of animals failed to respond (0 to 5 eggs).

### *c. Developmental Potential of Eggs from Superovulated Donors*

Eggs from FSH-treated (superovulated) rabbits have been extensively used for transfer. Experience has shown that their developmental potential compares favorably with that of eggs from untreated donors. Eggs from sexually immature donors, obtained by treatment with HAP, are also capable of developing into normal young.[13,14]

**B. The Recipient**

*1. Induction of Ovulation*

To induce ovulation in recipients, they are either mated with a vasectomized (sterile) male and/or treated with hCG or LHRH. In the rabbit the number of ovulations is positively correlated with live weight; data on Polish, Dutch, and Strain A stock are available.[18]

In a series of 39 does given 0.2 m$\ell$ (0.8 $\mu$g) LHRH i.v. (Receptal,® Hoechst A. G. Frankfurt [M], W. Germany), none failed to ovulate, and in 33 2-year-old does weighing 4150 ± 112 g, the mean number of ovulations at autopsy 27 to 30 hr after treatment with Receptal was 10.42 ± 0.49 (range 4 to 16).[19]

Rabbits become refractory to repeated treatment with hCG,[20,21] so that three or four treatments is the limit. This does not, however, apply to LHRH preparations.

*2. Age*

Within certain limits neither the age nor parity of the recipient exerts any marked effect on the outcome of egg transfer. However, in older does (4 to 5 years), pathological changes (cystic glandular hyperplasia) often affect the endometrium as a result of which egg survival is impaired.[8]

## III. SYNCHRONIZATION

Because ovulation is nonspontaneous, it is not necessary to synchronize donors and recipients by artificial means.

## IV. ANESTHESIA AND EUTHANASIA

Anesthesia is relatively easy to induce and manage in the rabbit. The following method is recommended for procedures involving the recovery and transfer of eggs. The first step is to inject intravenously pentobarbitone sodium, 60 mg/m$\ell$ (Sagatal,® May & Baker Ltd., Dagenham, Essex); approximately 1.3 m$\ell$ is needed for a rabbit weighing 3 to 4 kg, and it should be injected slowly (½ to 1 min) into a marginal ear vein. In the majority of rabbits the vein can be adequately visualized by applying a little warm water, and there is rarely any need to shave the injection area except when the hair is dark colored, e.g., in the Californian breed. For intravenous injection, which is much facilitated by the use of a good quality steel needle (26 guage), the rabbit should be placed on a rubber mat or on material such as sack cloth.

Anesthesia is completed by administering halothane (Fluothane,® I.C.I. Ltd., Macclesfield, Cheshire) delivered in oxygen: nitrous oxide (1:1, Entonox,® British Oxygen Company, Brentford, Middlesex) via a pediatric face mask. In the rabbit, the introduction of an endotracheal tube is difficult.[22] The required depth of anesthesia is quickly achieved and usually easily maintained. However, if respiratory failure occurs — this is more likely during the induction phase and with animals which have become excited — immediate action is necessary. The thorax is pressed repeatedly between the thumb and fingers, while the mouth is held open with the tongue extended till respiration is well established again. Staples[23] has described the indicators used in assessing the depth of anesthesia. For a comprehensive account of anesthesia, the reader is referred elsewhere.[24]

Surgical procedures, particularly those involving the oviducts, are facilitated by fasting rabbits for several hours, e.g., overnight; this much reduces the size of the cecum, permitting easier access to the ovaries and oviducts via a mid-line incision.

Before surgery recipients should be clipped to remove the "belly" fur, and then shaved. A bactericidal agent of 70% alcohol is applied to the operation site, which should be covered with a sterilized, fenestrated sheet. Surgical instruments should be sterilized. During the recovery phase, which usually takes 1 to 2 hr, rabbits should be kept warm.

Rabbits are most conveniently killed by means of an intravenous injection of pentobarbitone sodium, 200 mg/mℓ (Expiral, Abbott Laboratories Ltd., Queenborough, Kent). When about 1.2 mℓ is injected rapidly via a 23-guage needle, death is practically instantaneous.

## V. EGG RECOVERY

### A. Introduction

In the rabbit the eggs normally enter the uterus 72 to 84 hr p.c., i.e., approximately 60 to 72 hr after ovulation.[25-27] When ovulation is induced artificially by treatment with hCG, egg transport tends to be slightly accelerated.[27] The position of the eggs in the oviduct relative to the time elapsing from mating has been investigated.[28,29] Initially, passage through the ampulla is rapid, taking only a few minutes.[30] During the cleavage stages and up to about 72 hr p.c., the diameter of the rabbit egg is approximately 160 μm; at 96 hr p.c. the blastocyst measures 250 μm and at 144 hr p.c. about 5 mm.[31]

### B. At Autopsy

Immediately after killing, the rabbit should be bled out for a few minutes by severing the jugular veins. Then the rabbit is placed on its back and the ventral surface is swabbed with water; this helps to prevent hairs sticking to the dissecting instruments and contaminating the genital tract. A long mid-line incision is made through the skin and extended along the hind legs, after which the skin is pulled back. Next the peritoneal cavity is opened to permit removal of the genital tract, including the ovaries and the distal portion of the vagina.

#### *1. Oviducts*

To recover eggs from the oviducts, the ovaries are first removed and the oviducts separated from the uterus at the utero-tubal junction. Excess fat and connective tissue are then trimmed away from the oviducts, using Mayo scissors, whose curved points are directed away from the wall of the oviduct to avoid accidental nicking.

The eggs are recovered by flushing 2 mℓ medium, e.g., phosphate buffered saline, through the oviducts. The direction of flushing may be either from the fimbria towards the utero-tubal junction, or vice versa; in the former case a blunt-ended hypodermic needle is used, while for insertion into the isthmus, which requires more skill, a sharp pointed 26-g hypodermic needle is used. The latter technique has the advantage of requiring less pressure and reduces the risk of the tube coiling, which can lead to spillage of the flushing medium unless an assistant directs the isthmus over the collecting glass. In some animals stones may be present in the oviduct,[8] making flushing towards the utero-tubal junction either very difficult or impossible — sometimes the ampulla may even rupture as pressure builds up within the oviduct. The flushings are collected into two to four punted watch glasses (5-cm diameter); the majority of the eggs (90% + ) are usually found in the first glass. Therefore, if the volume of fluid entering the first glass is restricted, searching for the eggs is expedited. In punted glasses the eggs soon congregate in the flat central area: this can be accelerated by very gently swirling the fluid.

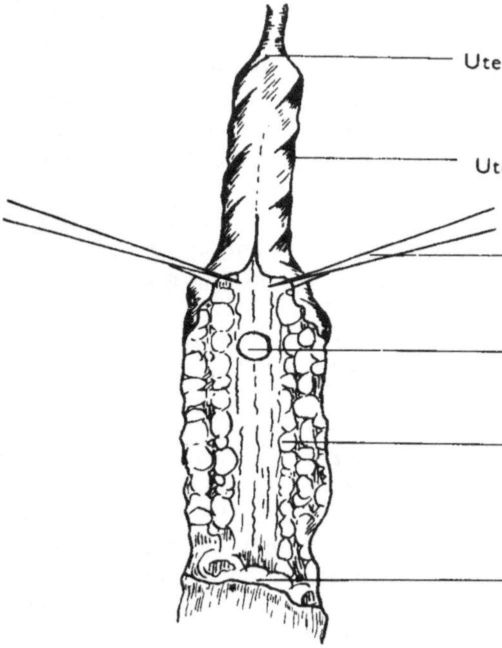


FIGURE 2. Method of opening the rabbit uterus for recovery of blastocysts. (From Adams, C. E., *J. Embryol. Exp. Morphol.*, 23, 21, 1970. With permission.)

*2. Uterus*

After entering the uterus the eggs, now early blastocysts, remain clustered for some time near the utero-tubal junction. They can be recovered by flushing. First the two uterine horns are separated, then each horn is "cleaned" by trimming along its mesometrial border. Any escaping blood should be taken up on a filter paper.

A 10-m$\ell$ syringe is used for flushing. If flushing is in the direction of the cervix, a short incision (1 cm) should be made through the cervix parallel with long axis of the uterus. A hypodermic needle is introduced into the lumen near the utero-tubal junction. For flushing towards the utero-tubal junction the tip of the uterus should be opened. Flushings are collected into watch glasses (7½- or 10-cm diameter).

Blastocysts may be flushed out from the uterus at least up to day 5½. To recover older blastocysts which become "spaced" along the horn, the following approach can be used. The horn is torn open along its antimesometrial border with two pairs of pointed forceps, starting from a slit in the cervix, as described by Lutwak-Mann et al.[32] and illustrated elsewhere[33] (Figure 2). The blastocysts can be removed with the aid of a small scoop.

Attachment of the blastocyst to the antimesometrial surface takes place about 6¾ days p.c.; at that time rupture of the blastocysts occurs if removal is attempted.

In certain experimental situations where both large blastocysts and small fertilized or unfertilized eggs may be present in the same uterine horn, a two-stage recovery technique can be applied.[33] After recovering the blastocysts by dissection, as described above, the uterine horn is held in the form of a semicircle with the endometrium facing downwards towards a watch glass containing about 6 m$\ell$ physiological saline or PBS, through which it is passed about 20 times (Figure 3). Only the endometrium is immersed.

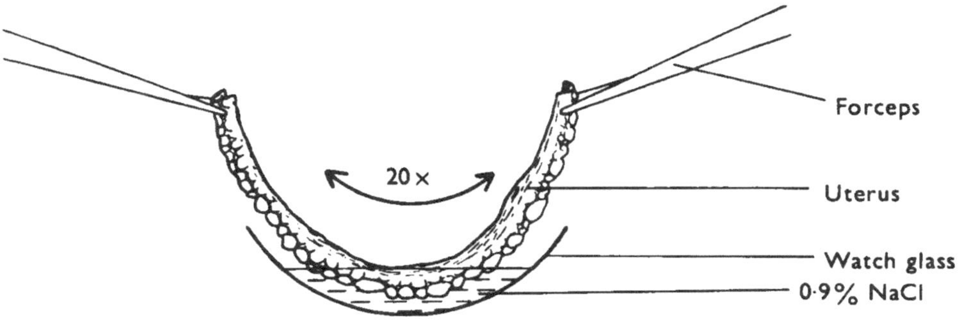


FIGURE 3. Recovery of eggs from the rabbit uterus. (From Adams, C. E., *J. Embryol. Exp. Morphol.*, 23, 21, 1970. With permission.)

*3. Vagina*

In certain experiments, it may be necessary to recover eggs from the vagina, either following their expulsion from the uterus[34] or direct transfer to the vagina.[12] Usually, the vagina will have been ligated to increase the chances of egg recovery, which proceeds as follows. The "cleaned" uterine horns and vagina are held over a 10-cm watch glass, the ligature is released, sometimes with the escape of fluid, and the vagina is flushed twice with physiological saline or PBS, using a 21 guage × 27 mm hypodermic needle and a 10-m*ℓ* syringe.

C. In Vivo

*1. Oviducts*

Techniques for the recovery of eggs from the oviduct in vivo have been described by Avis and Sawin[35] (Figure 4), Chang,[36] and Adams,[13] who reported results based on a series of immature rabbits. That technique, slightly modified, is still in use. It is carried out as follows: laparotomy is performed under general anesthesia. If both oviducts are to be flushed a mid-line approach is convenient, whereas if only one oviduct is to be involved a flank incision may be more satisfactory.

A special glass cannula, 2 to 3 mm o.d. and 8 to 10 cm long bent at an angle of 150°, and with fire polished ends, is introduced into the ampulla via the fimbria and held in place by an assistant. Various sizes of cannulae should be available to accommodate variation in the size of the ampullae. A sharp hypodermic needle (26 guage), connected to a 1- or 2-m*ℓ* Luer Lock syringe, is introduced into the lumen of the oviduct near the utero-tubal junction. In older rabbits, with much adipose tissue, exposing the oviduct and inserting the needle may present difficulty; in that case entry may be gained via the uterus as illustrated (Figure 5).[37] During flushing care should be taken to see that the ampulla does not become occluded by the cannula being held at the wrong angle. Flushing is facilitated by "tearing" the connective tissue around the fimbria and ampulla in order to straighten the oviduct.

The free end of the cannula is directed into a 2-m*ℓ* round bottomed glass tube. After standing for a few minutes the eggs gravitate to the bottom of the tube. They can be recovered by using a pipette to suck up the bottom layer of fluid either before or after gently removing the "supernatant".

Egg recovery from the oviduct works well up to about 60 hr p.c. In the hands of skilled, experienced personnel high recovery rates are possible, but overall recovery in vivo is not quite as reliable as flushing the excised oviduct.

*2. Uterus*

Venge[38] described a technique for the recovery of blastocysts from the uterus 5 to 6

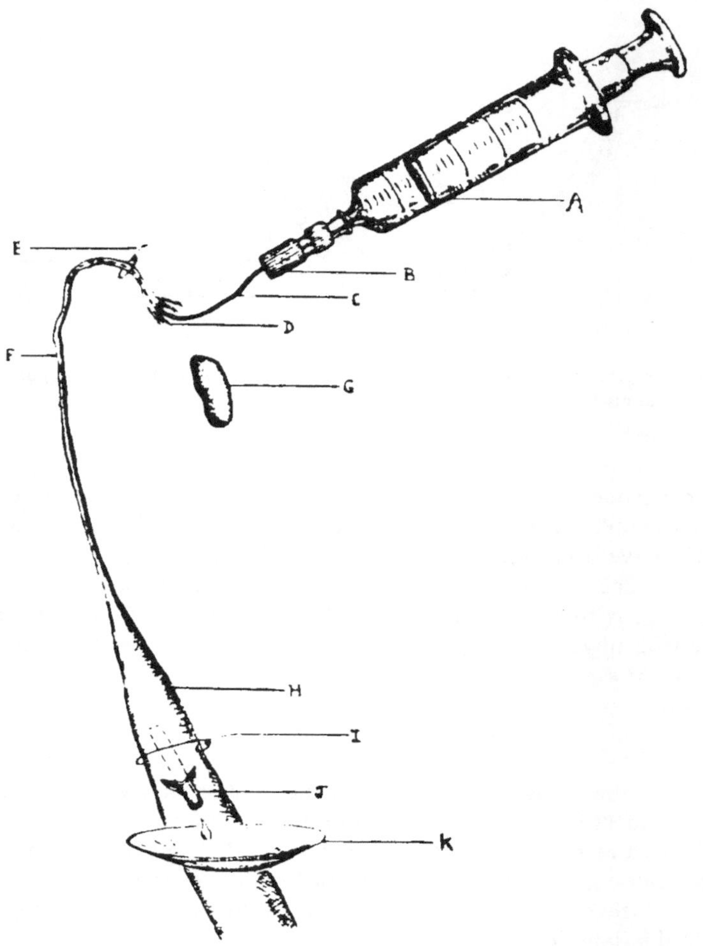


FIGURE 4. Technique for recovery of tubal eggs in the rabbit. (A) syringe; (B) boist adapter; (C) ureteral catheter; (D) fimbriae; (E) ligature; (F) right fallopian tube; (G) ovary; (H) right horn of uterus; (I) ligature; (J) gum rubber tubing; (K) watch glass. (From Avis, F. R. and Sawin, P. B., *J. Hered.*, 42, 259, 1951. With permission.)

days p.c. (Figure 6). The use of two types of indwelling device, permitting the nonsurgical collection of eggs, has also been investigated.[39] With the type B device a total of 149 eggs was collected in 32 attempts.

*3. Vagina*

Tsutsumi et al.[40] designed a two-way system for recovering eggs from the vagina by nonsurgical flushing (Figure 7). Results have been reported from its use in rabbits treated with estrogen,[40] prostaglandin $F_2\alpha$,[41-43] or after superovulation.[16]

## VI. EGG TRANSFER

### A. Nonsurgical

The anatomy of the female rabbit is such that the nonsurgical approach to egg transfer via the vaginal canal and cervix is difficult though not impossible. Testart[44] trans-

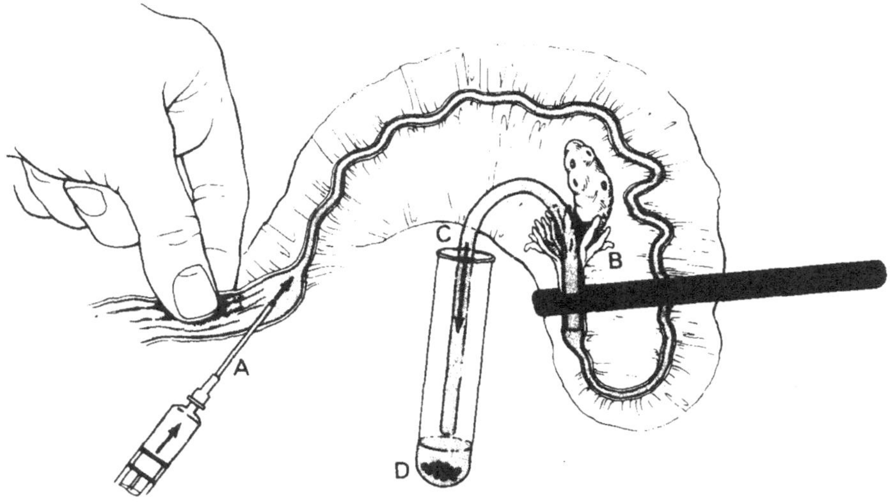


FIGURE 5. Recovery of rabbit eggs in vivo. (A) Flushing medium is injected into the uterus near the utero-tubal junction; (B) fimbriae; (C) polyethylene tube; (D) flushings containing the eggs. (From Hafez, E. S. E., *Methods in Mammalian Embryology,* Daniel, J. C., Jr., Ed., W. H. Freeman, San Francisco, 1971, 117. With permission.)

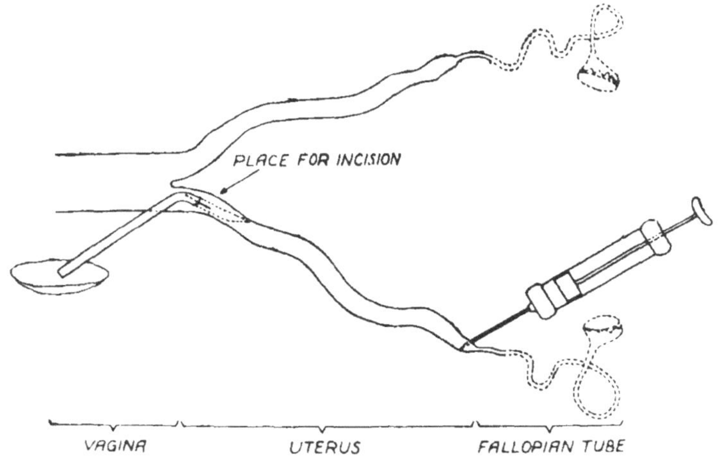


FIGURE 6. Recovery of rabbit blastocysts in vivo. (From Venge, O., *Nature (London),* 169, 590, 1952. With permission.)

ferred 4-day blastocysts via the cervix by means of an "inovulation" catheter. However, with this technique significantly fewer recipients became pregnant and fewer embryos implanted than after direct transfer to the uterus. On day 11 30% of the rabbits which had received eggs via the cervix had pyometritis, and only 47.5% had normal tracts compared with 91.4% after direct transfer.

The development of a technique for transferring eggs via the intraperitoneal route was considered by Hafez[17] but apparently not developed beyond an initial investigation. Dauzier[45] injected eggs i.p. by means of a trochar. One hundred and forty-one fertilized eggs, which he recovered 4 hr after ovulation, were transferred to 14 rabbits shortly after mating to a vasectomized male; at autopsy 24 hr later, only 13 (9.2%) of

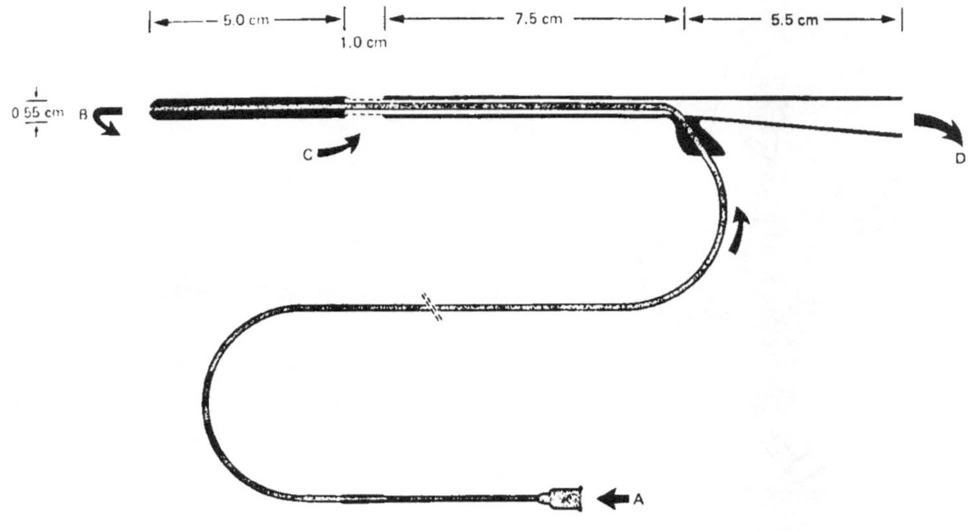


FIGURE 7. Diagram of the apparatus used for nonsurgical collection of rabbit eggs from the vagina. Warmed flushing fluid is infused into a narrow tube by an injector (A); the fluid flows through the catheter to the anterior part of the vaginal lumen (B). Then, the washings return to the catheter through the holes (C), and are recovered in the test tube (D). Arrows show the direction of flow. (From Tsutsumi, Y., Takeda, T., Yamanoto, K., and Tanabe, Y., *J. Reprod. Fert.*, 48, 393, 1976. With permission.)

the transferred eggs were recovered from the genital tract. Possibly the "pick-up" mechanism might have been more efficient had transfer taken place nearer the time of ovulation.

## B. Surgical

The choice of site of transfer, whether to the oviduct or uterus, is largely governed by the age of the egg (stage of development), but this may, of course, be overriden by experimental considerations. If the principal consideration is to provide optimal conditions for development, then two- to eight-cell eggs will be transferred to the oviducts via the fimbria, and later morulae (≥60 hr) and blastocysts will be transferred to the uterus. On balance it is probably better to transfer 2-day eggs (16 to 32 cells) to the oviducts than to the synchronous uterus; the alternative is to use 2½- to 3-day recipients.

### *1. Oviducts*

For two early accounts of egg transfer to the oviduct, the reader is referred to Heape[1] and Pincus.[46] Chang[47] used flank incisions for tubal transfer, and transferred the eggs in a fire-polished pipette (2-mm dameter), containing approximately 0.05 mℓ serum, with the tip of the pipette being inserted into the infundibulum to a depth of 5 to 10 mm. Most probably the pipette was fitted with a rubber bulb.[48] The author also uses a smooth-ended Pasteur pipette (o.d. 1 mm), bent at an angle of 20° 1 cm from the tip, and connected to a 2 mℓ syringe with pressure tubing. The eggs are expelled in approximately 20 μℓ fluid.

### *2. Uterus*

Early accounts of egg transfer to the uterus[47,49] provide few details of the technique, apart from noting the use of a mid-line approach and a pipette having a rough tip to penetrate the antimesometrial border of the uterine wall. However, a full description of a technique that is still in use has been provided by Adams,[50] who transferred 60-hr eggs to 72-hr uteri and observed that the rate of success was only marginally lower than in controls. Adams[51] also described the transfer of eggs to the uterus via the cervix and a short incision in the vaginal wall.

The direct transfer of blastocysts to the uterus has been reported.[23,45,52-54] Chang[52] transferred 6-day blastocysts in a 4-mm diameter pipette via a 5-mm incision in the uterine wall. The incision was closed with catgut sutures. Of 50 blastocysts transferred to 7-day six recipients, 21 developed to term. Shah[53] transferred 6-day blastocysts via the cervix through an incision in the vaginal wall. Four of five recipients, which had received a total of 32 blastocysts, became pregnant and bore 12 normal young. Staples,[23] in reviewing blastocyst transfer, noted that he had obtained very good results, particularly with smaller blastocysts, aged 4 to 4½ days, whereas the survival of 5¼- to 5¾-day blastocysts was not nearly so satisfactory. One third of the blastocysts transferred were not represented by implantations on day 12 and almost one third of those initiating implantation failed to survive to term. In the author's experience the increased levels of mortality are at least partly due to problems in "spacing" — when large blastocysts are transferred shortly before implantation contiguous attachment prejudices survival.

Hafez[54] compared the synchronous transfer of 3-day embryos "near the tip of the uterine horn" *circa* 1 in. below the utero-tubal junction), "approximately 1 in. above the cervix without penetration of the cervix," or via a vaginal incision a glass pipette was inserted for *circa* 1 in. through the cervix into one uterine horn. He concluded that penetration of the cervical canal had no harmful effects; approximately 60% of embryos implanted irrespective of the technique.

Dauzier[45] also compared two methods of transferring 4½-day blastocysts, each involving laparotomy; (1) after perforation of the vaginal wall, the pipette containing the blastocysts was guided through the cervix; and (2) a glass tube acting as a speculum was introduced via the vulva to the cervix, and the pipette was guided via the tube to and then through the cervix. In this way the vaginal wall was not perforated. Evaluation of the results took place 12days after transfer. With the first method 22 of 26 recipients (85%) became pregnant and 89/191 (47%) of the blastocysts implanted, whereas with the second method only 21/63 (33%) recipients became pregnant and 75/444 (17%) blastocysts implanted.

### C. Pregnancy Diagnosis

In the rabbit it is possible to diagnose pregnancy by palpation[55] as early as the 9th day of pregnancy with a high degree of accuracy. Palpation also enables one to follow the course of pregnancy, without resorting to laparotomy; this is particularly useful in experimental situations were resorption is anticipated.[18]

## VII. FACTORS AFFECTING THE SUCCESS OF EGG TRANSFER

### A. Media

Comparatively speaking, the rabbit egg is robust and quite tolerant of exposure to even simple salt solutions, e.g., physiological saline, 0.9% w/v sodium chloride, which has often been used for egg recovery and "immediate" transfer, i.e., within 1 hr of recovery. Nowadays, Dulbecco A® (phosphate buffered saline) is preferred. For longer periods of storage, it is necessary to pay more attention to the medium. When incubated at 37°C rabbit eggs will develop in simple salt solutions plus some organic component such as bovine plasma albumin, Fraction V,[56] or homologous plasma.[33]

The addition of plasma or serum to a medium has the possible disadvantage of causing rabbit eggs to adhere to glass or to each other, and tends to make them more

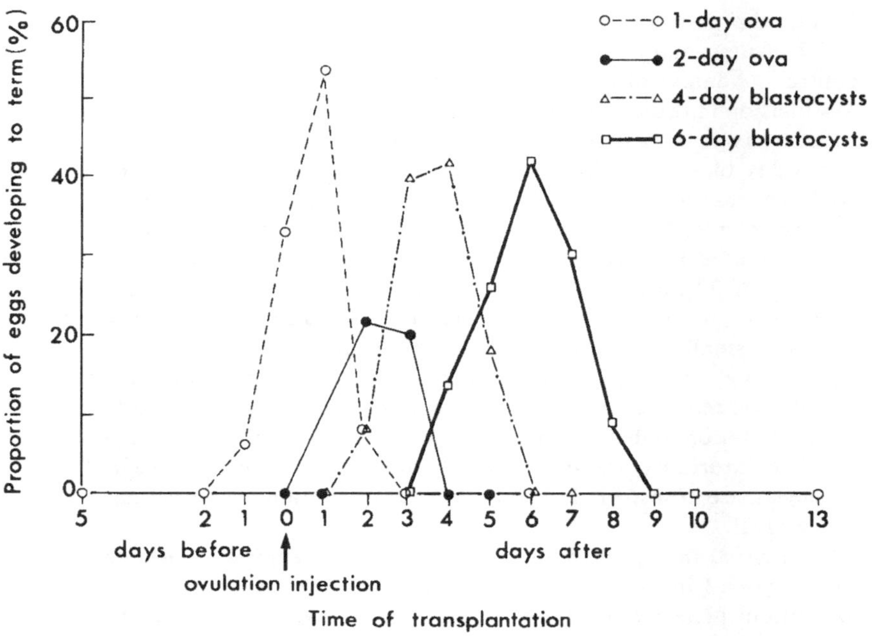


FIGURE 8. Development of transferred ova or blastocysts in relation to ovulation time of recipients. (From Chang, M. C., *J. Exp. Zool.*, 114, 197, 1950. With permission.)

difficult to handle. The problem is aggravated by the sticky mucin coat surrounding the egg.

Whatever the medium, it is most important that it should be sterilized because the luteal phase uterus is highly susceptible to infection; when infection occurs, the eggs do not survive long. Infections which are barely visible on day 6 can be severe by day 8.[57] Infection is probably the single most important cause of failure of egg transfer in the rabbit.

### B. Synchronization and Site of Transfer

The importance of synchronization of the donor and recipient rabbits was demonstrated by Chang,[47] whose findings are depicted in Figure 8. He concluded "there is little or no chance for a transferred ovum or blastocyst to develop into young when it is in the tracts one or two days before or after the corresponding corpus luteal stage." The site of transfer is also important. For example, the proportion of 2-day eggs surviving transfer to the day-2 uterus was quite low.[47,58] Transfer to the day-1 uterus was even less successful; in fact, only 13.5% of eggs could be recovered 24 hr after transfer.[59] Nearly all eggs transferred to the uterus prior to 36 hr are expelled into the vagina within 24 hr, but from 36 hr onwards an increasing proportion is retained *in utero,* though maximum retention is not until 76 to 84 hr p.c.[34] (Figure 9). When the uterine horns were ligated to prevent expulsion 66% of eggs transferred to the day-2 uterus implanted and 37/68 (54%) survived to term; even 23% (19/81) of eggs transferred at 36 hr p.c. implanted, and of those 42% survived to term.[34] The initial "survival" of rabbit eggs "auto-transferred" to ligated uteri 12 to 48 hr p.c. is shown in Figure 10.

When the recipient is in advance of the donor by 2 or more days, pregnancy fails.[47] Even limited exposure, 6 or 24 hr, to asynchronous conditions can affect embryo survival, as demonstrated by retransfer to a synchronized recipient[10] (Figure 11). The factor(s) present in the rabbit uterus on days 9 to 11 which can lead to the death of the embryo has yet to be identified.

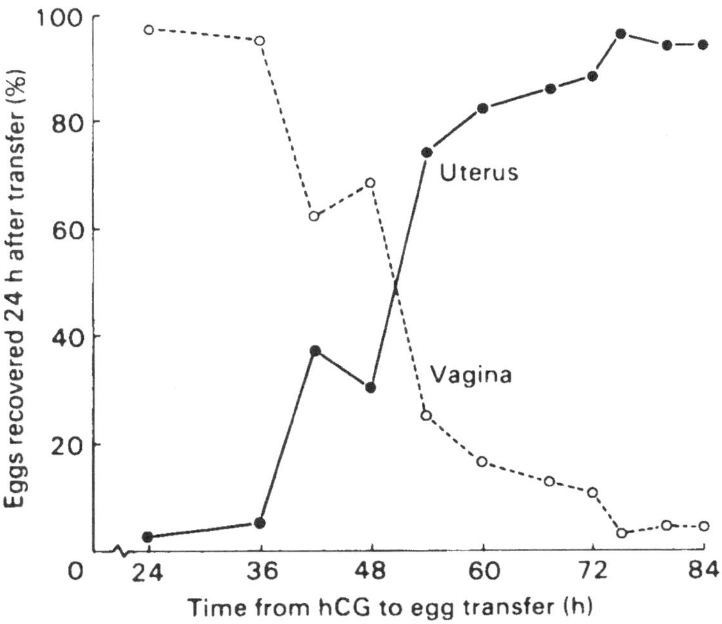


FIGURE 9. Recovery of eggs from the uterus and vagina of rabbits at autopsy 24 hr after transfer to the uterus. A total of 450 eggs was transferred to 43 rabbits. (From Adams, C. E., *J. Reprod. Fert.*, 55, 239, 1979. With permission.)

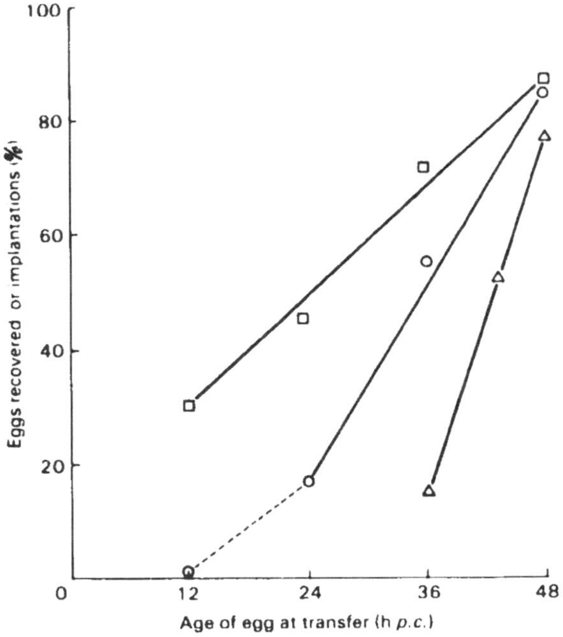


FIGURE 10. The survival of rabbit eggs relative to the time of transfer to the ligated uterus; □, 24 hr after transfer (16 rabbits:137 eggs); o, 72 hr after transfer (11 rabbits:96 eggs); △, implantation sites (13 rabbits:137 eggs). (From Adams, C. E., *J. Reprod. Fert.*, 55, 239, 1979. With permission.)

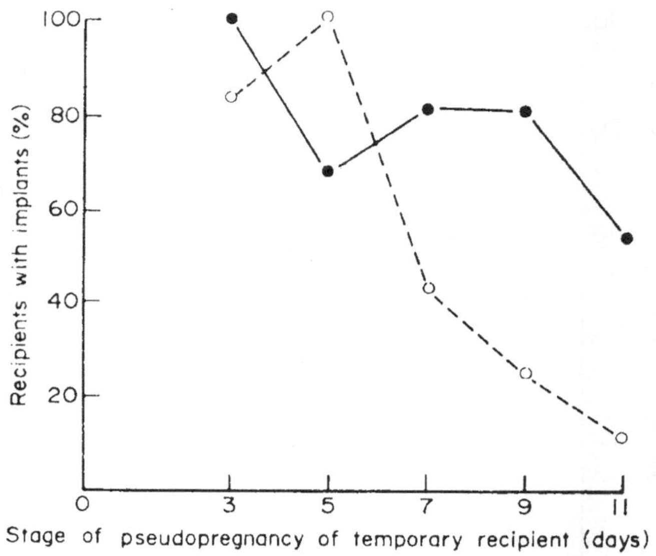


FIGURE 11. The proportion of recipients becoming pregnant (pregnancy rate) after receiving eggs which had spent either 6 hr (•) or 24 hr (o) in the uteri of pseudopregnant rabbits (temporary recipients). (From Adams, C. E., *J. Reprod. Fert.*, 26, 99, 1971. With permission.)

It has long been known that the progestational and decidual changes in the rabbit uterus which normally follow sterile mating can be inhibited by estrogen treatment,[60-65] and that if the estrogen treatment is suspended then the progestational changes will develop some days later.[63] More recently, it was shown[66,67] that the appearance of the uterine protein, uteroglobin, can be delayed (so-called delayed secretion) by two injections of 100 μg estradiol benzoate given 6 and 30 hr p.c. Thus, the secretory pattern of the day-8 uterus comes to resemble that of the normal day-4 stage. In turn, this affects synchronization requirements,[68,69] as illustrated in Figures 12 and 13.[70,71] Increasing the number of estradiol injections to four (6, 30, 48, and 72 hr p.c.) causes further delay so that 60-hr morulae can survive in "day-9 or -10" recipients. Because the corpora lutea are apparently unaffected by such treatment,[69] the technique affords the opportunity to permutate at will the three major components of pregnancy: corpus luteum, endometrium, and embryo.[71]

### C. Number of Eggs

Varying the number of eggs transferred within the range 5 to 35 had no significant effect on pregnancy rate, and neither did transfer to both horns as opposed to a single horn.[50] However, when increasing numbers of eggs were transferred the proportion surviving to implantation decreased significantly ($p < 0.001$) according to the formula: $y = 83.69 - 1.174x$, where y = % eggs implanted and x = number of eggs transferred ($r = -0.881$)[49] (Figure 14).[50] For example, in uteri receiving 5 eggs 82% implanted, whereas only 42% implanted of 34 eggs transferred. After implantation, embryonic mortality is much affected by the number of conceptuses, especially in the middle period, days 10 to 17[49] (Figure 15).

Within a recipient it is possible to have different levels of embryonic mortality in the two uterine horns, depending upon the numbers of eggs transferred.[50] This finding shows the importance of local conditions as factors in embryo mortality. There is no transcervical migration of eggs in the rabbit.

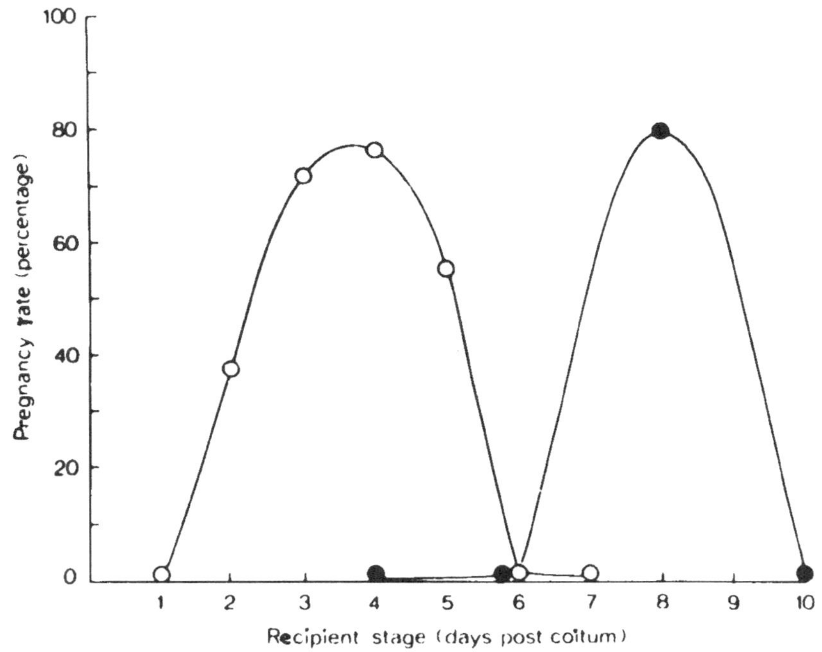


FIGURE 12. Pregnancy rate after transfer of 4-day blastocysts into normally synchronized recipient rabbits (Chang[47] o-o) or after transfer into estrogen-treated rabbits (Beier[67] •-•). (From Beier, H. M., *Advances in the Biosciences 13: Hormones and Embryonic Development*, Raspé, G., Ed., Pergamon Press, Oxford, 1974, 199. With permission.)

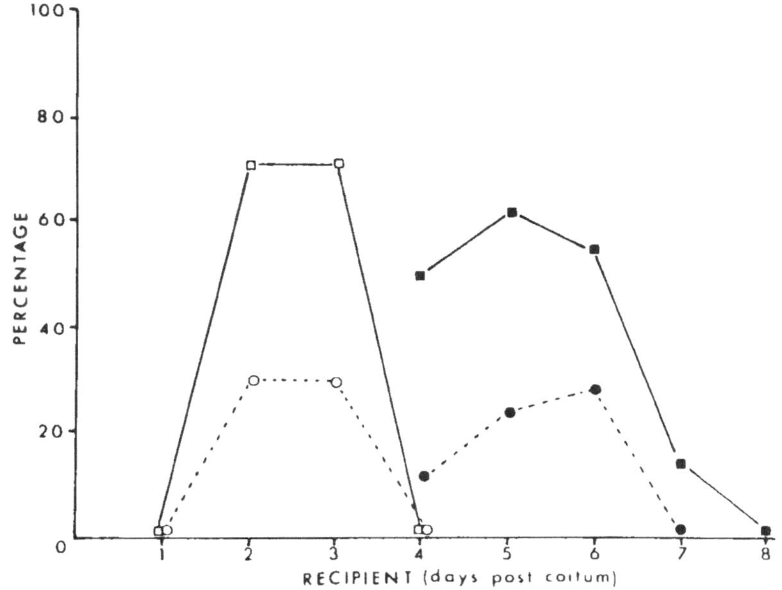


FIGURE 13. Transfer of 2-day eggs to the uterus of normally synchronized recipient rabbits (Chang[47]) and of 2½-day eggs to recipients treated with estradiol benzoate (Adams[69]). — pregnancy rate; --- eggs developing to term; o□Chang; •■Adams. (From Adams, C. E., *Ovum Transport and Fertility Regulation*, Harper, M. J. K., Pauerstein, C. J., Adams, C. E., Coutinho, E. M., Croxatto, H. B., and Paton, D. M., Eds., Scriptor, Copenhagen, 1976, 425. With permission.)

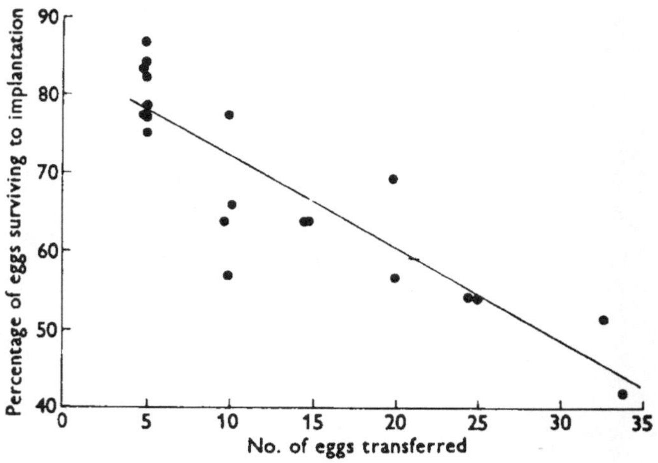


FIGURE 14. Relationship between the number of eggs transferred and proportion surviving to implantation. Formula for the regression line: y = 83.69 − 1.174 x, where y = % eggs implanting and x = no. of eggs transferred; r = 0.881. (From Adams, C. E., *J. Endocrinol.*, 24, 471, 1962. With permission.)

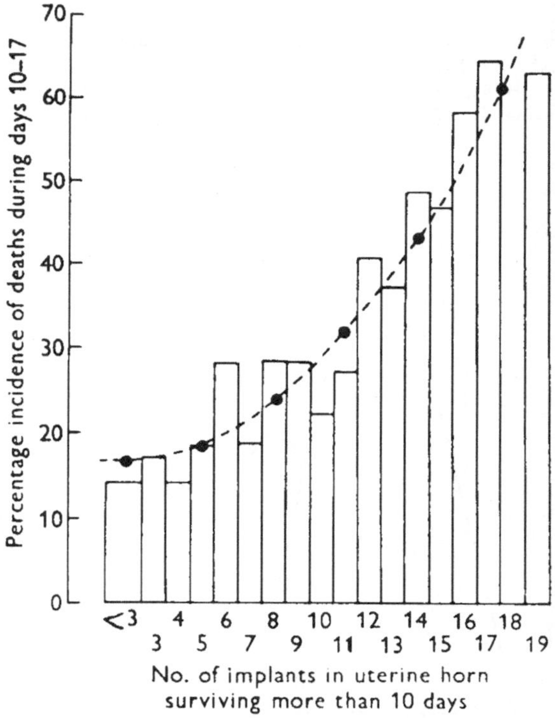


FIGURE 15. Relationship between "middle deaths", occurring from days 10 to 17, and number of implantations in uterine horn surviving more than 10 days. (From Adams, C. E., *J. Endocrinol.*, 24, 471, 1962. With permission.)

If the number of eggs transferred is restricted to one or two, then the outcome of pregnancy is seriously affected. Hafez[72] concluded that a "minimum of 2 implants seemed to be required for fetal survival and normal parturition." He recorded 67, 71, and 78%, respectively, of recipients with implantation following the transfer of one, two, or three eggs. Maximum embryonic mortality, up to 44%, occurred 10 to 12 days p.c. in recipients with one conceptus. Using Polish, Dutch, and a large breed as recipients, Adams[18] found that 53% of single-egg transfers led to implantation compared with 69 to 71% of two-egg transfers, and only 20% of does carrying one fetus maintained pregnancy to term compared with 87% of those with two conceptuses. Pregnancy failed most commonly between days 15 and 20.

A practical consideration arising from these observations concerns experiments or screening procedures to test the viability of eggs subjected to a possibly harmful treatment. To obviate failure due to the presence of too few "healthy" eggs, untreated eggs should be transferred to the contralateral horn, thereby helping to ensure pregnancy.

### D. Zona Pellucida and Mucin Layer

Rabbit eggs developing in culture not infrequently show rupture of the zona pellucida, as the early blastocyst attempts to expand,[33] and under certain conditions escape from the zona pellucida ("hatching") may occur. However, this must be regarded as an abnormal condition because the zona pellucida is normally retained up to implantation. Eggs with a ruptured zona pellucida may develop to term if transferred to a suitable recipient,[32] but in the absence of the zona pellucida, whether removed mechanically from 5-day blastocysts[73] or enzymically from 3-day eggs, development fails.[74]

The need for the zona pellucida hinders experiments concerned with the potential for development of individual blastomeres and the production of identical twins.[75]

Experiments designed to determine whether the mucin layer is essential for development and implantation of the rabbit blastocyst proved inconclusive insofar as the fate of eggs completely devoid of mucin was not established. However, it was found that eggs with less than 10 to 12% mucin could implant.[76,77]

## VIII. APPLICATIONS OF EGG TRANSFER IN THE RABBIT

The rabbit has proved a valuable model for a variety of studies which could only have been undertaken by means of egg transfer. These include the experimental shortening of the generation interval;[78] evaluation of egg development in ectopic sites, e.g., the vagina,[12] under abnormal endocrine conditions[10] or after culture in vitro;[33] assessment of the feasibility of fertilization *in utero*[59,79-81] and of the developmental potential of eggs fertilized by epididymal spermatozoa;[82] distinguishing between defects in the egg and in the uterine environment in relation to declining fertility in aging mothers;[83] evaluation of the action of a drug for a specific length of time on particular stages of development;[84] propagation of wild rabbits which would not breed under laboratory conditions;[85] and testing the capacity of the hormone-treated infantile uterus to support a pregnancy.[86]

## REFERENCES


1. **Heape, W.**, Preliminary note on the transplantation and growth of mammalian ova within a uterine foster mother, *Proc. R. Soc.*, 48, 457, 1891.
2. **Harper, M. J. K.**, The time of ovulation in the rabbit following the injection of luteinizing hormone, *J. Endocrinol.*, 22, 147, 1961.
3. **Harper, M. J. K.**, Ovulation in the rabbit: the time of follicular rupture and expulsion of the eggs, in relation to the injection of luteinising hormone, *J. Endocrinol.*, 26, 307, 1963.
4. **Pincus, G.**, Superovulation in rabbits, *Anat. Rec.*, 77, 1, 1940.
5. **Parkes, A. S.**, Induction of superovulation and superfecundation in rabbits, *J. Endocrinol.*, 3, 268, 1943.
6. **Adams, C. E.**, Artificial insemination in rodents, in *The Semen of Animals and Artificial Insemination*, Maule, J. P., Ed., Commonwealth Agricultural Bureaux, Farnham Royal, 1962, 316.
7. **Adams, C. E.**, Artificial insemination in the rabbit: the technique and application to practice, *J. Appl. Rabbit Res.*, 4, 10, 1981.
8. **Adams, C. E.**, Ageing and reproduction in the female mammal with particular reference to the rabbit, *J. Reprod. Fert. Suppl.*, 12, 1, 1970.
9. **Adams, C. E.**, unpublished observations, 1965.
10. **Adams, C. E.**, The fate of fertilized eggs transferred to the uterus or oviduct during advancing pseudopregnancy in the rabbit, *J. Reprod. Fert.*, 26, 99, 1971.
11. **Adams, C. E.**, Early embryonic mortality in the rabbit, *J. Reprod. Fert.*, 1, 315, 1960.
12. **Adams, C. E.**, The influence of maternal environment on preimplantation stages of pregnancy in the rabbit, in *Preimplantation Stages of Pregnancy*, Wolstenholme, G. E. W. and O'Connor, M., Eds., J. & A. Churchill Ltd., London, 1965, 345.
13. **Adams, C. E.**, Some aspects of ovulation, recovery and transplantation of ova in the immature rabbit, in *Mammalian Germ Cells*, Wolstenholme, G. E. W., Ed., J. & A. Churchill, London, 1953, 198.
14. **Adams, C. E.**, Studies on Growth and Reproduction in the Female Rabbit, Ph.D. thesis, University of Cambridge, 1953.
15. **Kennelly, J. J. and Foote, R. H.**, Superovulatory response of pre- and post-pubertal rabbits to commercially available gonadotrophins, *J. Reprod. Fert.*, 9, 177, 1965.
16. **Tsutsumi, Y., Terami, Y., Takeda, T., Suzuki, M., and Matsui, S.**, In vivo egg recovery from the vagina and the pattern of egg distribution in superovulated rabbits, *Jpn. J. Anim. Reprod.*, 26, 6, 1980.
17. **Hafez, E. S. E.**, An experimental study of ova reception and some related phenomena in the rabbit *J. Morphol.*, 108, 327, 1961.
18. **Adams, C. E.**, Maintenance of pregnancy relative to the presence of few embryos in the rabbit, *J. Endocrinol.*, 48, 243, 1970.
19. **Adams, C. E.**, unpublished data, 1981.
20. **Adams, C. E.**, Artificial insemination in the rabbit, *J. Reprod. Fert.*, 2, 521, 1961.
21. **Greenwald, G. S.**, Development of ovulatory refractoriness in the rabbit to cyclic injections of HCG, *Fert. Steril.*, 21, 163, 1970.
22. **Murdoch, H. R.**, Anaesthesia in the rabbit, *Fed. Proc.*, 28, 1510, 1969.
23. **Staples, R. E.**, Blastocyst transplantation in the rabbit, in *Methods in Mammalian Embryology*, Daniel, J. C., Jr., Ed., W. H. Freeman, San Francisco, 1971, 290.
24. **Green, C. J.**, *Animal Anaesthesia*, Laboratory Animals, London, 1979.
25. **Assheton, R.**, A re-investigation into the early stages of the development of the rabbit, *J. Microsc. Sci.*, 37, 113, 1894.
26. **Gregory, P. W.**, The eary embryology of the rabbit, *Contr. Embryol. Carnegie Inst.*, 21, 141, 1930.
27. **Hodgson, B. J. and Pauerstein, C. J.**, Comparison of oviductal transport of fertilized and unfertilized ova after HCG or coitus induced ovulation in rabbits, *Biol. Reprod.*, 14, 377, 1976.
28. **Chang, M. C.**, Fertility and sterility as revealed in the study of fertilization and development of rabbit eggs, *Fert. Steril.*, 2, 205, 1951.
29. **Greenwald, G. S.**, A study of the transport of ova through the rabbit oviduct, *Fert. Steril.*, 12, 80, 1961.
30. **Harper, M. J. K.**, Egg movement through the ampullar region of the fallopian tube of the rabbit, *Proc. 4th Int. Congr. Anim. Reprod. A.I. (The Hague)*, 2, 275, 1975.
31. **Adams, C. E.**, Egg development in the rabbit: the influence of post coital ligation of the uterine tube and of ovariectomy, *J. Endocrinol.*, 16, 283, 1958.
32. **Lutwak-Mann, C., Boursnell, J. C., and Bennett, J. P.**, Blastocyst uterine relationships: uptake of radioactive ions by the early rabbit embryo and its environment, *J. Reprod. Fert.*, 1, 169, 1960.

33. **Adams, C. E.**, The development of rabbit eggs after culture in vitro for 1 to 4 days, *J. Embryol. Exp. Morphol.*, 23, 21, 1970.
34. **Adams, C. E.**, Retention and development of eggs transferred to the uterus at various times after ovulation in the rabbit, *J. Reprod. Fert.*, 60, 309, 1980.
35. **Avis, F. R. and Sawin, P. B.**, A surgical technique for the reciprocal transplantation of fertilized eggs, *J. Hered.*, 42, 259, 1951.
36. **Chang, M. C.**, Fertilization of rabbit ova and the effects of temperature in vitro on their subsequent fertilization and activation in vivo, *J. Exp. Zool.*, 121, 351, 1952.
37. **Hafez, E. S. E.**, Egg storage, in *Methods in Mammalian Embryology*, Daniel, J. C., Jr., Ed., W. H. Freeman, San Francisco, 1971, 117.
38. **Venge, O.**, A method for continuous chromosome control of growing rabbits, *Nature (London)*, 169, 590, 1952.
39. **Yoshinaga, K., Stocker, J. F., and Pincus, G.**, Non surgical collection of rabbit eggs, *Proc. 6th Int. Congr. Anim. Reprod. A. I. (Paris)*, 1, 585, 1968.
40. **Tsutsumi, Y., Takeda, T., Yamanoto, K., and Tanabe, Y.**, Non-surgical recovery of fertilized eggs from the vagina of oestrogen treated rabbits, *J. Reprod. Fert.*, 48, 393, 1976.
41. **Takeda, T., Tsutsumi, Y., Tanabe, Y., and Yamanoto, K.** Administration of prostaglandin $F_{2}\alpha$ for the recovery of fertilized eggs from the vaginas of rabbits, *Fert. Steril.*, 28, 759, 1977.
42. **Takeda, T., Suzuki, H., Terami, Y., and Tsutsumi, Y.**, Egg recovery from the vagina of the rabbit treated with prostaglandin $F_{2}\alpha$, *Jpn. J. Anim. Reprod.*, 25, 23, 1970.
43. **Tsutsumi, Y., Terami, Y., Suzuki, H., and Sarto, S.**, Effects of prostaglandin $F_{2}\alpha$ on egg recovery from the vagina and egg transport in superovulated rabbit, *J. Fac. Agric. Hokkaido Univ.*, 59, 312, 1979.
44. **Testart, J.**, Comparision de différentes techniques de transplantation des blastocystes chez la lapine, *Ann. Biol. Anim. Biochem. Biophys.*, 9, 351, 1969.
45. **Dauzier, L.**, Nouvelles données sur la transplantation des oeufs, chez la lapine, pas voie vaginale où intrapéritoneale, *Ann. Biol. Anim. Biochem. Biophys.*, 2, 17, 1962.
46. **Pincus, G.**, *The Eggs of Mammals*, Macmillan, New York, 1936.
47. **Chang, M. C.**, Development and fate of transferred rabbit ova or blastocyst in relation to the ovulation time of recipients, *J. Exp. Zool.*, 114, 197, 1950.
48. **Chang, M. C.**, The effects of low temperature on fertilized rabbit ova in vitro, and the normal development of ova kept at low temperature for several days, *J. Gen. Physiol.*, 31, 385, 1948.
49. **Dowling, D. F.**, Problems of the transplantation of fertilized ova, *J. Agric. Sci.*, 39, 394, 1949.
50. **Adams, C. E.**, Studies on prenatal mortality in the rabbit, Oryctolagus cuniculus, the effect of transferring varying numbers of eggs, *J. Endocrinol.*, 24, 471, 1962.
51. **Adams, C. E.**, A study of fertilization in the rabbit: the effect of post-coital ligation of the fallopian tube or uterine horn, *J. Endocrinol.*, 13, 296, 1956.
52. **Chang, M. C.**, Transplantation of rabbit blastocysts at late stage: probability of normal development and viability at low temperature, *Science (New York)*, 111, 544, 1950.
53. **Shah, M. K.**, Reciprocal egg transplantations to study the embryo-uterine relationship in heat induced failure of pregnancy in rabbits, *Nature (London)*, 177, 1134, 1956.
54. **Hafez, E. S. E.**, Effect of progestational stage of the endometrium on implantation, fetal survival and fetal size in the rabbit, Oryctolagus cuniculus, *J. Exp. Zool.*, 151, 217, 1962.
55. **Suitor, A. E.**, Palpating Domestic Rabbits to Determine Pregnancy, Bull. No. 245, U.S. Department of Agriculture, Washington, D.C., 1946.
56. **Adams, C. E.**, Egg transfer and fertility in the rabbit, *Proc. 3rd Int. Congr. Anim. Reprod. A. I. (Cambridge)*, 2, 5, 1956.
57. **Adams, C. E.**, unpublished observations, 1978.
58. **Adams, C. E.**, Consequences of accelerated ovum transport, including a re-evaluation of Estes' operation, *J. Reprod. Fert.*, 55, 239, 1979.
59. **Chang, M. C.**, Développement de la capacité fertilisatrice des spermatozoides du lapin à l'interieure du tractus génital femelle et fécondabilité des oeufs de lapine, in *La Fonction Tubaire et ses Troubles*, Funck-Bretano, P., Ed., Masson, Paris, 1955, 40.
60. **Brouha, A.**, Action empêchante de la folliculine sur le déciduome chez la lapine, *C. R. Soc. Biol. Paris*, 110, 1023, 1932.
61. **Brouha, A.**, Études des rapports entre les modifications prégravidiques de la muqueuse utérine et les hormones ovariennes, *Arch. Biol. Paris*, 45, 571, 1934.
62. **Courrier, R. and Kehl, R.**, La déciduome experimental chez la lapine gestante, *C. R. Soc. Biol. Paris*, 104, 1180, 1930.
63. **Courrier, R. and Kehl, R.**, Sur la suspension expérimentale de la phase luteinique, *C. R. Soc. Biol. Paris*, 109, 877, 1932.
64. **Courrier, R. and Kehl, R.**, Sur la existence de seuils differentiel endocriniens dans les réactions utérines de la phase luteinique, *C. R. Soc. Biol. Paris*, 113, 607, 1933.

65. **Leonard, S. L., Hisaw, F. L., and Fevold, M. L.**, Further studies of the follicular-corpus luteum hormone relationship in the rabbit, *Am. J. Physiol.*, 100, 111, 1932.
66. **Beier, H. M., Kühnel, W., and Petri, G.**, Uterine secretion proteins as extrinsic factors in preimplantation development, in *Advances in the Biosciences 6: Intrinsic and Extrinsic Factors in Early Mammalian Development*, Raspé, G., Ed., Pergamon Press, Oxford, 1971, 165.
67. **Beier, H. M.**, Die hormonelle Steuerung der Uterussekretion und frühen Embryonalentwicklung des Kaninchens, Habil-Schrift Med. Fakult. (Thesis) Univ. Kiel, 1972.
68. **Beier, H. M., Mootz, U., and Kühnel, W.**, Asynchrone Eitransplantationen während der verzögerten Uterussekretion beim Kaninchen, *Proc. 7th Int. Congr. Anim. Reprod. A. I. (Munich)*, 3, 1891, 1972.
69. **Adams, C. E.**, Asynchronous egg transfer in the rabbit, *J. Reprod. Fert.*, 35, 613, 1973.
70. **Beier, H. M.**, Ovarian steroids in embryonic development before nidation, in *Advances in the Biosciences 13: Hormones and Embryonic Development*, Raspé, G., Ed., Pergamon Press, Oxford, 1974, 199.
71. **Adams, C. E.**, Egg survival relative to maternal endocrine status, in *Ovum Transport and Fertility Regulation*, Harper, M. J. K., Pauerstein, C. J., Adams, C. E., Coutinho, E. M., Croxatto, H. B., and Paton, D. M., Eds., Scriptor, Copenhagen, 1976, 425.
72. **Hafez, E. S. E.**, Some factors causing postimplantation mortality in the rabbit, *Proc. 6th Int. Congr. Anim. Reprod. A. I. (Paris)*, 1, 425, 1968.
73. **Adams, C. E.**, unpublished data, 1966.
74. **Rottmann, O. J. and Lampeter, W. W.**, Development of early mouse and rabbit embryos without zona pellucida, *J. Reprod. Fert.*, 61, 303, 1981.
75. **Moore, N. W., Adams, C. E., and Rowson, L. E. A.**, Developmental potential of single blastomers of the rabbit egg, *J. Reprod. Fert.*, 17, 527, 1968.
76. **Adams, C. E.**, Development of the rabbit egg with special reference to the mucin layer, *Proc. 1st Int. Congr. Endocrinol. (Copenhagen)*, p. 687, 1960.
77. **Greenwald, G. S.**, The role of the mucin layer in development of the rabbit blastocyst, *Anat. Rec.*, 142, 407, 1962.
78. **Adams, C. E.**, The experimental shortening of the generation interval, *Proc. Br. Soc. Anim. Prod.*, p. 97, 1954.
79. **Adams, C. E.**, Egg-uterus interrelationships, in *Advances in the Biosciences 4: Mechanisms Involved in Conception*, Raspé, G., Ed., Pergamon Press, Oxford, 1970, 149.
80. **Bedford, J. M.**, The influence of oestrogen and progesterone on sperm capacitation in the reproduction tract of the female rabbit, *J. Endocrinol.*, 46, 191, 1970.
81. **Glass, R. H.**, Fate of rabbit eggs fertilized in the uterus, *J. Reprod. Fert.*, 31, 139, 1972.
82. **Overstreet, J. W. and Bedford, J. M.**, Embryonic mortality in the rabbit is not increased after fertilization by young epididymal spermatozoa, *Biol. Reprod.*, 15, 54, 1976.
83. **Adams, C. E.**, The influence of advanced maternal age on embryo survival in the rabbit, *Proc. 5th Int. Congr. Anim. Reprod. A.I. (Trento)*, 2, 305, 1964.
84. **Adams, C. E., Hay, M. F., and Lutwak-Mann, C.**, The action of various agents upon the rabbit embryo, *J. Embryol. Exp. Morphol.*, 9, 468, 1961.
85. **Adams, C. E.**, Stimulation of reproduction in captivity of the wild rabbit Oryctolagus cuniculus, *J. Reprod. Fert.*, 43, 97, 1975.
86. **Abbott, M.**, Pregnancy in the infant rabbit, *J. Reprod. Fert.*, 35, 620, 1973.

Chapter 4

# EGG TRANSFER IN CARNIVORES AND RODENTS, BETWEEN SPECIES, AND TO ECTOPIC SITES

**C. E. Adams**

## TABLE OF CONTENTS

# I. INTRODUCTION

Among the small laboratory animals the most commonly used species for studies on egg transfer have been the mouse, rat, and rabbit. Other small animals in which egg transfer has been performed, though only to a limited extent, include four carnivores — the cat *(Felis catus)*, the dog *(Canis familiaris)*, the ferret *(Mustela putorius furo)*, and the mink (*Mustela vison*) — and two rodents — the golden hamster *(Mesocricetus auratus)* and the Mongolian gerbil *(Meriones unguiculatus)*. Observations on these six species will be presented in this chapter, which will also include transfer between species and to ectopic sites.

# II. CARNIVORES

## A. Cat

There are only two accounts of egg transfer in the cat,[1,2] and both involve the same material. From a total of 17 attempted collections, 6 days p.c., 9 were successful, yielding 47 embryos which were transferred surgically to nine recipients. Of the four recipients which became pregnant, two gave birth to four live young, one litter was stillborn, and the remaining recipient either aborted or experienced resorption. The first kitten was born on April 17, 1978. The embryos were recovered surgically, using Ham's F10 medium containing 25 m*M* Hepes buffer, and were kept at 32°C for approximately 30 min pending transfer. Synchrony was within ±1 day and ovulation was induced by either hCG or mating with a vasectomized tom.[2]

Ovulation is nonspontaneous, taking place 24 to 30 hr p.c. or some comparable stimulus.[3,4] It can also be induced in anestrous queens by treatment with follicle stimulating hormone, given daily i.m. until estrous behavior is observed. Treatment with PMS proved unsatisfactory due to the rupture of follicles prior to estrus.[5] Concerning the time of entry of eggs into the uterus, reports vary somewhat, e.g., 4 to 5 days p.c.,[6] "some time during the fifth day after ovulation,"[4] or as late as the 6th or 7th day.[7,8] Fertilization in vitro has been reported.[9,10]

## B. Dog

The three reports of egg transfer in the dog[1,2,11] all emanate from the same laboratory at Texas A & M University where four puppies have been born as a result of egg transfer, the first on October 1, 1978.

The following methods were used: at the onset of proestrous bleeding, bitches were exposed daily to a vasectomized male. The first day that a bitch allowed the male to mount was designated day 1 of estrus; copulation was not permitted. Prospective donors were artificially inseminated either on days 1, 3, and 5 or days 2, 4, and 6, and eggs were collected on days 14 and 15. Under anesthesia each uterine horn was flushed separately with 50 mℓ Ham's F10 medium containing 12.5% heat-treated fetal calf serum plus 50 units penicillin and 50 μg streptomycin/mℓ. The medium was introduced into the uterine lumen at the tip of the horn by means of a plastic catheter and collected via a cannula inserted "distal to the body of the uterus".

Within 45 min of recovery the eggs in 1 to 5 μℓ medium were transferred surgically to the uteri of recipients which had shown estrus within ± 4 days of the onset of estrus in the donors. Altogether 72 embryos were recovered from 26 attempted collections. Thirty-seven of the embryos were transferred to seven recipients, three of which were diagnosed pregnant. The low rate of survival to term (4/37) could have been due to a lack of synchrony between the donors and recipients.

In the dog the early stages of development were first described by Bischoff[12] as early as 1845, but for a comprehensive account of preimplantation events reference should

**Table 1**
**TEMPORAL SEQUENCE OF EVENTS IN THE PREIMPLANTATION DEVELOPMENT OF THE BEAGLE[13]**

| Days elapsed from the end of vaginal cornification | Stage of development |
|---|---|
| 0—1 | Cleavage, morula formation |
| 2—3 | Growth to 16-cell morula |
| 4—5 | Entry into uterus and formation of blastocyst |
| 6—7 | Growth of blastocyst to 1200 μm, migration through uterus |
| 8—11 | Growth to 2500 μm, elongation, establishment of definitive implantation sites |
| 11—12 | Local uterine edema and beginning primitive streak. |

be made to a recent report[13] based on purebred beagles. The temporal sequence of events in the preimplantation development of the beagle is shown in Table 1. Ovulation, which is spontaneous, occurs 1 to 2 days after the bitch's first acceptance of the male. The tubal phase is unusually long, such that the eggs do not enter the uterus until 8 to 12 days p.c., between the 16-cell and early blastocyst stages. The free-floating blastocyst stage lasts about 1 week, by which time it measures 2.6 mm in diameter compared with 118 to 135 μm before cleavage.

According to Holst and Phemister,[13] the vaginal smear provides the best clinical index for timing embryonic development in the bitch; they noted that the end of vaginal cornification is easily observed. The relationship between reproductive behavior, serum luteinizing hormone, and time of ovulation has been investigated.[14]

### C. Ferret

Egg transfer in the ferret has been described by Chang.[15,16] From synchronous, uterine transfers of 51 6-, 7-, and 8-day eggs, he observed 31 implantations, of which 17 developed normally into fetuses, including 4 to term.[15] Chang[16] subsequently demonstrated the importance of synchronizing donor and recipient in order to obtain normal development.

In the ferret ovulation is nonspontaneous, occurring 30 hr p.c.;[17] it may also be induced by giving 90 i.u. hCG i.p.[15] The majority of the eggs enter the uterus 4¼ to 7 days after hCG. The early stages of development, from fertilization to the formation of the prochordal plate, have been described.[18]

### D. Mink

There are only two published accounts of egg transfer in the mink.[15,19] Chang transferred 11 eggs to two recipients, but only a single regressing implantation site was present at autopsy. In synchronous transfers of early uterine blastocysts, 25% survival to term was obtained, using 0.9% sodium chloride as the recovery medium.[20] Because the uterine wall is very tough it is necessary to first pierce it with a hypodermic needle to facilitate entry of the Pasteur pipette containing the eggs.

As in the ferret, ovulation is normally induced by coitus, taking place after 36 to 37 hr,[21] or, according to Enders,[22] at 42 to 52 hr p.c.; considerable variation, 33 to 72 hr, has been noted within and between animals.[23] Ovulation is also very reliabily induced by giving 50 i.u. hCG s.c.,[24] or 90 i.u. i.p., when ovulation occurs after 35 to 40 hr.[15] The eggs enter the uterus as unexpanded blastocysts about 8 days p.c.,[21] and the earliest implantation may occur is about 2 or 3 days later. Normally, however, implantation is delayed 5 to 25 days, the longer periods of delay being associated with matings early in the season. Classic accounts of the biology and physiology of reproduction of the mink are available;[21,22] the more recent literature has been extensively reviewed.[25]

Hansson[21] used the technique of "massaging" the excised oviduct to recover eggs between 3 and 8 days p.c.,; however, he did not report on the efficiency of the technique. Although the mink oviduct is relatively small, eggs may be recovered by flushing as described elsewhere,[26] but the technique demands patience and skill. Using a stereomicroscope with incident lighting, each oviduct is dissected free from the ovary, which it surrounds, and then straightened out. A curved, blunt-ended hypodermic needle is introduced into the fimbria and held in place with forceps. About 0.5 $m\ell$ fluid is then carefully flushed through the oviduct and collected into a watch glass.

It was found that when flushing was attempted 5 to 6 days p.c., the fluid could not be forced through the oviduct, even under considerable pressure, unless the utero-tubal junction was first removed. However, on days 7 or 8 p.c. the junction offered no resistance to flushing. The oviducts are very robust.

The uterine horns are very easily flushed by inserting a sharp-pointed hypodermic needle (25 g) into the lumen near the utero-tubal junction; about 0.5 to 1 $m\ell$ of fluid is used for each horn.

## III. RODENTS

### A. Golden Hamster

In 1972 Sato and Yanagimachi[27] noted that few studies had been made on egg transfer in the golden hamster, citing only three reports — those of Blaha,[28] Orsini and Psychoyos,[29] and Yang.[30] Subsequently, one additional study has been reported.[31]

Several accounts are available of the preimplantation stages in the golden hamster;[32-34] the main events are illustrated in Figure 1. According to Ward[35] the peak of ovulatory activity is about 01.00 hr. The eggs begin to enter the uterus during day 3 and entry is completed on day 4 (day 1 = day spermatozoa found in vaginal smear). Blaha[36] noted that at 3 days 15 hr after ovulation egg recovery is difficult because adherence of the blastocyst and uterus may already be occurring. The blastocyst is zona free at 2 days 15 to 20 hr after ovulation.[36]

Superovulation can be induced in adult hamsters[37,38] by treatment with 30 i.u. PMSG on the day of ovulation followed by 25 i.u hCG 76 hr later;[38] heat occurred within 5 hr of giving the hCG. Immature hamsters (5 to 6 weeks old and 60 to 80 g in weight) have been induced to superovulate with 25 i.u. PMSG i.p. and 25 i.u. hCG given i.p. 48 to 56 hr later.[31] In adult females killed 2 to 3 days after mating, 41 to 45 normally developing eggs were recovered.[38]

The procedure for egg transfer described by Sato and Yanagimachi[27] was as follows: hamsters, which had been kept on a 12-hr light cycle (06.00 to 18.00 hr), were used at 90 to 120 g live weight, i.e., 2 to 3 months of age. The day of post estrous vaginal discharge was taken as the first day of the estrous cycle. On the fourth day of the cycle, between 21.00 and 22.00 hr, prospective recipients were subjected to mechanical stimulation of the vagina and cervix[39] in order to induce pseudopregnancy, while donors were mated with fertile males.

To recover eggs, the oviducts and uteri were excised from the donors and each flushed with 0.2 $m\ell$ sterile TC 199 medium. The embryos were rinsed twice with TC-199 and stored no longer than 20 min pending transfer. The recipients were anesthetized, and the oviducts and uteri exposed via dorso-lateral incisions. Eggs were transferred to the ampulla through the infundibulum, or into the uterine lumen, inserting the pipette through a needle puncture near the utero-tubal junction. The total volume of fluid containing the eggs was about 20 $\mu\ell$.

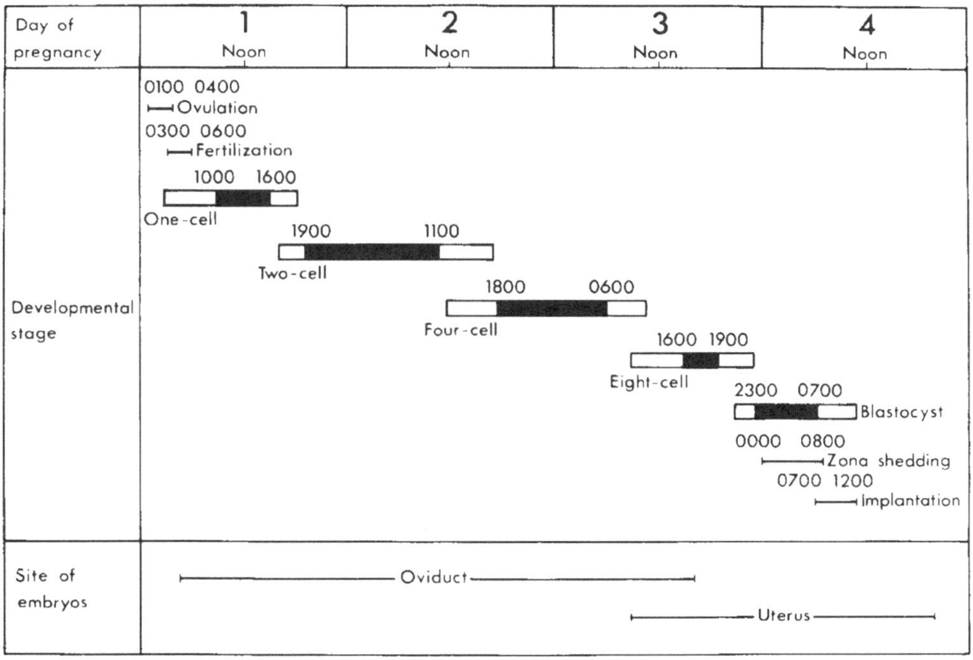


FIGURE 1. Major events and developmental stages of golden hamster embryos during the first 4 days of pregnancy. Columns illustrate times during which respective embryos can be recovered; the optimal recovery times are indicated in black. (Reproduced from Sato, A. and Yanagimachi, R., *J. Reprod. Fert.*, 30, 329, 1972. With permission.)

None of 15 females receiving one- and two-cell eggs transferred to the oviducts became pregnant, whereas 50 to 100% of those receiving four-cell or older eggs (uterine transfers) did so. The best results were obtained with blastocysts transferred synchronously; over 80% implanted successfully of which approximately 70% developed normally to day 14. The proportions of four-cell and eight-cell eggs and of blastocysts developing to term were 31%, 37%, and 36%, respectively.[27]

The transferred one- and two-cell eggs failed to undergo cleavage and started to degenerate in the oviducts.[27] That this may have been due to the medium is suggested by the observations of Whittingham and Bavister,[31] who showed that hamster eggs are highly sensitive to the medium, especially pH. Using PBI medium (a modified Dulbecco's phosphate buffered medium with a stable pH of 7.2), they reported that 59% of two-cell eggs had developed into blastocysts when recovered 48 hr after transfer. Blaha[28] recorded 49% survival to term of six- to eight-cell eggs, and Yang[30] obtained 49% and 54% survival to fetal stages of day 3 or 4 eggs transferred synchronously; his asynchronous transfers (day 4 → 3 and day 3 → 4) were less successful, with 14% and 35%, respectively, surviving. To separate the effects of aging of the female reproductive system from intrinsic defects of the egg, Blaha[28] used the egg transfer technique, which is particularly suitable for this type of study. Further observations on this subject are discussed elsewhere.[36] Recently, Fleming and Yanagimachi[38] have used egg transfer to determine the developmental ability of superovulated hamster ova.

## B. Mongolian Gerbil

There is only one account of egg transfer in the Mongolian gerbil,[40] and even that is not primarily concerned with homologous transfers, but rather with the transfer of gerbil blastocysts in delay to the rat uterus. For the control, gerbil → gerbil transfers, blastocysts were recovered from the excised uteri of nursing mothers on days 7 to 10. Each horn was flushed with 0.5 mℓ phosphate buffered saline (PBS) at 37°C. Recipients were anesthetized with methoxyflurane (Penthrane),[41] and one uterine horn was exposed via a flank incision. The blastocysts were transferred with a fine Pasteur pipette connected to a Clinac micropipette (La Pine Sci. Co., Chicago, Illinois), and to facilitate entry into the lumen the wall of the uterus was first pierced with a 25-g hypodermic needle. Since the recipients were killed 2 to 5 days after transfer in order to evaluate blastocyst survival, it appears that no-term young have been produced as a result of egg transfer in the gerbil.

Ovulation, which is spontaneous, starts about 21.00 hr on the day of mating (day 0) and is completed by 01.00 hr on day 1, some 6 to 10 hr p.c. Egg transport through the oviducts is comparatively slow, as uterine entry does not normally occur until days 5 to 6, when the majority of eggs are at the blastocyst stage.[42] Superovulation has been induced in the mature female gerbil[42-44] and in the immature female,[45] in which treatment with 10 i.u. PMS followed by 20 i.u. hCG 50 hr later gave an average response of 33 eggs. In adults, the maximum response, 31.4 ± 3.4 eggs/female, was obtained in 2-month-old animals given 20 i.u. PMS followed by 20 i.u hCG 54 hr later.[44]

## IV. INTERSPECIES EGG TRANSFER

Reference has been made already in Chapter 1 to some of the early observations on interspecies egg transfer, a subject previously reviewed in 1969.[46] Most work in this field has been undertaken since 1950.

Briones and Beatty[47] transferred eggs between guinea pigs, mice, rats, and rabbits in 9 of the 12 possible combinations. On recovery of the eggs 1 or at most 2 days following transfer, they found that only 6% of the rodent eggs and 28% of the rabbit eggs had continued to develop, and in three of the combinations — rat → mouse, rat → rabbit, guinea pig → mouse — there was a lack of success, which Tarkowski[48] later ascribed to technical inadequacies. In contrast, a contemporary report[49] revealed that sheep eggs could survive up to 5 days in the genital tract of the rabbit.

Initially, because of the known synchronization requirements for transfers within species, synchronization of the recipients and donors was attempted, but due to variations between species in the onset of estrus, time of ovulation, rate of egg development, and egg transport, this is not as straightforward as within species. Moreover, though the eggs may have been transferred to the oviducts, recovery, even in short-term experiments, was often from the uterus to which the alien eggs were transported along with the natives. However, egg recovery from the uterus not only takes longer but is less efficient than from the oviducts. For this reason egg development was tested and found to be equally satisfactory in the ligated oviduct.[50] Subsequently, follicular phase rabbits were used, thereby avoiding the need for synchronization. In such animals, ligation is essential to prevent loss of eggs due to accelerated tubal transport and its consequence, uterine expulsion.[51] Under the influence of estrogen the mucin layer secreted by the rabbit oviduct is much reduced (Figure 2).

In three extensive series of transfers involving cow,[52] sheep,[53] and pig eggs,[54] it was established that fertilized eggs of each of these species could develop to the early blastocyst stage in the ligated rabbit oviduct; as a rule the younger the egg at transfer the longer the period of survival. Of further interest are the two findings that the rate of development remained characteristic of the egg and that the hatching process could be initiated in the rabbit oviduct (Figure 2), thereby emphasizing the significance of the intrinsic mechanism.

Brinster and Tenbroeck[55] showed that the stage of the egg at transfer is important in the transfer of mouse eggs to the rabbit; thus, no blastocyst developed from 730

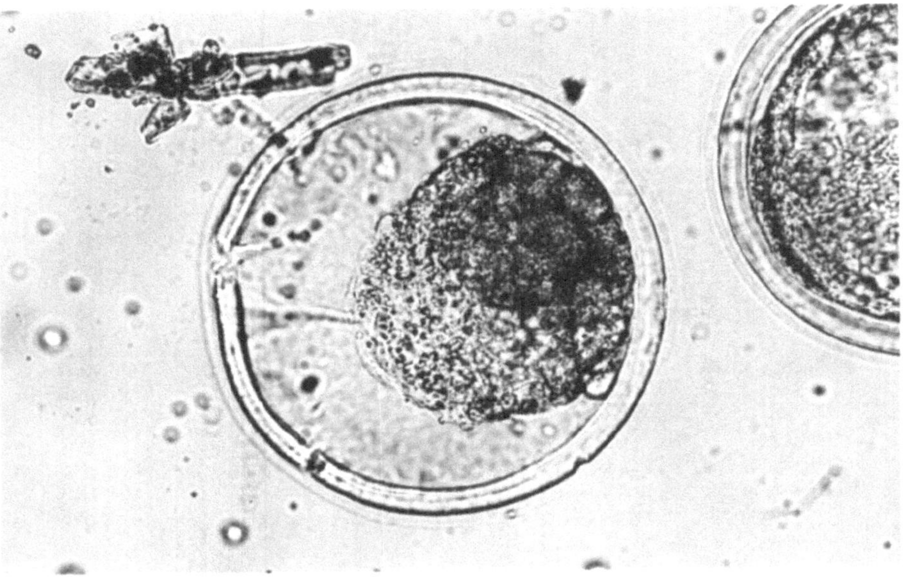

FIGURE 2. Two blastocysts recovered 3 days after transfer of day-3 cow eggs to the ligated oviduct of a follicular phase rabbit. Note the rupture (hatching) of the zona pellucida and the very thin mucin layer.

one-cell eggs, whereas a total of 141 blastocysts developed from 540 two-cell eggs transferred. They, therefore, implied that the secretions of the oviducts of different species are not functionally interchangeable.

A list of species which have served as donors and recipients is detailed in Table 2.[15,47-49,52-71] Altogether at least ten different species have been used as recipients for the eggs of 14 species, though many of the possible combinations still remain unexplored. In general, it appears that from early cleavage up to the formation of the blastocyst the mammalian egg is quite tolerant of a foreign environment, but, thereafter, with certain notable exceptions, e.g., ferret and mink, sheep, and goat, it quickly succumbs, thus signaling its dependence upon highly specific conditions. For that reason, therefore, the transferred eggs are generally not allowed to remain in the foreign host for more than a few days simply because of the limited period of survival. Table 2 also provides information on whether the eggs were retransferred. The principal reason for retransfer, now to a synchronous host, is to test egg viability and capacity to develop to term, which is not possible solely on the basis of morphological criteria.

Rowson[72] described the occurrence of dead twin calves with abnormal limb development, which he attributed to the fact that as eggs they had spent too long — 4 days as opposed to 3 — in a rabbit's oviduct prior to retransfer to a recipient cow. The same author also noted that "storing of sheep eggs prior to retransfer for too long, namely 5 days or more, leads in many cases to abnormal development of the conceptus." However, in such cases it should be emphasized that the embryos died and no abnormal young were born. Except for the cattle twins, no other examples of abnormal development have been directly attributed to interspecies transfer.

In practice, the rabbit has been much used as a "temporary incubator", particularly for the eggs of species which could not be induced to develop in vitro. Even today this approach is still valid because of failure to develop satisfactory, alternative techniques. Another use is to test the viability of eggs following a particular treatment, e.g., deep freezing.

## Table 2
## TRANSFER OF EGGS BETWEEN SPECIES

| Donor | Recipient | Site of transfer | Stage of development or age of eggs at transfer (days) | Interval transfer-examination (days) | Whether eggs retransferred | Ref. |
|---|---|---|---|---|---|---|
| Mouse | Rabbit | Oviduct | | 3 | No | 56 |
| Mouse | Rabbit | Oviduct | 1 and 2 cells | 4—5 | No | 55 |
| Mouse | Rabbit | Oviduct, uterus | 1½, 2½, 4½ | 1—2 | No | 47 |
| Mouse | Rat | Ovarian capsule, uterus | 1½, 2½, 3½ | 1—2 | No | 47 |
| Mouse | Rat | Oviduct, uterus | 2 cell, blastocyst | 5—11 | Yes | 48 |
| Mouse | Guinea pig | Uterus | 3½ | 1—2 | No | 47 |
| Rat | Mouse | Oviduct | 2—8 cell | 7—13 | Yes | 48 |
| Rat | Mouse | Uterus | 114 hr | 1—1½ | No | 57 |
| Rat | Mouse | Uterus | 3½, 4½ | 1—2 | No | 47 |
| Rat | Rabbit | Uterus | 4½ | 1—2 | No | 47 |
| Rat | Rabbit | Oviduct[a] | 5, 7—11 | 2 | Yes | 58 |
| Rat | Golden hamster | Uterus | 4—5 | ? | No | 59 |
| Rat | Golden hamster | Ovarian capsule, uterus | Pronuclear, morula | ? | No | 60 |
| Golden hamster | Rat | Uterus | 4—5 | ? | No | 59 |
| Golden hamster | Rat | Ovarian capsule | Pronuclear, morula | 2—3 | No | 60 |
| Golden hamster | Rat | Uterus | Morula | ? | No | 60 |
| Field vole | Mouse | Oviduct | 2 and 8 cell | | No | 48 |
| Guinea pig | Mouse | Uterus | 5½ | 1—2 | No | 47 |
| Rabbit | Rat | Ovarian capsule, uterus | 1, 2, and 3 | 1—2 | No | 47 |
| Rabbit | Guinea pig | Uterus | 1 | 1—2 | No | 47 |
| Rabbit | Mouse | Uterus | ¼, 3 | 1—2 | No | 47 |
| Rabbit | Ferret | Ovarian capsule, uterus | 1 cell, morula, blastocyst | 2—4 | No | 61 |
| Rabbit | Cow | Uterus | 2—3 | ½ or 2 | No | 62 |

| | | | | | | |
|---|---|---|---|---|---|---|
| abbit | Rat | Uterus | 4 | 2 | No | 58 |
| owshoe are | Rabbit | Oviduct | Pronuclear, 2—4 cell | 6 or 13 | No | 63 |
| rret | Rabbit | Oviduct | 4—6 | 2 or 3 | Yes | 64 |
| rret | Rabbit | Oviduct, uterus | 2—4 cell, morula, blastocyst | 2—4 | No | 61 |
| rret | Mink | | 6—7 | 5—23 | No | 15 |
| ink | Ferret | Uterus | 6—8 | 6—25 | No | 15 |
| ow | Rabbit | Oviduct | Follicular eggs | ? | ? | 65 |
| ow | Rabbit | Oviduct, uterus | 3—5 days | 2 | No | 62 |
| ow | Rabbit | Oviduct | 1—8 cell | 2—4 | Yes | 52 |
| ow | Rabbit | Oviduct | Follicular oocytes | ½—7 | No | 66 |
| eep | Goat | | | 30—term | No | 67 |
| eep | Goat | | 3 or 4 | | No | 68 |
| eep | Rabbit | Oviduct | 2½—3 (2—12 cell) | 4—5 | Yes | 49 |
| eep | Rabbit | Oviduct[a] | 1—8 cell | 4.2—5.3 | Yes | 69 |
| eep | Rabbit | Oviduct[a] | 2—16 cell | 3—7 | Yes | 53 |
| oat | Sheep | | | 30—term | No | 67 |
| oat | Sheep | | | Near term | No | 70 |
| oat | Sheep | | 5 | | No | 68 |
| g | Rabbit | Oviduct[a] | 8, 32, 56, or 80 hr after fertilization | 1, 2, or 3 | Yes | 54 |
| uirrel ionkey | Rabbit | Oviduct[a] | Follicular oocytes 10—12 hr after hCG + 24 hr culture | 2 or 3 | No | 71 |

[a] Ligated.

## V. ECTOPIC TRANSFER

Whereas most emphasis has been placed on the transfer of eggs to the genital tract, the capacity of eggs to develop ectopically, i.e., in extrauterine sites, has also been subjected to experimental investigation. Initially, the preferred sites of transfer were the peritoneal cavity[47,73-76] and the anterior chamber of the eye,[74,75,77-79] probably because of the ease of access and known occurrence of abdominal pregnancies and, in the case of the eye, the facility offered for direct observation through the transparent cornea. Subsequently, other sites including the kidney,[80-83] spleen,[84] testis,[85] brain,[86] and abdominal muscle[87] were employed. Apart from Nicholas[73] and Ahlgren and Bengtsson,[79] who used rat and rabbit eggs respectively, all the other workers used mouse eggs. Most studies have focused on the development of the transferred eggs, but a few have examined the effect of an extrauterine pregnancy on the host[88,89] and immunological aspects.[90-93]

Ahlgren and Bengtsson[79] described a technique for transplanting two- to four-cell rabbit eggs to the anterior chamber of the eye. After 2 to 4 days egg recovery was successful in 90% of cases; development proceeded normally up to the 72-hr morula stage, but, thereafter, growth was retarded and degenerative changes occurred. By means of reciprocal transfers Kirby[82] showed that the mouse blastocyst was four times more successful than that of the rat in implanting under the kidney capsule; moreover, the mouse blastocyst also produced a more "flourishing" trophoblast than that of the rat. On the contrary, mouse blastocysts failed to implant in the rat testis, whereas rat blastocysts developed in the mouse testis.[94]

Kirby's[95] suggestion that a mouse blastocyst to realize totipotency must be subjected to the environment of a hormonally receptive uterus was later refuted.[96] The extrauterine mouse egg as an experimental model was reviewed in 1969 by Kirby,[86] whose early death later that year was a major loss to this field of investigation.

In women, the incidence of tubal pregnancy, which accounts for 96 to 98% of ectopics, is 0.25 to 1% of all pregnancies.[97] However, tubal pregnancy has never been induced experimentally in any animal model, where the eggs fail to survive beyond the early blastocyst stage, e.g., in the rabbit[98] and sheep.[99] The question of the signal for maternal recognition in an ectopic pregnancy has been discussed recently.[100]

## REFERENCES

1. **Kraemer, D. C., Flow, B. L., Schriver, M. D., Kinney, G. M., and Pennycook, J. W.,** Embryo transfer in the non human primate, feline, and canine, *Theriogenology,* 11, 51, 1979.
2. **Kraemer, D. C., Kinney, G. M., and Schriver, M. D.,** Embryo transfer in the domestic canine and feline, *Arch. Androl.,* 5, (Abstr. 104), 111, 1980.
3. **Greulich, W. W.,** Artificially induced ovulation in the cat *(Felis domesticus), Anat. Rec.,* 58, 217, 1934.
4. **Amoroso, E. C.,** Placentation, in *Marshall's Physiology of Reproduction,* Vol. 2, 3rd ed., Parkes, A. S., Ed., Longmans, Green & Co., London, 1952, chap. 15.
5. **Wildt, D. E., Kinney, G. M., and Seagar, S. W. J.,** Gonadotropin induced reproductive cyclicity in the domestic cat, *Lab. Anim. Sci.,* 28, 301, 1978.
6. **Herron, M. A. and Sis, R. F.,** Ovum transport in the cat and the effect of estrogen administration, *Am. J. Vet. Res.,* 35, 1277, 1974.
7. **Hill, J. P. and Tribe, M.,** Early development of the cat *(Felis domesticus), Q. J. Micros. Sci.,* 68, 513, 1924.
8. **Manwell, E. J. and Wickens, P. G.,** The mechanism of ovulation and implantation in the domestic cat, *Anat. Rec.,* 38 (Abstr. 54), 1928.

9. **Hamner, C. E., Jennings, L. L., and Sojka, N. J.**, Cat (*Felis Catus* L.) spermatozoa require capacitation, *J. Reprod. Fert.*, 23, 477, 1970.
10. **Bowen, R. A.**, Fertilization in vitro of feline ova by spermatozoa from the ductus deferens, *Biol. Reprod.*, 17, 144, 1977.
11. **Kinney, G. M., Pennycook, J. W., and Schriver, M. D.**, Surgical collection and transfer of canine embryos, *Biol. Reprod.*, 20 (Suppl. 2) (Abstr. 170), 96A, 1979.
12. **Bischoff, T. L. W.**, *Entwickelungsgeschiechte des Hundeies*, F. Vieweg, Braunschweig, 1845.
13. **Holst, P. A. and Phemister, R. D.**, The prenatal development of the dog: preimplantation events, *Biol. Reprod.*, 5, 194, 1971.
14. **Wildt, D. E., Chakraborty, P. K., Panko, W. B., and Seager, S. W. J.**, Relationship of reproductive behaviour, serum luteinizing hormone and time of ovulation in the bitch, *Biol. Reprod.*, 18, 561, 1978.
15. **Chang, M. C.**, Reciprocal insemination and egg transfer between ferrets and mink, *J. Exp. Zool.*, 168, 49, 1968.
16. **Chang, M. C.**, Development of transferred ferret eggs in relation to the age of corpora lutea, *J. Exp. Zool.*, 171, 459, 1969.
17. **Hammond, J. and Walton, A.**, Notes on ovulation and fertilization in the ferret, *J. Exp. Biol.*, 11, 307, 1934.
18. **Hamilton, W. J.**, The early stages in the development of the ferret fertilization to the formation of the prochordal plate, *Trans. R. Soc. Edin.*, 58, 251, 1933 and 1934.
19. **Zheleznova, A. I. and Golubitsa, A. N.**, Transplantation of blastocysts in the mink, *Dokl. Akad. Nauk S.S.S.R.*, 238, 462, 1978; *Anim. Breed. Abstr.*, 47 (Abstr. 4970), 545, 1979.
20. **Adams, C. E.**, unpublished data, 1975.
21. **Hansson, A.**, The physiology of reproduction in mink (Mustela vison, schreb.) with special reference to delayed implantation, *Acta Zool.*, 28, 1, 1947.
22. **Enders, R. K.**, Reproduction in the mink *(Mustela vison)*, *Proc. Am. Phil. Soc.*, 96, 691, 1952.
23. **Venge, O.**, Reproduction in the fox and mink, *Anim. Breed. Abstr.*, 27, 129, 1959.
24. **Adams, C. E.**, unpublished data, 1974.
25. **Venge, O.**, Reproduction in the mink, *R. Vet. Agric. Univ. Copenhagen*, Yearbook, 95, 1973.
26. **Adams, C. E.**, Observations on the induction of ovulation and expulsion of uterine eggs in the mink, *Mustela vison*, *J. Reprod. Fert.*, 63, 241, 1981.
27. **Sato, A. and Yanagimachi, R.**, Transplantation of preimplantation hamster embryos, *J. Reprod. Fert.*, 30, 329, 1972.
28. **Blaha, G. C.**, Effect of age of the donor and recipient on the development of transferred golden hamster ova, *Anat. Rec.*, 150, 413, 1964.
29. **Orsini, M. W. and Psychoyos, A.**, Implantation of blastocysts transferred into progesterone treated virgin hamsters previously ovariectomized, *J. Reprod. Fert.*, 10, 300, 1965.
30. **Yang, W. H., cited by Chang, M. C. and Pickworth, S.**, Egg transfer in the laboratory animal, in *The Mammalian Oviduct: Comparative Biology and Methodology*, Hafez, E. S. E. and Blandau, R. J., Eds., University of Chicago Press, Chicago, 1969, chap. 16.
31. **Whittingham, D. G. and Bavister, B. D.**, Development of hamster eggs fertilized in vitro or in vivo, *J. Reprod. Fert.*, 38, 489, 1974.
32. **Venable, J. H.**, Preimplantation stages in the golden hamster *(Cricetus auratus)*, *Anat. Rec.*, 94, 105, 1946.
33. **Hamilton, W. J. and Samuel, D. M.**, The early development of the golden hamster *(Cricetus auratus)*, *J. Anat.*, 90, 395, 1956.
34. **Austin, C. R.**, Ovulation, fertilization and early cleavage in the hamster *(Mesocricetus auratus)*, *J. R. Microsc. Soc.*, 75, 141, 1956.
35. **Ward, M. C.**, A study of the oestrous cycle and the breeding of the golden hamster, *Cricetus auratus*, *Anat. Rec.*, 94, 139, 1946.
36. **Blaha, G. C.**, Egg transfer between young and old animals, in *Aging Gametes Their Biology and Pathology*, Blandau, R. J., Ed., S. Karger, Basel, 1975, 219.
37. **Greenwald, G. S.**, Distinction between developing and reserve follicles in the cyclic hamster, *Ann. Biol. Clin. Biochem. Biophys.*, 13, 199, 1973.
38. **Fleming, A. D. and Yanagimachi, R.**, Developmental ability of superovulated hamster ova, *Arch. Androl.*, 5 (Abstr. 76), 83, 1980.
39. **Diamond, M. and Yanagimachi, R.**, Induction of pseudopregnancy in the golden hamster, *J. Reprod. Fert.*, 17, 165, 1968.
40. **Norris, M. L.**, Studies on Reproduction in the Female Gerbil, Ph.D. thesis, National Council for Academic Awards, London, 1979.
41. **Norris, M. L.**, Portable anaesthetic apparatus designed to induce and maintain surgical anaesthesia by methoxyfluorane inhalation in the Mongolian gerbil *(Meriones unguiculatus)*, *Lab. Anim.*, 15, 153, 1981.

42. **Marston, J. H. and Chang, M. C.**, The morphology and timing of fertilization and early cleavage in the Mongolian gerbil and deer mouse, *J. Embryol. Exp. Morphol.*, 15, 169, 1966.
43. **Noske, I. G.**, In vitro fertilization of the Mongolian gerbil egg, *Experientia*, 28, 1348, 1972.
44. **Ishijima, Y., Nonaka, S., Mori, C., and Higa, H.**, Induction of superovulation in the Mongolian gerbil, *Jpn. J. Anim. Reprod.*, 25, 146, 1979.
45. **Fischer, T. V. and Fisher, D. L.**, Effect of gonadotropins on ovulation and ovarian histology in the immature Mongolian gerbil, *Am. J. Anat.*, 142, 391, 1975.
46. **Chang, M. C. and Pickworth, S.**, Egg transfer in the laboratory animal, in *The Mammalian Oviduct: Comparative Biology and Methodology*, Hafez, E. S. E. and Blandau, R. J., Eds., University of Chicago Press, Chicago, 1969, chap. 16.
47. **Briones, H. and Beatty, R. A.**, Interspecific transfers of rodent eggs, *J. Exp. Zool.*, 125, 99, 1954.
48. **Tarkowski, A. K.**, Inter-specific transfers of eggs between rat and mouse, *J. Embryol. Exp. Morphol.*, 10, 476, 1962.
49. **Averill, R. L. W., Adams, C. E., and Rowson, L. E. A.**, Transfer of mammalian ova between species, *Nature (London)*, 176, 167, 1955.
50. **Adams, C. E. and Rowson, L. E. A.**, unpublished data, 1960.
51. **Adams, C. E.**, The influence of maternal environment on preimplantation stages of pregnancy in the rabbit, in *Preimplantation Stages of Pregnancy*, Wolstenholme, G. E. W. and O'Connor, M., Eds., Churchill Livingstone, London, 1965, 345.
52. **Lawson, R. A. S., Rowson, L. E. A., and Adams, C. E.**, The development of cow eggs in the rabbit oviduct and their viability after re-transfer to heifers, *J. Reprod. Fert.*, 28, 313, 1972.
53. **Lawson, R. A. S., Adams, C. E., and Rowson, L. E. A.**, The development of sheep eggs in the rabbit oviduct and their viability after re-transfer to ewes, *J. Reprod. Fert.*, 29, 105, 1972.
54. **Polge, C., Adams, C. E., and Baker, R. D.**, Development and survival of pig embryos in the rabbit oviduct, *Proc. 7th Int. Congr. Anim. Reprod. A.I.*, Munich, 1972, 513.
55. **Brinster, R. L. and Tenbroeck, J. T.**, Blastocyst development of mouse pre-implantation embryos in the rabbit fallopian tube, *J. Reprod. Fert.*, 9, 417, 1969.
56. **Pincus, G.**, *The Eggs of Mammals*, Macmillan, New York, 1936, 89.
57. **Potts, D. M., Wilson, I. B., and Smith, M. S. R.**, Observations on rat eggs transplanted to the mouse uterus, *J. Reprod. Fert.*, 22, 425, 1970.
58. **Yoshinaga, K. and Adams, C. E.**, Reciprocal transfer of blastocysts between rat and rabbit, *J. Reprod. Fert.*, 14, 325, 1967.
59. **Blaha, G. C. and De Feo, V. J.**, Interspecies ova transfer between hamsters and rats, *Anat. Rec.*, 148 (Abstr.), 261, 1964.
60. **Yang, W. H., cited by Chang, M. C. and Pickworth, S.**, Egg transfer in the laboratory animal, in *The Mammalian Oviduct: Comparative Biology and Methodology*, Hafez, E. S. E. and Blandau, R. J., Eds., University of Chicago Press, Chicago, 1969, chap. 16.
61. **Chang, M. C.**, Reciprocal transplantation of eggs between rabbit and ferret, *J. Exp. Zool.*, 161, 297, 1966.
62. **Hafez, E. S. E. and Sugie, T.**, Reciprocal transfer of cattle and rabbit embryos, *J. Anim. Sci.*, 22, 30, 1963.
63. **Chang, M. C.**, Artificial insemination of snowshoe hares *(Lepus americanis)* and the transfer of their fertilized eggs to the rabbit *(Oryctolagus cuniculus)*, *J. Reprod. Fert.*, 10, 447, 1965.
64. **Chang, M. C., Casas, J. H., and Hunt, D. M.**, Development of ferret eggs after 2 to 3 days in the rabbit fallopian tube, *J. Reprod. Fert.*, 25, 129, 1971.
65. **Umbaugh, R. E.**, Synchronization and ovum transfer in cattle, *Am. J. Vet. Res.*, 10, 295, 1949.
66. **Trounson, A. O., Willadsen, S. M., and Rowson, L. E. A.**, Fertilization and development capability of bovine follicular oocytes matured in vitro and in vivo and transferred to the oviducts of rabbits and cows, *J. Reprod. Fert.*, 51, 321, 1977.
67. **Warwick, B. L. and Berry, R. O.**, Inter-generic and intra-specific embryo transfers, *J. Hered.*, 40, 297, 1949.
68. **Hancock, J. L. and McGovern, P. T.**, Transfer of goat x sheep hybrid eggs to sheep and reciprocal transfer of eggs between sheep and goats, *Res. Vet. Sci.*, 9, 411, 1968.
69. **Hunter, G. L., Bishop, G. P., Adams, C. E., and Rowson, L. E. A.**, Successful long-distance aerial transport of fertilized sheep ova, *J. Reprod. Fert.*, 3, 33, 1962.
70. **Lopyrin, A. I., Loginova, N. V., and Karpov, P. L.**, The effect of changed conditions during embryogenesis on growth and development of lambs, *Sov. Zootekh.*, 6, 83, 1951; cited in *Anim. Breed. Abstr.*, 20 (Abstr. 729), 1952.
71. **Kuehl, T. J., Ariga, S., and Dukelow, W. R.**, The development of squirrel monkey and mouse zygotes following transfer to the rabbit oviduct, *Fed. Proc.*, 36 (Abstr. 544), 372, 1977.
72. **Rowson, L. E. A.**, Prolonged storage of gametes in relation to fertility and pregnancy characteristics in farm animals, in *Aging Gametes Their Biology and Pathology*, Blandau, R. J., Ed., S. Karger, Basel, 1975, 249.

73. **Nicholas, J. S.**, Experiments on developing rats. I. Limits of foetal regeneration; behaviour of embryonic material in abnormal environments, *Anat. Rec.*, 58, 387, 1934.
74. **Fawcett, D. W., Wislocki, G. B., and Waldo, C. M.**, The development of mouse ova in the anterior chamber of the eye and in the abdominal cavity, *Am. J. Anat.*, 81, 413, 1947.
75. **Jones, I. C.**, Endocrine effects of the virgin female host of fertilized mouse ova developing in extra uterine sites, *J. Endocrinol.*, 7, 63, 1951.
76. **McLaren, A. and Tarkowski, A. K.**, Implantation of mouse eggs in the peritoneal cavity, *J. Reprod. Fert.*, 6, 385, 1963.
77. **Runner, M. N.**, Development of mouse eggs in the anterior chamber of the eye, *Anat. Rec.*, 98, 1, 1947.
78. **Shapiro, H. and Harvey, T. S.**, Ocular implantation of the mammalian ovum, *Proc. Am. Philos. Soc.*, 101, 164, 1957.
79. **Ahlgren, M. and Bengtsson, L. Ph.**, Transplantation of fertilized rabbit eggs to the anterior chamber of the eye, *J. Reprod. Fert.*, 3, 89, 1962.
80. **Fawcett, D. W.**, The development of mouse ova under the capsule of the kidney, *Anat. Rec.*, 108, 71, 1950.
81. **Kirby, D. R. S.**, Development of mouse eggs beneath the kidney capsule, *Nature (London)*, 187, 707, 1960.
82. **Kirby, D. R. S.**, Reciprocal transfer of blastocysts between rats and mice, *Nature (London)*, 194, 785, 1962.
83. **Kirby, D. R. S.**, The influence of the uterine environment on the development of mouse eggs, *J. Embryol. Exp. Morphol.*, 10, 496, 1962.
84. **Kirby, D. R. S.**, Development of the mouse blastocyst transplanted to the spleen, *J. Reprod. Fert.*, 5, 1, 1963.
85. **Kirby, D. R. S.**, The development of mouse blastocysts transplanted to the scrotal and cryptorchid testis, *J. Anat. London*, 97, 119, 1963.
86. **Kirby, D. R. S.**, The extra-uterine mouse egg as an experimental model, in *Advances in the Biosciences 4*, Raspé, G., Ed., Pergamon Press, Oxford, 1970, 255.
87. **Bland, K. P. and Donovan, B. T.**, Experimental ectopic implantation of eggs and early embryos in guinea pigs, *J. Reprod. Fert.*, 10, 189, 1965.
88. **Kirby, D. R. S.**, Endocrinological effects of experimentally induced extra-uterine pregnancies in virgin mice, *J. Reprod. Fert.*, 10, 403, 1965.
89. **Kirby, D. R. S.**, The difference in response by the mouse adrenal X zone to trophoblast derived from transplanted tubal eggs and uterine blastocysts, *J. Endocrinol.*, 36, 85, 1966.
90. **Simmons, R. L. and Russell, P. S.**, The antigenicity of mouse trophoblast, *Ann. N.Y. Acad. Sci.*, 99, 717, 1962.
91. **Simmons, R. L. and Russell, P. S.**, Histocompatibility antigens in transplanted mouse eggs, *Nature (London)*, 208, 698, 1965.
92. **Kirby, D. R. S., Billington, W. D., and James, D. A.**, Transplantation of eggs to the kidney and uterus of immunised mice, *Transplantation*, 4, 713, 1966.
93. **Kirby, D. R. S.**, The immunological consequences of extra uterine development of allogeneic mouse blastocysts, *Transplantation*, 6, 1005, 1968.
94. **Mayer, G. and Dulac, A.-J.**, Homotransplantation et heterotransplantation intratesticulaires chez le rat et la souris de blastocystes en phase de preimplantation et de lethargie, *C. R. Acad. Sci. (Paris)*, 263, 1111, 1966.
95. **Kirby, D. R. S.**, The role of the uterus in the early stages of mouse development, in *Preimplantation Stages of Pregnancy*, Wolstenholme, G. E. W. and O'Connor, M., Eds., Churchill Livingstone, London, 1965, 325.
96. **Billington, W. D., Graham, C. F., and McLaren, A.**, Extra-uterine development of mouse blastocysts cultured in vitro from early cleavage stages, *J. Embryol. Exp. Morphol.*, 20, 391, 1968.
97. **Adams, C. E.**, Bibliography with review on ectopic pregnancy. Literature 1963 to 1976, bibliography no. 102, *Bibl. Reprod.*, 30, 97, 1977.
98. **Adams, C. E.**, Egg development in the rabbit; the influence of post coital ligation of the uterine tube and of ovariectomy, *J. Endocrinol.*, 16, 283, 1958.
99. **Wintenberger-Torres, S.**, Les rapports entre l'oeuf en segmentation et le tractus maternel chez la brebis, *Proc. 2nd Int. Congr. Anim. Reprod. A. I.*, Cambridge, Sect. 1, 62, 1956.
100. **Anon.**, Final general discussion, in Maternal Recognition of Pregnancy, *Excerpta Medica*, Amsterdam, 1979, 402.

Chapter 5

# SYNCHRONIZATION OF ESTRUS AND SUPEROVULATION IN CATTLE

I. Gordon

## TABLE OF CONTENTS

# I. SYNCHRONIZATION OF ESTRUS

## A. Introduction

Embryo transfer studies in the farm mammals have clearly shown that the survival of embryos depends on their stage of development corresponding closely with the developmental stage of the recipient's uterus. However, much remains in determining the precise nature of many factors involved in the survival of a bovine embryo after transfer; it may not always be a question of simply providing exact synchrony of estrus between donor and recipient. In the early days of surgical embryo transfer, recipient cattle were almost invariably selected on the basis of a naturally occurring estrus. More recently, however, selection of recipients has been widened to include cattle that have their estrous cycles controlled by hormonal means. The synchronizing agents employed must clearly not result in an adverse effect on pregnancy rates in such recipients.

Experiments specifically aimed at elucidating the importance of exact synchrony between donor and recipient to the successful establishment of pregnancy in cattle have been rather few; a considerable body of less controlled evidence has accumulated from the efforts of those engaged in the commercial application of embryo transfer. Taking as our starting point the first successful establishment of cattle pregnancies by surgical embryo transfers, Rowson et al.[1] showed that exact synchrony resulted in the highest pregnancy rate and that a degree of variation between donor and recipient cycles of ±2 days could be tolerated; the start of estrus in donor and recipient is the reference point, although it might be better matching on the basis of the end of heat rather than the beginning, in view of the fact that it is ovulation that is the crucial event requiring "alignment". The Cambridge workers subsequently provided evidence of substantial decreases in pregnancy rates in recipients for each day they were out of exact synchrony with donors over the range ±3 days; there were strong indications in these data that synchronization requirements for successful embryo transfer in cattle may be more acute than in the sheep and should not vary by more than ±1 day.[2] Although this work showed evidence of a substantial reduction in pregnancy rate between recipients at exact synchrony and those at ±1 day (91% vs. 52 to 57%), there was an exceptionally and unusually high proportion of pregnancies recorded at exact synchrony. The contemporary work of Sugie et al.[3] in Japan showed little difference in pregnancy rates between cattle at exact synchrony and those out-of-phase by ±1 day, although their rates were on a much lower level than those of the Cambridge workers. In a further report in which the Cambridge group reported a less exceptional pregnancy rate at exact synchrony, differences between exact synchrony and ±1 day out-of-phase proved no more than about 10%.[4]

Workers in Ireland, where high pregnancy rates were obtained by surgical transfers, reported a higher proportion of viable embryos in exactly synchronized recipients than in those ±1 day out-of-phase.[5,6] Although limited evidence by Trounson et al.[7] reported recipients ±1 day out-of-phase to have higher pregnancy rates when the donor was more advanced than the recipient, the extensive data of Shea et al.[8] showed a 63% pregnancy rate after exact synchrony in 1126 recipients, 60% pregnancy rate in 334 recipients in estrus the day before the donors (−1), and 49% pregnancies in 556 recipients exhibiting estrus the day after the donors (+1). Further evidence along these lines is in the large-scale data generated in commercial embryo transfer operations by Schneider et al.[9] in Texas; on the basis of 2556 transfers, recipient cows at exact synchrony showed a pregnancy rate of 67%, with a similar figure (66%) in recipients showing estrus 12 hr earlier than donors, but a significantly lower proportion (61%) among recipients in estrus 12 hr after donors. These authors note that in large-scale transfer programs, recipients that come in estrus 12 hr after a donor are suitable for the next donor to be used; selection of recipients is therefore in favor of animals in estrus at the same time or before the donor. There is some evidence about the possible merit of asynchrony in the report on nonsurgical transfers in Brahman cattle of Halley et al.;[10] recipients in estrus 1 day earlier than donor animals had a significantly higher pregnancy rate (42%) than those at exact synchrony (26%). However, these authors note possible metabolic and developmental differences between the embryos of *Bos taurus* and *Bos indicus* cattle species which may explain such results.

On the basis of present evidence for *Bos taurus* cattle, the aim for those engaged in embryo transfer should be exact synchrony between donor and recipient animal. With increasing knowledge and expertise in the freezing of cattle embryos, it is now possible to arrange exact synchrony on the basis of thawing the frozen embryo and carrying out the transfer when the recipient reaches the appropriate stage of the estrous cycle. In the early surgical transfer work in cattle, when early cleavage stage embryos (3- to 4-day-old) were employed, the hazards of asynchrony were likely to be much greater than they currently are when 7-day embryos and older are commonly employed in transfers. Early cleavage-stage embryos are known to be more susceptible to adverse effects, whether this takes the form of reduced temperatures[11] or premature exposure of the early embryo to the uterine environment.[4] Newcomb and Rowson[4] record that the transfer of day-4 embryos to the uterus of day-3 recipients resulted in a significantly higher pregnancy rate than when the day-3 donor was synchronous with the day-3 recipient.

Much remains to be examined in the synchronization of donor and recipient cattle, which may yield information contributing towards optimal conception rates after embryo transfer. It may be that embryos at certain stages of their development or after certain culture treatments tolerate some degree of asynchrony more readily than others. With the increasing scale of commercial cattle embryo transfer operations, especially in the U.S., there is ample opportunity to examine many of these points from the figures that become available.

### B. Synchronization of Estrus Using Progestins

There are numerous reports dealing with the hormonal control of estrus in cattle, dating from the use of injections of progesterone by Casida's group at Wisconsin more than 30 years ago.[12] Emphasis and interest during the 1960s centered almost entirely around the orally active progestagens, especially medroxy progesterone acetate (MAP), chlormadinone acetate (CAP), and melengestrol acetate (MGA); treatment at the appropriate dose level for a period approximating to one cycle interval (18 to 21 days) was shown to be effective in controlling estrus, so that a majority of animals showed heat within the space of 3 days. Apart from the fact that the response to the oral treatments could be described more accurately as "grouping" of estrus rather than "synchronization", first service conception rate was generally found to be unacceptably low. Extensive data from studies conducted over a 5-year period with MGA, for example, showed acceptable results in "grouping" heat periods, but a conception rate that was only 70% of that obtained in the controls.[13]

Towards the end of the 1960s a new approach to the use of progesterone and progestagens, administered by one or other of several routes, became available when Wiltbank and Kasson[14] reported successful estrus control after a relatively short (9-day) oral progestagen regime combined with an estrogen injection on day 2 of treatment. Previously, studies had shown that regression of the mature bovine corpus luteum can be induced readily by the administration of estradiol.[15,16] Estradiol administered at the start of progestagen treatment was also shown to result in the enhancement of corpus luteum regression.[17] The lack of immediate luteal regression after estradiol is taken to indicate that estrogens act through an intermediate target, presumably the uterus, rather than directly on the corpus luteum. As well as achieving satisfactory control of estrus, short-term progestagen treatments were shown to result in a first service conception rate comparable to that in control animals. When this short-term progestagen treatment was initiated in the early luteal phase of the estrous cycle, however, it was evident from several studies that heats in cattle were less well synchronized than when treatments commenced later.[18,19] Fortunately, Wiltbank and Gonzalez-Padilla[20] were able to show that this could be overcome by modifying treatments to bring the cyclic corpus luteum under closer control; the use of a combined progestagen/estrogen injection rather than estrogen alone at the time of commencing progestagen treatment permits adequate control in those cattle that are in the late stages of the estrous cycle or have recently ovulated (a consequence of the antiluteotrophic property of substantial progestagen dosage).

In explaining why short-term (9 days) rather than long-term (18 days) progestagen treatments can result in normal fertility at the controlled estrus, there is some evidence that short-term progestagen treatments may not result in the same degree of estrogenic activity as is associated with long-term progestagen treatment.[21,22] Adverse estrogenic effects could involve changes in cervical mucus and in normal sperm and egg transport.

### C. Methods in Applying Short-Term Progestagen Treatments

There are three main ways in which the short-term progestagen treatment may be applied to potential recipient cattle; these are by intravaginal sponge pessary, by the coil, or by the ear implant. A substantial amount of research in Ireland over the years has been with the progestagen-impregnated intravaginal sponge pessary, and it is clear that a high degree of synchronization and normal fertility can follow short-term (9 to 10 days) intravaginal treatment in conjunction with a progestagen/estrogen dose at the time of initiating the application.[23-27] At the present time, this form of intravaginal sponge treatment, employing doses of 3 g progesterone or 200 mg fluorogestone acetate (FGA), although highly effective, is not commercially available. There is, however, an alternative intravaginal treatment, the Progesterone-Releasing Intravaginal Device (PRID) produced by Abbott Laboratories and now commercially available in several countries. The PRID was first reported on as an effective method for administering progesterone by Mauer et al.,[28] the device consisting of a metal spiral coated with progesterone impregnated silicone rubber. The development of the device as a satisfactory short-term treatment has been largely due to the efforts of Roche[29-32] in Ireland. Currently recommended procedure with the PRID is to insert it with an attached estrogen capsule (estradiol benzoate) and leave it in the vagina for 12 days. In contrast to what is required with the intravaginal sponge, additional progesterone or progestagen is not necessary at the start of treatment, because of the initial high rate of release of progesterone from the PRID.[33]

The third method involves the use of a subcutaneous implant, impregnated with the highly potent progestagen SC21009 (Norgestamet®), which is inserted in the ear for 9 days, the start of treatment being preceded by an injected dose of progestagen and estrogen (usually 5 mg estradiol valerate and 3 mg Norgestamet®). Removal of the implant is followed by a satisfactory estrous response and normal fertility.[34-36] With this form of treatment, as when using intravaginal sponges and coils, the majority of cattle can be expected to exhibit estrus 24 to 48 hr after removal of the implant. The estrous response in one study reported by Wishart[37] was 83% in the 350 heifers observed closely every 4 hr. Regardless of the method used in applying short-term progestagen treatment, the precision of the estrous response in cattle and the proportion showing heat are generally lower than in sheep, in which an estrous response in excess of 95% may be anticipated among cyclic sheep after progestagen treatment.[38]

In sheep, it has been established for some time that pregnancy rate to embryo transfer is not adversely affected by progestagen employed in the synchronization of recipients,[39-41] and similar data have been provided for cattle by Sreenan et al.[6] and Sreenan and Beehan.[42]

### D. Synchronization of Estrus Using a Prostaglandin

An alternative to the short-term progestagen treatment is available in the form of prostaglandin (PG)$F_2$ alpha. Administration by intramuscular injection of a luteolytic dose of $PGF_{2\alpha}$ (20 to 30 mg) or one of the several synthetic analogues, e.g., 500 mcg of cloprostenol (I.C.I. Ltd.), will induce regression of the corpus luteum in heifers between days 5 and 16 after estrus.[43-48] When the intrauterine mode of administration is used rather than the intramuscular, substantially lower dose levels of PG may be employed,[49] but the procedure is tedious and little used for that reason. The fact that prostaglandin is ineffective in cattle between estrus and day 5 of the cycle means that a single-dose treatment is only applicable among cattle at a known stage of the cycle. When dealing with animals at unknown stages, two injections (10 to 12 days apart) are required. The onset of estrus, relative to treatment withdrawal, is less rapid with PG than with progestagen due to the fact that regression of the corpus luteum must first occur; the onset of estrus can be expected at 48 to 72 hr after PG, rather than the 24 to 48 hr usual for progestagens. In cyclic heifer cattle, normal fertility levels can be expected after prostaglandin treatment, and those who have used this as the synchronizing agent in their recipients have recorded a pregnancy rate no different from that in heifer cattle showing a natural estrus.[6,49] In view of evidence provided in several reports that the endocrine changes following PG injection are essentially similar to those occurring in the heifer around the time of normal estrus, there is no reason why pregnancy rates in PG-treated recipients should be adversely affected.

### E. Progestagen in Combination with Prostaglandin $F_{2\alpha}$

One way of dealing with the luteal period in which the *corpus luteum* is insensitive to $PGF_{2\alpha}$ is to inhibit estrus and ovulation for at least 5 days before administering the dose of prostaglandin; various authors have reported on this possibility using either the Norgestamet® implant or the PRID for a 7-day period and administering a luteolytic dose of $PGF_{2\alpha}$ 24 hr before or at the time of progestagen withdrawal.[50-52] Using such combinations avoids the need for estrogen or estrogen/progestagen doses at the commencement of the short progestagen treatment.

### F. General Considerations in Estrus Synchronization Programs

Quite apart from what hormonal means are employed to achieve estrus synchronization in recipient cattle, it must be recognized that the pregnancy rates obtained in such animals may be substantially affected by their nutritional status and general body condition. Studies on the effect of body condition and nutrition on conception rate at first service have provided plenty of evidence that fertility can be influenced by the level of feeding provided over the service period.[27,35,53] When compared with the usual farm ration, supplementation to provide an additional 20 M/J metabolizable energy/day over the service period increased calving rates to heat periods synchronized by prostaglandin or short-term progestagen treatments from 50.0% to 68.9% ($p<0.0001$) in heifer cattle.[54] The present commercial application of estrus synchronization procedures (prostaglandin and PRID) will probably serve to focus attention more sharply on the effect of nutritional and other forms of stress on first service conception rates in cattle; hopefully, this may accelerate the development of management routines that can be employed among recipient cattle, whether they are synchronized or not, to achieve optimal pregnancy rates after embryo transfer.

### G. Other Possibilities

An alternative to using hormonally-induced synchronization, and one which has been common among commercial cattle embryo transfer units, is to keep a sufficiently large herd of potential recipients on hand so that at any one time the required number can be found whose heat period closely coincides with that of the donor. The usual rule is to have 10 to 20 potential recipients available and in synchrony with each donor cow.[55] Seidel and Seidel[55] draw attention to the fact that the major cost in operating a cattle embryo transfer unit is in maintaining hundreds of potential recipients in prime reproductive condition as well as the recipients that have already received embryos and are waiting until perhaps 90 days of gestation before finally being dispatched to the customer. Currently, some units in the U.S. operate with up to 2000 recipients on hand at all times; other units are mobile and conduct their embryo recoveries and transfers wholly on the farm.

There are obvious practical difficulties for cattle embryo transfer units in recruiting and maintaining large groups of recipients in prime reproductive condition. Quite apart from ensuring that recipients are of acceptable quality and that their feeding and management are of a sufficiently high order, it is possible that cattle held in such large groups may not always be capable of functioning at full reproductive efficiency. Williams[56] drew attention some years ago to a temporary form of subfertility which can occur in cattle grouped together in unfamiliar surroundings.

Although the availability of frozen cattle embryos will permit exact synchrony between donor and recipient to be readily obtained in future years, there may be limited scope for use of an alternative procedure involving embryos not stored at freezing temperatures. Cooling below 15°C can be employed to halt mitosis and development of the bovine embryo.[57] When the recipient reaches the luteal stage corresponding to the developmental stage of the embryo, it is possible that the embryo could be brought to body temperature and transferred.

## II. SUPEROVULATION IN CATTLE

### A. Introduction

Earliest reports on superovulation in cattle came from the Wisconsin group;[58,59] subsequently, the induction of a high superovulatory response in sexually mature heifers and cows using pituitary and placental gonadotrophins has been investigated by many workers employing a wide variety of techniques. Much of this work has already been covered in several comprehensive reviews.[60-62] The aim in most cattle superovulation treatments is to administer an appropriate FSH-type preparation several days in advance of estrus and ovulation in doses sufficient to grow 10 to 20 or more follicles to maturity. Although it is possible to stimulate the release of well in excess of 100 eggs in some individual cattle with exogenous gonadotrophins, problems in recovery and fertilization generally mean that the average number of useful embryos may be less than ten. In an early survey of commercial cattle transfer clinics in North America,[63] a figure of less than four fertilized eggs per donor treated was quoted; some current reports[9,62,64] would put this figure at eight rather than four, but the fact remains that the yield of fertilized eggs per donor is one of the main limiting factors in cattle embryo transfer. Part of the difficulty has been the fact that quantitative information on many of the hormonal factors involved in the bovine estrous cycle and in follicular growth has been severely limited. This situation is being rectified with the widespread use of radioimmunoassays for steroid and polypeptide hormones, but for those attempting to superovulate cattle, there is particular need for information on the factors determining ovarian response to exogenous gonadotrophins. Factors affecting the complement of oocytes within the available vesicular (Graafian) follicles are of special interest in superovulation work, since it is only the oocytes in these structures that are able to undergo ovulation. Available information shows the number of vesicular follicles to remain fairly constant in cattle from 2 months of age until 8 to 10 years, after which it slowly declines.[65] Most of these follicles are lost as a result of atresia, and some understanding of the complex cellular interactions that exist within the specialized microenvironment of each follicle is only now beginning to emerge;[66] investigators are still some way from understanding why 99% of vesicular follicles regress in the normal course of events.

### B. Gonadotrophins Used in Superovulation

The ability of pregnant mare serum gonadotrophin (PMSG) to induce superovulation has led to the widespread use of this placental gonadotrophin in cattle embryo transfer work. The outcome of many investigations shows that the larger the dose of PMSG the greater the superovulatory response and the greater the variability in the number of ovulations.[67-72] For use in commercial applications, PMSG has the virtue of being available in quantity and at low cost (relative to FSH and LH preparations); administering the agent in multiple injections does not enhance the superovulatory response beyond that found with a single dose.[70] There is evidence to show that PMSG is much less easily inactivated in the body than pituitary FSH preparations and that measurable amounts are still present in the cow 10 days after injection.[73]

Pituitary extracts of FSH were employed in many early cattle superovulation studies, and reports of comparative studies showed them to be as effective or more so than PMSG in producing yields of fertilized eggs.[67] However, cost and availability are considerations that are much in favor of PMSG. A crude horse pituitary extract of FSH (HAP) was employed by Moore[74] in comparative studies with PMSG, but no clear difference in ovarian response was recorded. Commercially available FSH and LH preparations were used by Seidel et al.[75] and Elsden et al.[76] in work showing that a FSH-LH regimen was clearly superior to a PMSG regimen in terms of the yield of "transferable" embryos. For the repeated superovulation of donor cattle, there may be some advantage in alternating treatment between PMSG and other gonadotrophic preparations. Using human menopausal gonadotrophin (hMG) in a comparison with PMSG, Newcomb[77] found response after a second superovulation treatment to be comparable (3000 i.u. PMSG, $11.1 \pm 2.0$; hMG, $10.6 \pm 1.7$), and notes that this gonadotrophin may be a useful alternative treatment on occasions for inducing multiple ovulations in cattle.

Regardless of the source of the FSH preparation, several early studies on superovulation in cattle showed that the timing of gonadotrophin administration, relative to the stage of the estrous cycle, could be critical. Attempts to employ FSH in the main luteal phase were seldom effective and the greatest success in inducing multiple ovulations and obtaining an acceptable yield of "transferable" embryos followed application of gonadotrophin treatment in the follicular phase (from day 16 onwards). A certain measure of control over the bovine estrous cycle in early superovulation studies was achieved by enucleation of the *corpus luteum*,[67,78] but there was evidence that the fertilization of eggs released after PMSG treatment could be influenced by the completeness or otherwise of the enucleation;[70] much more effective than enucleation is the use of $PGF_{2\alpha}$ (or one of its analogues) to induce luteolysis.

### C. Prostaglandin in Combination with Gonadotrophin

Part of the variation in the response of cattle to a PMSG superovulation regimen in early studies was probably due to an inability to anticipate with sufficient accuracy the time at which estrus would occur after gonadotrophin treatment, regardless of whether the preparation was given in the follicular phase of the cycle or after expressing the corpus luteum by manipulation *per rectum.* The solution to this particular difficulty came with the finding that $PGF_{2\alpha}$ is luteolytic in cattle. Currently, the most common method of superovulating cattle is a single dose of 2000 to 2500 i.u. PMSG during the midluteal phase of the estrous cycle, followed 48 to 72 hr later by a luteolytic dose of $PGF_{2\alpha}$ (e.g., 30 mg Lutalyse®; 0.5 to 1.0 mg cloprostenol).[6,62,64,75] With prostaglandin given at 48 hr, donors can be expected to exhibit estrus two days later, thus providing an interval of 4 days between gonadotrophin injection and the start of estrus. In aligning donor and recipient cattle, it is recognized that the interval between PG and estrus is substantially shortened in PMSG-treated cattle, estrus starting on the 2nd rather than the 3rd day after the prostaglandin.[79] Evidence that the day of administering the PMSG may be critical, with lower ovulatory responses being obtained when gonadotrophin is given in the early luteal phase (days 3 to 8) than in the midluteal stage (days 9 to 12), is documented in several studies.[62,80-82] Regardless of whether FSH or PMSG is employed, the interval between gonadotrophin administration and ovulation can markedly influence the superovulatory response.[71] Follicles in the bovine ovary may require 4 or 5 days to mature; the use of $PGF_{2\alpha}$ and other synchronizing agents can be the means of adjusting the PMSG to estrus interval, thus exercising a degree of control over the number of follicles which ovulate. Newcomb and Rowson[81] found that decreasing the PMSG to estrus interval from 4 to 3 days resulted in a statistically significant decline in the superovulatory response.

### D. Dose Level and Batch of PMSG

Dose levels of PMSG appropriate for inducing high levels of superovulation were examined in the early years of cattle embryo transfer research; the general conclusion reached was that 3000 i.u. was about the upper limit;[70,83] in practice, most superovulation regimens in which PMSG is used employ doses of 2000 to 2500 i.u.,[62] although there may be justification on occasions to go to 3000 i.u.[77] Apart from considerations of dose level, reports in the literature have occasionally suggested that the superovulatory effect of PMSG may vary with the particular batch employed.[84-86] The PMSG molecule possesses both FSH and LH properties, and this has led some to suggest that the ratio of these components in different batches of PMSG may influence ovarian response.[72] However, batches of PMSG have now been analyzed for their FSH and LH activity by various groups without significant differences being evident,[73,87] and controlled comparisons of batches used in superovulation have also failed to reveal evidence of significant differences in ovarian response.[61,88,89]

### E. Progestagen, Estrogen, and Ovulating Hormones in Cattle Superovulation

An alternative to relying on natural or prostaglandin-induced luteolysis occurring a few days after PMSG administration in cattle superovulation treatments is to time the gonadotrophin injection to precede the withdrawal of an appropriate progestagen treatment. Several investigators have reported on this approach using various forms of progesterone or progestagen pretreatment. Scanlon[90] employed progesterone-impregnated intravaginal sponges, inserted on day 13 of the cycle and withdrawn on day 20, 4 days after PMSG administration; the intravaginal progesterone treatment resulted in some increase in the proportion of follicles ovulating, compared with the response of heifers receiving PMSG on day 16, but the yield of fertilized eggs was not improved. Testart et al.[91] used Norgestamet® implants, inserted during the luteal phase of the cycle, with an estrogen injection to accelerate the demise of the corpus luteum; implants were removed after 8 or 10 days and PMSG administered 2 days before their withdrawal. Moore[74] employed injections of progesterone (40 mg/day) for periods of 17 to 30 days, with PMSG administered on the day prior to the last injection. Probably the most comprehensive work is that reported by Sreenan et al.,[92] who compared prostaglandin, Norgestamet® implants, and intravaginal sponge pessaries (impregnated with progesterone or cronolone) as the means of controlling estrus in donors being treated with PMSG, the gonadotrophin being given 2 days before the end of progestagen treatment. Although mean ovulation and fertilization rates were similar, regardless of the synchronizing agent employed, the estrous response was higher and the ovulatory response less variable with the PMSG-PG regime than with the progestagen treatments. Available evidence suggests that progestagen as the synchronizing agent offers no advantage over prostaglandin.

Estrogen has been used in conjunction with gonadotrophins in superovulation studies. Hafez et al.[93] administered estradiol before or after PMSG in beef heifers; their most promising regimen, yielding an average of 11.3 fertilized eggs from eight donors, involved 3000 i.u. PMSG on day 16, estradiol on days 19 and 20, and 2000 i.u. human chorionic gonadotrophin (hCG) on day 21. With modifications, this treatment has been reported on by Sugie et al.[94] and employed by Scanlon[90] and Gordon[95] to produce average yields of fertilized eggs comparable to those achieved with other regimens, including that in which $PGF_{2\alpha}$ is employed in combination with the PMSG.[96] Although it is now evident that hCG administered at the onset of estrus in donor cattle does not increase ovulatory response to PMSG by reducing the incidence of unovulated follicles,[74,77,96] there is a suggestion of a favorable effect on the yield of embryos recovered, which may indicate an influence on the transport or retention of embryos.[77] The use of a gonadotrophin-releasing hormone (Gn-RH) preparation at the time of estrus in donors has been reported on to a limited extent and without evidence of a response to a single injection treatment.[77]

### F. Breed and Age Differences

Evidence exists to show that cattle breeds, even strains within a breed, can differ markedly in sensitivity to a given dose level of PMSG. As shown in Table 1, data from several studies show a tendency for Friesians to give lower yields of eggs than some beef breeds, at the same PMSG dose levels. Work by the Cambridge group in Friesian heifers and lactating cows indicates that this breed will respond satisfactorily to PMSG if given at an appropriate dosage;[77,100] their data suggest that for Friesians 2000 i.u. PMSG is probably the minimum dose level required for a satisfactory superovulatory response.

Moore[74] examined response to superovulation regimes in relation to age and found a greater ovulatory response in heifers than in cows. It might be expected that there would be a decline in ovarian response with increasing age due to the effects of follicular atresia. In rabbits, decreased litter size in aged animals is paralleled by a decline in ovulatory response to a standard dose of gonadotrophin.[101] Although similar evidence is not available for cattle, Erickson[65] did report a general decrease in the number of vesicular follicles from the 4th year of life to a virtual absence in some cows at 5 to 20 years of age.

The physiological status of cattle, whether they are "dry" or lactating or perhaps even nursing calves, may be expected to influence superovulatory responses, but as yet comparative data on which to assess these effects are not available.

### G. Seasonal and Nutritional Effects

In Finland, Rajakoski[102] found vesicular follicles in cattle to be more numerous in the winter and spring periods than in the summer and autumn; in South Africa[103] and in North America[104] marked seasonal differences in the response of cattle to PMSG have been reported. The reason for such differences may not be readily apparent in view of the complex of nutritional and environmental factors which may be associated with a particular season.

There is clear evidence that fasting the cow in the period immediately after PMSG treatment can markedly reduce superovulatory response.[72] The effect of undernutrition

**Table 1**
**SUPEROVULATORY RESPONSE IN DIFFERENT BREEDS**

| PMSG dose (i.u.) | Breed | No. cattle | Mean ovulation rate | Ref. |
|---|---|---|---|---|
| 3000 | Friesian | 31 | 7.3 | 90 |
| | Hereford/Angus | 203 | 12.2 | |
| 3000 | Friesian | 18 | 13.7 | 97 |
| | Hereford/Angus | 34 | 32.0 | |
| 1600 | Friesian | 78 | 2.6 | 98 |
| | Charolais | 52 | 6.2 | |
| 1600 | Friesian | 15 | 2.5 | 99 |
| | Charolais | 17 | 6.2 | |
| 2000 | Friesian | 16 | 5.0 | |
| | Charolais | 14 | 10.4 | |

on the development of vesicular follicles in cattle has been reported on by Hill et al.;[105] low plane feeding showed no evident effect on vesicular follicles up to day 15 of the cycle, but beyond that time it did influence the number of large follicles. Those engaged in cattle embryo transfer work are unlikely to be dealing with donors other than those on an adequate level of nutrition, but there could be occasions when cattle are in poor body condition because of disease, injury, or old age. There may be subtle nutritional effects operating under some conditions, not necessarily exerting an influence on the ovulatory response but perhaps on fertilization rates; Segerson et al.,[106] using PMSG in beef cattle, found evidence of significant beneficial effects on fertilization rates in animals receiving supplemental selenium/vitamin E and maintained on an adequate plane of nutrition.

### H. Ovarian Follicle Population

Some evidence in cattle indicates that the superovulatory response after gonadotrophin treatment may be related to the number of follicles in the ovaries.[98] It is well established that the number of primordial follicles and the number, size, and viability of vesicular follicles can vary markedly among cattle.[65,107,108] In studies with beef heifers, Scanlon[90] found ovulatory response to be markedly influenced by the vesicular follicle content of the donors' ovaries.

It would be of considerable value to learn much more about hormonal and other factors influencing the vesicular follicle population, both long term and short term; it may eventually prove possible to devise an evaluation procedure (hormonal or otherwise) by which the superovulatory response of a donor animal can be accurately predicted.

### I. Breeding Superovulated Donor Cattle

After completing the hormonal treatment for the induction of superovulation, donor cattle are checked for estrus and inseminated. Regardless of the treatment regimen, 10% or more of donor animals may not exhibit estrus.[61] Although in early superovulation studies it tended to be assumed that normal breeding and insemination procedures were adequate to effect fertilization in donor animals, the current widespread use of $PGF_{2\alpha}$ among embryo transfer units to control the timing of estrus permits more precise recommendations to be made. McKenzie and Kenney[109] reported that donors inseminated with fresh semen yielded a higher proportion of fertilized eggs than those inseminated with frozen semen (67% vs. 20%), but subsequent reports clearly show that it is possible to obtain satisfactory results with frozen semen. Controlling the occurrence of estrus in donors with prostaglandin does permit standard insemination procedures to be adopted. Donors can be expected in estrus about 2 days after PG, and it is usual to inseminate in the morning and again in the evening if they exhibit heat early on the day; if heat occurs later in the day, they would be inseminated that afternoon and again the next morning. Insemination with frozen semen is generally on the basis of using two straws (20 million total spermatozoa/straw) in the initial insemination after the onset of estrus and an additional one or two straws after a further interval of 12 to 24 hr. When fresh semen is employed, one insemination rather than two during estrus appears sufficient, presumably because fresh spermatozoa tend to survive longer than those which are frozen, and because ovulations in donors may occur over several hours, there is a greater chance of fertilization.[110]

The yield of fertilized eggs after insemination with frozen semen has been shown to be markedly influenced by the particular bull used;[77,100] although there is normally variation from bull to bull in fertility, such differences may be accentuated in the superovulated cow. A further point is to avoid inseminating donors late after the end of standing estrus, as this may adversely affect the yield of embryos subsequently recovered.[77]

Dealing with data from superovulation regimens employed before prostaglandins became available, Gordon[61] reported marked differences in fertilization rates between donors exhibiting estrus and those that did not, even though the latter animals were induced to ovulate by hCG and inseminations with fresh semen were subsequently performed at 12-hr intervals. The use of $PFG_{2\alpha}$ to control estrus and ovulation does afford the opportunity to inseminate donors at fixed time intervals relative to the time of administering prostaglandin (e.g., at 48, 60, 72 hr) so that "nonestrous" cattle can be dealt with in much the same way as those that do show a heat period.

### J. Inducing Superovulation on More Than One Occasion

Conflicting evidence appears in the early literature on the outcome of attempts to induce superovulation in donor cattle on more than one occasion. Reduced ovarian response, often substantial, was recorded by some authors,[111-113] but not to the same extent by others.[114,115] More recently, Saumande and Chupin[116] and Saumande et al.[99] employed PMSG regimens on eight occasions at 7- to 9-week intervals; response to repeat superovulation was variable and such that they concluded that problems in this area are a major obstacle to the possibilities of bovine embryo transfer. More encouraging evidence has been reported by Christie et al.,[117] who took a small group of heifers and treated them repeatedly at 6-week intervals with 2000 i.u. PMSG plus $PGF_{2\alpha}$ after 48 hr. Most animals responded satisfactorily to the repeat treatments, even up to the tenth time; there was no significant decline in the average ovulatory response between the second and subsequent PMSG-PG applications. There are several reports showing that ovarian response after a second gonadotrophin treatment can be comparable to response after the first one,[77,89,96] given minor adjustments in the superovulation regimen. In considering repeat gonodotrophin treatments, the normal assumption is that embryos will be collected by nonsurgical procedures. Even so, there are those employing surgical techniques who report satisfactory superovulatory responses after four successive treatments.[64] The fact that acceptable ovarian responses can be maintained after several PMSG treatments accords with the conclusions of Schams et al.[73] that the production of antibodies against PMSG poses little problem in cattle, even after repeated injections with doses usually given in superovulation studies. Saumande and Chupin,[116] in reporting their general lack of success with repeat superovulation, found that ovulation rate decreased after the second treatment, remained low until the sixth one, but then appeared to improve. This variable form of ovarian response presumably was the result of factors (hormonal or ovarian) other than an immunological response.

**K. Gonadotrophin Treatment of the Prepuberal Calf**

Superovulation of the prepuberal calf could have merit in providing a supply of embryos for research. In breeding programs there could be some genetic advantage in reducing the generation interval and initiating early progeny testing of females. Heifer calves are born with many vesicular follicles already evident in their ovaries; Erickson[65] records the highest number at about 4 months of age with a decline as the heifer approaches puberty. PMSG and other follicle-stimulating agents have been employed in many studies by workers at Cambridge[118] and in the U.S.[119-128] to induce multiple ovulations in the young calf. Considerable variability in the superovulatory response with low recovery and fertilization rates were a common feature of these reports; calves are found to respond very readily to gonadotrophins and usually show a much larger number of ovulations than postpuberal cattle. Fertility and egg recovery results, on the other hand, have been unsatisfactory.

Work by Testart[129-131] involved treating calves with intravaginal progestagen to obtain greater ovarian uniformity prior to administering gonadotrophin. Mickelsen et al.,[132] using an LH preparation after treatment with either PMSG or FSH, report a substantial improvement in the total number of ovulations and in subsequent egg recovery and fertilization rates. As yet, however, reports of pregnancies after transfer of embryos derived from prepuberal animals have been very few,[62,126] and the degeneration of bovine embryos recovered from the prepuberal reproductive tract has been observed.[124,126] In sheep, embryos from prepuberal ewe-lambs do not have the same potential for continued development when cultured in vitro as those from adult ewes,[133] and egg transfer studies by Quirke and Hanrahan[134] have shown substantial differences in the survival of embryos from adult sheep and ewe-lambs (73% vs. 33%) when transferred to the uteri of adult ewes. In using prepuberal animals it may be that questions of embryo normality arise which are not found among postpuberal cattle and sheep. Other than for research purposes, the practical utilization of calves remains very limited at this time.

**L. Use of Follicular Oocytes**

An eventual alternative to superovulation as a source of embryos for research and other purposes in cattle embryo transfer may prove to be the maturation and fertilization of the follicular oocyte (using either in vivo or in vitro techniques). Induction of twinning in cattle by embryo transfer procedures may be facilitated by the utilization of oocytes from heifers slaughtered for meat. Cattle ovaries contain numerous vesicular follicles; examination of 258 heifer tracts by Scanlon[90] showed an average of 45 per animal. Sreenan[97] confirmed an earlier report of Edwards[135] that nuclear maturation can be achieved by culture of the bovine follicular oocyte in a suitable medium; others have reported likewise.[136-139] It is now clear from the work of Thibault[140] and others that nuclear maturation is not necessarily accompanied by cytoplasmic maturation of the oocyte, as this occurs within the preovulatory follicle in the live animal. The growing awareness that there is a complex regulatory system operating within the Graafian follicle, involving hormonal and other factors, has resulted in the intact follicle rather than the oocyte itself being cultured, using suitable media and gas phases.[141] There is evidence that a normal pattern of cytoplasmic protein synthesis occurs in the oocytes matured in the intact follicle, but not necessarily in the oocytes liberated from the follicle.[142] However, one report is available on the birth of two calves after surgical transfer of embryos derived from such liberated oocytes fertilized in vivo after in vitro maturation;[143] although the success rate was low, refinements to technique can possibly make it more effective.

Studies with cattle and sheep oocytes matured within intact preovulatory follicles have achieved both normal nuclear and cytoplasmic maturation, the developmental capacity of the oocytes being confirmed subsequently by their fertilization in the oviduct of the species in question and in the development of normal young.[144,145] In cattle studies, follicular oocytes were recovered 6 to 24 hr after injection of hCG in animals previously treated with PMSG, and placed in the oviducts of inseminated estrous heifers. Although the oocytes were fertilized in the heifer's oviduct, the possibility exists that this will eventually be achieved in vitro. Capacitation of bull spermatozoa may be achieved in the rabbit uterus;[138] artificially matured extrafollicular bovine oocytes were readily penetrated by such spermatozoa in vitro. Capacitation of bull spermatozoa would not appear to be species specific, but may be accomplished in the reproductive tract of sheep[97] and pigs,[137] as well as in the rabbit.[138] Eventually, the in vitro capacitation of bull spermatozoa by appropriate culture treatment will become available.

### M. Use of Isolated Blastomeres

The possibility exists of using isolated blastomeres as a source of embryos for use in transfers. There is good evidence from a number of species to indicate that single blastomeres are capable of further cleavage and development, even to term. Single blastomeres from early-stage fertilized eggs of the rat, mouse, rabbit, pig, and sheep have been shown to be capable of further development,[146-149] and single blastomeres of two-, four-, and eight-cell rabbit and sheep eggs have developed into apparently normal young after transfer to recipient females.[147,150] The production of identical twins in sheep has recently been achieved by separating the blastomeres of the two-cell egg by micromanipulation techniques;[151] this raises the possibility that isolated blastomeres may be useful in cattle embryo transfer, especially for research purposes when there is a need for identical twins or multiples.

## REFERENCES


1. **Rowson, L. E. A., Moor, R. M., and Lawson, R. A. S.,** Fertility following egg transfer in the cow: effect of method medium and synchronization of oestrus, *J. Reprod. Fertil.*, 18, 517, 1969.
2. **Rowson, L. E. A., Lawson, R. A. S., Moor, R. M., and Baker, A. A.,** Egg transfer in the cow: synchronization requirements, *J. Reprod. Fertil.*, 28, 427, 1972.
3. **Sugie, T., Soma, T., Fukumitsu, S., and Otsuki, K.,** Studies on the ovum transfer in cattle with special reference to transplantation of fertilized ova by means of non-surgical techniques, *Natl. Inst. Anim. Ind. Bull.*, 25, 35, 1972.
4. **Newcomb, R. and Rowson, L. E. A.,** Conception rate after uterine transfer of cow eggs in relation to synchronization of oestrus and age of eggs, *J. Reprod. Fertil.*, 43, 539, 1975.
5. **Sreenan, J. M. and Beehan, D.,** Egg transfer in the cow; pregnancy rate and egg survival, *J. Reprod. Fertil.*, 41, 497, 1974.
6. **Sreenan, J. M., Beehan, D., and Mulvehill, P.,** Egg transfer in the cow; factors affecting pregnancy and twinning rates following bilateral transfers, *J. Reprod. Fertil.*, 44, 77, 1975.
7. **Trounson, A. O., Willadsen, S. M., and Rowson, L. E. A.,** The influence of in vitro culture and cooling on the survival and development of cow embryos, *J. Reprod. Fertil.*, 47, 367, 1976.
8. **Shea, B. F., Hines, D. J., Lightfoot, D. E., Ollis, G. W., and Olson, S. M.,** The transfer of bovine embryos, in *Egg Transfer in Cattle*, Rowson, L. E. A., Ed., Commission of the European Communities, Luxemborg, 1976, 145.
9. **Schneider, H. J., Jr., Castleberry, R. S., and Griffin, J. L.,** Commercial aspects of bovine embryo transfer, *Theriogenology*, 13, 73, 1980.
10. **Halley, S. D., Rodes, R. C., McKellar, L. D., and Randel, R. D.,** Successful superovulation, nonsurgical collection and transfer of embryos from Brahman cows, *Theriogenology*, 12, 97, 1979.
11. **Trounson, A. O., Willadsen, S. M., Rowson, L. E. A., and Newcomb, R.,** The storage of eggs at room temperature and at low temperatures, *J. Reprod. Fertil.*, 46, 173, 1976.

12. **Christian, R. E. and Casida, L. E.**, The effects of progesterone in altering the estrus cycle of the cow, *J. Anim. Sci.*, 7, 540, 1948.
13. **Zimbleman, R. G., Lauderdale, J. W., Sokolowski, J. H., and Schalk, T. G.**, Safety and pharmacologic evaluations of melengestrol acetate in cattle and other animals: a review, *J. Am. Vet. Med. Assoc.*, 157, 1528, 1970.
14. **Wiltbank, J. N. and Kasson, C. W.**, Synchronization of estrus in cattle with an oral progestational agent and an injection of an estrogen, *J. Anim. Sci.*, 27, 113, 1968.
15. **Wiltbank, J. N., Ingalls, J. E., and Rowden, W. W.**, Effect of various forms and levels of oestrogen alone or in combination with gonadotrophins on the oestrous cycle in beef heifers, *J. Anim. Sci.*, 20, 341, 1961.
16. **Kaltenbach, C. C., Niswender, G. D., Zimmerman, D. R., and Wiltbank, J. N.**, Alteration of ovarian activity in cycling, pregnant and hysterectomized heifers with exogenous oestrogens, *J. Anim. Sci.*, 23, 995, 1964.
17. **Rich, T. D., Johnson, C. L., and Randel, R. D.**, Plasma LH, progestins and corticoids in heifers injected with estradiol valerate and FSH during estrous synchronization with MGA, *J. Anim. Sci.*, 35, 90, 1972.
18. **Smith, L. E., Jr. and Vincent, C. K.**, Stage of cycle effect on bovine estrus control, *J. Anim. Sci.*, 36, 216, 1973.
19. **Woody, C. O. and Pierce, R. A.**, Influence of day of estrous cycle at treatment on response to estrous cycle regulation by norethandrolone implants and estradiol valerate injections, *J. Anim. Sci.*, 39, 903, 1974.
20. **Wiltbank, J. N. and Gonzalez-Padilla, E.**, Synchronization and induction of estrus in heifers with a progestagen and estrogen, *Ann. Biol. Anim. Biochim. Biophys.*, 15, 225, 1975.
21. **Moseley, W. M., Forrest, D. W., Kaltenbach, C. C., and Dunn, T. G.**, Effect of Norgestamet on peripheral levels of progesterone and estradiol-$17_\beta$ in beef cows, *Theriogenology*, 11, 331, 1979.
22. **Smith, J. F., Fairclough, R. J., and Peterson, A. J.**, Plasma levels of progesterone, Provera, oestradiol-$17_\beta$ and prostaglandin F in cows treated with Provera impregnated intravaginal sponges, *J. Reprod. Fertil.*, 55, 359, 1979.
23. **Sreenan, J. M.**, Effect of long- and short-term intravaginal progestagen treatments on synchronization of oestrus and fertility in heifers, *J. Reprod. Fertil.*, 45, 479, 1975.
24. **Sreenan, J. M. and Mulvehill, P.**, Synchronization of oestrus in cattle. II. The effect of short-term (10-day) progestagen-oestrogen administration on oestrous response and fertility in cattle, *J. Reprod. Fertil.*, 45, 1975.
25. **Sreenan, J. M.**, The effects of progesterone and oestrogen treatment in heifers on oestrous cycle control and plasma progesterone levels, *Vet. Rec.*, 101, 13, 1977.
26. **Mulvehill, P. and Sreenan, J. M.**, Field-scale evaluation of short-term progestagen synchronization treatments in beef cow herds, *Ir. J. Agric. Res.*, 17, 241, 1978.
27. **Mulvehill, P. and Sreenan, J. M.**, Oestrous cycle control and fertility in beef cattle following short-term progestagen treatments, in *Control of Reproduction in the Cow*, Sreenan, J. M., Ed., Martinius Nijhoff, The Hague, 1978, 511.
28. **Mauer, R. E., Webel, S. K., and Brown, M. D.**, Ovulation control in cattle with progesterone intravaginal device (PRID) and gonadotropin releasing hormone (GnRH), *Ann. Biol. Anim. Biochim. Biophys.*, 15, 291, 1975.
29. **Roche, J. F.**, Synchronization of oestrus in cows using intravaginal silastic coils containing progesterone, *Ann. Biol. Anim. Biochim. Biophys.*, 15, 301, 1975.
30. **Roche, J. F.**, Retention rate in cows and heifers of intravaginal silastic coils impregnated with progesterone, *J. Reprod. Fertil.*, 46, 253, 1976.
31. **Roche, J. F.**, Calving rate of cows following insemination after a 12-day treatment with silastic coils impregnated with progesterone, *J. Anim. Sci.*, 43, 164, 1976.
32. **Roche, J. F.**, Control of estrus and progesterone levels in heifers given intravaginal progesterone coils and injections of progesterone and oestrogen, *J. Anim. Sci.*, 44, 1026, 1977.
33. **Mawhinney, S. and Roche, J. F.**, Factors involved in oestrous cycle control in the bovine, in *Control of Reproduction in the Cow*, Sreenan, J. M., Ed., Martinius Nijhoff, The Hague, 1978, 511.
34. **Wishart, D. F. and Young, I. M.**, Artificial insemination of progestin (SC21009)-treated cattle at predetermined times, *Vet. Rec.*, 95, 503, 1974.
35. **Wishart, D. F., Young, I. M., and Drew, B.**, Fertility of Norgestamet treated dairy heifers, *Vet. Rec.*, 100, 417, 1977.
36. **Spitzer, J. C., Burrell, W. C., Lefever, D. G., Whitman, R. W, and Wiltbank, J. N.**, Synchronization of estrus in beef cattle. I. Utilization of a Norgestamet implant and injection of estradiol valerate, *Theriogenology*, 10, 181, 1978.
37. **Wishart, D. F.**, Controlled breeding of progestin treated cattle, in *Egg Transfer in Cattle*, Rowson, L. E. A., Ed., Commission of the European Communities, Luxembourg, 1976, 276.

38. **Gordon, I.**, Hormonal control of reproduction in sheep, *Proc. Br. Soc. Anim. Prod.*, 4, 79, 1975.
39. **Hunter, G. L., Adams, C. E., and Rowson, L. E. A.**, Interbreed ovum transfer in sheep, *J. Agric. Sci. (Camb.)*, 46, 143, 1955.
40. **Moore, N. W., Rowson, L. E. A., and Short, R. V.**, Egg transfer in sheep. Factors affecting the survival and development of transferred eggs, *J. Reprod. Fertil.*, 1, 332, 1960.
41. **Shelton, J. N. and Moore, N. W.**, Survival of fertilized eggs transferred to ewes after progesterone treatment, *J. Reprod. Fertil.*, 11, 149, 1966.
42. **Sreenan, J. M. and Beehan, D.**, Methods of induction of superovulation in the cow and transfer results, in *Egg Transfer in Cattle*, Rowson, L. E. A., Ed., Commission of the European Communities, Luxembourg, 1976, 19.
43. **Lauderdale, J. W.**, Effects of $PGF_{2\alpha}$ on pregnancy and oestrous cycle of cattle, *J. Anim. Sci.*, 35, 246, 1972.
44. **Lauderdale, J. W., Chenault, J. R., Sequin, B. E., and Thatcher, W. W.**, Fertility of cattle after $PGF_{2\alpha}$ treatment, *J. Anim. Sci.*, 37, 319, 1973.
45. **Lauderdale, J. W.**, The use of prostaglandins in cattle, *Ann. Biol. Anim. Biochim. Biophys.*, 15, 419, 1975.
46. **Cooper, M. J.**, Control of oestrous cycles of heifers with a synthetic prostaglandin analogue, *Vet. Rec.*, 95, 200, 1974.
47. **Roche, J. F.**, Synchronization of oestrus with prostaglandins, *Vet. Sci. Commun.*, 1, 121, 1977.
48. **Waters, R. J. and Ball, R.**, Commercial ovulation control and fixed-time artificial insemination in cattle, *Vet. Rec.*, 103, 585, 1978.
49. **Rowson, L. E. A., Tervit, R., and Brand, A.**, The use of prostaglandins for synchronization of oestrus in cattle, *J. Reprod. Fertil.*, 29, 145, 1972.
50. **Wishart, D. F.**, Synchronization of oestrus in cattle using a potent progestin (SC21009) and $PGF_{2\alpha}$, *Theriogenology*, 1, 87, 1974.
51. **Thimonier, J., Chupin, D., and Pelot, J.**, Synchronization of oestrus in heifers and cyclic cows with progestagens and prostaglandin analogues alone or in combination, *Ann. Biol. Anim. Biochim. Biophys.*, 15, 437, 1975.
52. **Heersche, G., Jr., Kiracse, G. H., de Benedetti, R. C., Wen, S., and McKee, R. M.**, Synchronization of estrus in beef heifers with a Norgestamet implant and prostaglandin $F_{2\alpha}$, *Theriogenology*, 11, 197, 1979.
53. **Leaver, J. D.**, Rearing of dairy cattle. VII. Effect of level of nutrition and body condition on the fertility of heifers, *Anim. Prod.*, 25, 219, 1977.
54. **Drew, B.**, Management factors in oestrous cycle control, in *Control of Reproduction in the Cow*, Sreenan, J. M., Ed., Martinius Nijhoff, The Hague, 1978, 475.
55. **Seidel, G. E., Jr. and Seidel, S. M.**, Bovine embryo transfer: costs and success rates, *Adv. Anim. Breed.*, 24, 6, 1978.
56. **Williams, G. L.**, Fertility in cattle, *Vet. Rec.*, 72, 197, 1960.
57. **Sreenan, J., Scanlon, P., and Gordon, I.**, Storage of fertilized cattle ova in vitro, *J. Agric. Sci. (Camb.)*, 74, 593, 1970.
58. **Casida, L. E., Nalbandov, A., McShan, W. H., Meyer, P. J., and Wisnicky, W.**, Potential fertility of artificially matured and ovulated ova in cattle, *Proc. Am. Soc. Anim. Prod.*, 303, 1940.
59. **Casida, L. E., Meyer, R. K., McShan, W. H., and Wisnicky, W.**, Effects of pituitary gonadotrophins on the ovaries and the induction of superfucundity in cattle, *Am. J. Vet. Res.*, 4, 76, 1943.
60. **Foote, R. H. and Onuma, H.**, Superovulation, ovum collection culture and transfer — a review, *J. Dairy Sci.*, 53, 1681, 1970.
61. **Gordon, I.**, Problems and prospects in cattle egg transfer, *Ir. Vet. J.*, 29, 21, 1975.
62. **Betteridge, J. J., Ed.**, Embryo Transfer in Farm Animals, Monograph 16, Canada Department of Agriculture, Ottawa, 1977, 1.
63. **Graham, E. F.**, Ova transfer, in *Proc. 5th Tech. Conf. A.I. Reprod.*, Natl. Assoc. Anim. Breed., Columbia, 1974, 21.
64. **Marshall, D. P. J. and Struthers, G. A.**, Commercial surgical embryo transfer in cattle, *N. Z. Vet. J.*, 26, 287, 1978.
65. **Erickson, B. H.**, Development and senescence of postnatal bovine ovary, *J. Anim. Sci.*, 25, 800, 1966.
66. **Richards, J. A.**, Hormonal control of ovarian follicular development: a 1978 perspective, *Recent Prog. Horm. Res.*, 35, 343, 1979.
67. **Dowling, D. F.**, Problems of the transplantation of fertilized ova, *J. Agric. Sci. (Camb.)*, 39, 1, 1949.
68. **Rowson, L. E. A.**, Methods of inducing multiple ovulation in cattle, *J. Endocrinol.*, 7, 260, 1951.
69. **Gordon, I., Williams, G., and Edwards, J.**, The use of serum gonadotrophin (PMS) in the induction of twin pregnancy in the cow, *J. Agric. Sci.*, 59, 143, 1962.

70. **Hafez, E. S. E., Sugie, T., and Gordon, I.,** Superovulation and related phenomena in the beef cow. I. Superovulatory responses following PMS and HCG injections, *J. Reprod. Fertil.*, 5, 359, 1963.
71. **Scanlon, P. F., Sreenan, J., and Gordon, I.,** Hormonal induction of superovulation in cattle, *J. Agric. Sci. (Camb.)*, 70, 179, 1968.
72. **Lamond, D. R.,** The effect of pregnant mare serum gonadotrophin (PMS) on ovarian function of beef heifers, as influenced by progestins, plane of nutrition and fasting, *Aust. J. Agric. Res.*, 21, 153, 1970.
73. **Schams, D., Menzer, Ch., Schallenberger, E., Hoffman, B., Hahn, J., and Hahn, R.,** Some studies on pregnant mare serum gonadotrophin (PMSG) and on endocrine responses after application for superovulation in cattle, in *Control of Reproduction in the Cow*, Sreenan, J. M., Ed., Martinius Nijhoff, The Hague, 1978, 122.
74. **Moore, N. W.,** The control of time of oestrus and ovulation and the induction of superovulation in cattle, *Aust. J. Agric. Res.*, 26, 295, 1975.
75. **Seidel, G. E., Jr., Elsden, R. P., Nelson, L. D., and Brown, R. A.,** Superovulation of cattle with pregnant mare's serum gonadotrophin and follicle stimulating hormone, in *Control of Reproduction in the Cow*, Sreenan, J. M., Ed., Martinius Nijhoff, The Hague, 1978, 159.
76. **Elsden, R. P., Nelson, L. D., and Seidel, G. E., Jr.,** Superovulation of cows with follicle stimulating hormone and pregnant mare's serum gonadotrophin, *Theriogenology*, 9, 17, 1978.
77. **Newcomb, R.,** Investigation of factors affecting superovulation and non-surgical embryo recovery from lactating British Friesian cows, *Vet. Rec.*, 106, 48, 1980.
78. **Avery, T. L., Faning, M. L., and Graham, E. F.,** Investigations associated with the transplantation of bovine ova. II. Superovulation, *J. Reprod. Fertil.*, 3, 212, 1962.
79. **Tervit, H. R., Rowson, L. E. A., and Brand, A.,** Synchronization of oestrus in cattle using a prostaglandin $F2_{\alpha}$ analogue (ICI 79939), *J. Reprod. Fertil.*, 34, 179, 1973.
80. **Phillippo, M. and Rowson, L. E. A.,** Prostaglandins and superovulation in the bovine, *Ann. Biol. Anim. Biochim. Biophys.*, 15, 233, 1975.
81. **Newcomb, R. and Rowson, L. E. A.,** Multiple ovulation, egg transplantation: towards twinning, in *Principles of Cattle Production*, Swan, H. and Broster, W. H., Eds., Butterworths, London, 1976, 89.
82. **Sreenan, J. M. and Gosling, J. P.,** The effect of cycle stage and plasma progesterone level on the induction of multiple ovulations in heifers, *J. Reprod. Fertil.*, 50, 367, 1977.
83. **Folley, S. J. and Malpress, F. H.,** The response of the bovine ovary to pregnant mare's serum and horse pituitary extract, *Proc. R. Soc. London B*, 132, 164, 1944.
84. **Baker, A. A.,** Ovum transfer in the cow, *Aust. Vet. J.*, 49, 424, 1973.
85. **Bowen, J. M.,** Problems associated with ovum transplants, *Vet. Rec.*, 92, 17, 1973.
86. **Rowson, L. E. A.,** The use of egg transfer in practice and research, *Br. Cattle Breed. Club Dig.*, 23, 19, 1973.
87. **Stewart, F., Allen, W. R., and Moor, R. M.,** Pregnant mare serum gonadotrophin: ratio of follicle stimulating hormone and luteinizing hormone activities measured by radio-receptor assay, *J. Endocrinol.*, 71, 371, 1976.
88. **Beehan, D. and Sreenan, J. M.,** Factors affecting ovarian response to superovulation in the synchronized bovine oestrous cycle, in Stud. Fertil., *J. Reprod. Fert.*, 51, 519, 1977.
89. **Newcomb, R., Christie, W. B., Rowson, L. E. A., Walters, D. E., and Bousfield, W. E. D.,** Influence of dose, repeated treatment and batch of hormone on ovarian response in heifers treated with PMSG, *J. Reprod. Fertil.*, 56, 113, 1979.
90. **Scanlon, P. F.,** Studies in Reproduction in the Cow, Ph.D. thesis, National University of Ireland, Dublin, 1969.
91. **Testart, J., Goddard-Siour, C., and du Mesnil du Buisson, F.,** Transvaginal transplantation of an extra egg to obtain twinning in cattle, *Theriogenology*, 4, 163, 1975.
92. **Sreenan, J. M., Beehan, D., and Gosling, J. P.,** Ovarian responses in relation to endocrine status following PMSG stimulation in the cow, in *Control of Reproduction in the Cow*, Sreenan, J. M., Ed., Martinius Nijhoff, The Hague, 1978, 144.
93. **Hafez, E. S. E., Sugie, T., and Hunt, W. L.,** Superovulation and related phenomena in the beef cow. II. Effect of oestrogen administration on production of ova, *J. Reprod. Fertil.*, 5, 381, 1963.
94. **Sugie, T., Soma, T., Fukumitsu, S., and Otsuki, K.,** Studies on the ovum transfer in cattle, with special reference to collection of ova by means of non-surgical technique, *Natl. Inst. Anim. Ind. Bull.*, 25, 27, 1972.
95. **Gordon, I.,** Cattle twinning by the egg transfer approach, in *Egg Transfer in Cattle*, Commission of the European Communities, Luxembourg, 1976, 305.
96. **Gordon, I. and Boland, M. P.,** Cattle twinning by non-surgical egg transfer, in *Control of Reproduction in the Cow*, Sreenan, J. M., Ed., Martinius Nijhoff, The Hague, 1978, 336.
97. **Sreenan, J. M.,** Studies Related to Ovum Transfer in Cattle, Ph.D. thesis, National University of Ireland, Dublin, 1969.

98. **Mariana, J. G., Mauleon, P., Benoit, M., and Chupin, D.**, Variability and repeatability of the number of ovulations obtained after injection of 1,600 i.u. PMSG and 1,500 i.u. HCG, *Ann. Biol. Anim. Biochim. Biophys.*, 10, 47, 1970.
99. **Saumande, J., Chupin, D., Mariana, J. C., Ortavant, R., and Mauleon, P.**, Factors affecting the variability of ovulation rates after PMSG stimulation, in *Control of Reproduction in the Cow*, Sreenan, J. M., Ed., Martinius Nijhoff, The Hague, 1978, 195.
100. **Newcomb, R., Rowson, L. E. A., and Trounson, A. O.**, The Sacrewell project: an on-farm demonstration of the potential of egg transfer, *Vet. Rec.*, 130, 415, 1978.
101. **Adams, C. E.**, Ageing and reproduction in the female mammal with particular reference to the rabbit, *J. Reprod. Fertil.*, Suppl. 12, 1, 1970.
102. **Rajakowski, E.**, Ovarian follicular system in sexually mature heifers with special reference to the seasonal cyclical and left right variations, *Acta Endocrinol.*, 34 (Suppl. 52), 1, 1960.
103. **Denny, J. E. F. M.**, Ovulation control as a first step in the induction of twin births in dairy cattle, *Proc. S. Afr. Soc. Anim. Prod.*, 57, 1964.
104. **Church, R.**, Some aspects of bovine embryo transfer, in *Egg Transfer in Cattle*, Rowson, L. E. A., Ed., Commission of the European Communities, Luxembourg, 1976, 73.
105. **Hill, J. R., Lamond, D. R., Henricks, D. M., Dickey, J. F., and Niswender, G. D.**, Effect of undernutrition on ovarian function and fertility in beef heifers, *Biol. Reprod.*, 2, 78, 1970.
106. **Segerson, E. C., Jr., Murray, F. A., Moxon, A. L., Redman, D. R., and Conrad, H. R.**, Selenium/vitamin E: role in fertilization of bovine ova, *J. Dairy Sci.*, 60, 1001, 1977.
107. **Choudary, J. B., Gier, H. T., and Marion, G. B.**, Cyclic changes in bovine vesicular follicles, *J. Anim. Sci.*, 27, 468, 1968.
108. **Summers, P. M. and Campbell, R. S. F.**, A histological study of ovarian function in infertile beef cows, *Res. Vet. Sci.*, 17, 131, 1974.
109. **McKenzie, B. E. and Kenney, R. M.**, In vitro culture of bovine embryos, *Am. J. Vet. Res.*, 34, 1271, 1973.
110. **Rowson, L. E. A.**, Embryo transfer, *Br. Cattle Breed. Club Dig.*, 34, 77, 1979.
111. **Willett, E. L., Buckner, P. J., and McShan, W. P.**, Refractoriness of cows repeatedly superovulated with gonadotrophins, *J. Dairy Sci.*, 36, 1083, 1953.
112. **Dziuk, P. J., Donker, F. D., Nichols, J. P., and Petersen, W. E.**, Problems associated with the transfer of ova between cattle, *Univ. Minn. Agric. Exp. Stn. Tech. Bull.*, 222, 1958.
113. **Hafez, E. S. E., Jainudeen, M. R., and Lindsay, D. R.**, Gonadotrophin-induced twinning and related phenomena in beef cattle, *Acta Endocrinol. Copenh.*, Suppl. 102, 5, 1965.
114. **Umbaugh, R. E.**, Superovulation and ovum transfer in cattle, *Am. J. Vet. Res.*, 10, 295, 1949.
115. **Scanlon, P. F.**, Ovarian response of cows following PMSG treatment during two successive estrous cycles, *J. Dairy Sci.*, 55, 527, 1972.
116. **Saumande, J. and Chupin, D.**, Superovulation: a limit to egg transfer in cattle, *Theriogenology*, 7, 141, 1977.
117. **Christie, W. B., Newcomb, R., and Rowson, L. E. A.**, Ovulation rate and egg recovery in cattle treated repeatedly with pregnant mare serum gonadotrophin and prostaglandin, *Vet. Rec.*, 104, 281, 1979.
118. **Marden, W. G. W.**, The hormonal control of ovulation in the calf, *J. Agric. Sci. (Camb.)*, 43, 381, 1953.
119. **Black, W. G., Ulberg, L. G., Christian, R. E., and Casida, L. E.**, Ovulation and fertilization in the hormone-stimulated calf, *J. Dairy Sci.*, 36, 274, 1953.
120. **Howe, G. R., Black, D. L., Foley, R. C., and Black, W. G.**, Ovarian activity in prepuberal dairy calves, *J. Anim. Sci.*, 21, 82, 1962.
121. **Jainudeen, M. R., Hafez, E. S. E., and Lineweaver, J. A.**, Superovulation in the calf, *J. Reprod. Fertil.*, 12, 149, 1966.
122. **Onuma, H. and Foote, R. H.**, Superovulation in prepuberal calves on two levels of nutrition intake, *J. Anim. Sci.*, 28, 771, 1969.
123. **Onuma, H. and Foote, R. H.**, In vitro development of ova from prepuberal cattle, *J. Dairy Sci.*, 52, 1085, 1969.
124. **Onuma, J., Hahn, J., and Foote, R. H.**, Factors affecting superovulation, fertilization and recovery of superovulated ova in prepuberal cattle, *J. Reprod. Fertil.*, 21, 119, 1970.
125. **Onuma, H., Hahn, J., Maurer, R. R., and Foote, R. H.**, Repeated superovulation in calves, *J. Anim. Sci.*, 28, 634, 1969.
126. **Seidel, G. E., Larson, L. L., Spilman, G. H., Hahn, J., and Foote, R. H.**, Transfer of superovulated calf ova, *J. Anim. Sci.*, 31, 230, 1970.
127. **Seidel, G. E., Larson, L. L., Spilman, C. H., Hahn, J., and Foote, R. H.**, Culture and transfer of calf ova, *J. Dairy Res.*, 54, 923, 1971.
128. **Seidel, G. E., Larson, L. L., and Foote, R. H.**, Effects of age and gonadotrophin treatment on superovulation in the calf, *J. Anim. Sci.*, 33, 617, 1971.

129. **Testart, J.**, Follicular response of immature heifers to various treatments with serum gonadotrophin, with or without progestin, *Ann. Biol. Anim. Biochim. Biophys.*, 12, 397, 1972.
130. **Testart, J.**, Synchronization of induced ovulation in the prepuberal female calf, *Proc. 7th Int. Congr. Anim. Reprod. A. I. (Munich)*, 1, 493, 1972.
131. **Testart, J. and Arrau, J.**, Oocyte maturation following follicle stimulation in the calf, *Ann. Biol. Anim. Biochim. Biophys.*, 13 (Suppl. 157), 1973.
132. **Mickelsen, W. D., Wright, R. W., Jr., Menino, A. R., Zamora, C. S., and Paisley, L. G.**, Superovulation, fertilization and embryo recovery in gonadotropin treated prepuberal calves, *Theriogenology*, 10, 167, 1978.
133. **Wright, R. W., Jr., Anderson, G. B., Cupps, P. T., Drost, M., and Bradford, G. E.**, In vitro culture of embryos from adult and prepuberal ewes, *J. Anim. Sci.*, 42, 912, 1976.
134. **Quirke, J. F. and Hanrahan, J. P.**, Comparison of the survival in the uteri of adult ewes of cleaved ova from adult ewes and ewe lambs, *J. Reprod. Fertil.*, 51, 487, 1977.
135. **Edwards, R. G.**, Maturation in vitro of mouse, sheep, cow, pig, rhesus monkey and human ovarian oocytes, *Nature (London)*, 206, 349, 1965.
136. **Hunter, R. H. F., Lawson, R. A. S., and Rowson, L. E. A.**, Maturation transplantation and fertilization of ovarian oocytes in cattle, *J. Reprod. Fertil.*, 30, 325, 1972.
137. **Shea, B. F., Bedirian, K. N., and Baker, R. D.**, Sperm penetration of bovine follicular oocytes following in vitro and in vivo maturation, *Biol. Reprod.*, 9, 84, 1973.
138. **Iritani, A. and Niwa, K.**, Capacitation of bull spermatozoa and fertilization in vitro of cattle follicular oocytes matured in culture, *J. Reprod. Fertil.*, 50, 119, 1977.
139. **Sato, E., Iritani, A., and Nishikawa, Y.**, Factors involved in maturation of pig and cattle follicular oocytes cultured in vitro, *Jpn. J. Anim. Reprod.*, 23, 12, 1977.
140. **Thibault, C.**, Are follicular maturation and oocyte maturation independent processes?, *J. Reprod. Fertil.*, 51, 1, 1977.
141. **Thibault, C., Gerard, M., and Menezo, Y.**, In vitro acquired ability of rabbit and cow oocyte to ensure sperm nucleus decondensation during fertilization (MPGF), *Ann. Biol. Anim. Biochim. Biophys.*, 15, 705, 1975.
142. **Warnes, G. M., Moor, R. M., and Johnson, M. H.**, Changes in protein synthesis during maturation of sheep oocytes in vivo and in vitro, *J. Reprod. Fertil.*, 49, 331, 1977.
143. **Newcomb, R., Christie, W. B., and Rowson, L. E. A.**, Birth of calves after in vivo fertilization of oocytes removed from follicles and matured in vitro, *Vet. Rec.*, 102, 461, 1978.
144. **Moor, R. M. and Trounson, A. O.**, Hormonal and follicular factor affecting maturation of sheep oocytes in vitro and their subsequent developmental capacity, *J. Reprod. Fertil.*, 49, 101, 1977.
145. **Trounson, A. O., Willadsen, S. M., Rowson, L. E. A., and Newcomb, R.**, Fertilization and development capacity of bovine follicular oocytes matured in vitro and in vivo and transferred to the oviducts of rabbits and cows, *J. Reprod. Fertil.*, 51, 321, 1977.
146. **Nicholas, J. S. and Hall, B. V.**, Experiments on developing rats. II. The development of isolated blastomeres and fused eggs, *J. Exp. Zool.*, 90, 441, 1942.
147. **Seidel, F.**, The development potential of an isolated blastomere in the 2-cell stage mammalian egg, *Naturwissenschaft*, 15, 355, 1952.
148. **Tarkowski, A. K.**, Experiments on the development of isolated blastomeres of mouse eggs, *Res. Vet. Sci.*, 17, 131, 1959.
149. **Moore, N. W., Polge, C., and Rowson, L. E. A.**, The survival of single blastomeres of pig eggs transferred to recipient gilts, *Aust. J. Biol. Sci.*, 22, 979, 1969.
150. **Trounson, A. O. and Moore, N. W.**, Attempts to produce identical offspring in the sheep by mechanical division of the ovum, *Aust. J. Biol. Sci.*, 27, 505, 1974.
151. **Willadsen, S.**, A method for culture of micro-manipulated sheep embryos and its use to produce monozygotic twins, *Nature (London)*, 277, 298, 1979.

Chapter 6

# EGG RECOVERY AND TRANSFER IN CATTLE

R. Newcomb

## TABLE OF CONTENTS

## I. INTRODUCTION

Egg transfer techniques have been commercially exploited in cattle to a considerable extent in recent years, far more than in any other species. In 1969, Rowson et al.[1] reported excellent results following the surgical transfer of cow eggs, and this coincided with a boom in the prices paid for European breeds of cattle, "exotic" to the English speaking countries. As a result, commercial egg transfer groups were rapidly established throughout the English speaking world. Although the position was to change very rapidly as the value of these exotic cattle fell, the enthusiasm which had been generated was not to waver, for a reservoir of expertise had been established and research stimulated. In a decade the technique had advanced from being applied only surgically to nonsurgical techniques being available for both egg recovery and transfer.

The concepts currently being applied are, however, by no means novel. As early as 1949, devices were available for the nonsurgical recovery of cattle eggs,[2,3] and in 1951 the first pregnancy was obtained from a transferred cow egg.[4] An enthusiastic article by Hervey[5] appeared in 1949 entitled "Thirty Calves a Year from Your Best Cow".

Nevertheless, the initial enthusiasm was tempered by 20 years without substantial progress. One name — that of L. E. A. Rowson — must be singled out as the person most involved with the progress of egg transfer in cattle. His nonsurgical egg recovery device was commercially available in 1949, and it was his group at Cambridge which in 1969 reported the first consistent success with the technique. At a course held in Cambridge in June 1972 under Rowson's direction, the seeds were sown which were to result in the rapid growth of commercial egg transfer. Rowson retired in 1980, having seen many of his aspirations reach commercial fruition within a working lifetime, a tribute to his great optimism.

## II. SOURCE OF SUPPLY OF EGGS

Currently the main source of supply of cow eggs for transfer is through the superovulation of donors to produce fertilized eggs. However, because of the economic value of cattle, exploitation of other sources of supply, e.g., ovarian oocytes, may become commercially viable.

Ultimately, the costs of producing eggs might be reduced by obtaining them as a by-product of a slaughterhouse production line, either from inseminated, superovulated donors, or by maturing, fertilizing, and culturing ovarian oocytes in vitro. Obviously, eggs produced in this terminal manner could not be from animals of the highest genetic merit, but provided animal health restrictions were not such as to preclude or render nonviable such a system, they could be used for such purposes as inducing twin pregnancy or beef production from the dairy herd.

Current knowledge of methods for obtaining cow eggs is outlined below.

### A. Ovarian Oocytes

#### *1. Stimulation and Maturation of Oocytes in Vivo*

Trounson et al.[6] obtained pregnancies by maturing oocytes in donor heifers, fertilizing the eggs in inseminated recipient heifers, and retransferring the fertilized eggs.

The donor heifers were prepared with 2000 i.u. PMSG IM in the midluteal phase of the estrous cycle followed after 48 hr with 750 μg cloprostenol IV and 48 hr later with 2500 i.u. hCG (i/v). Donors were slaughtered 24 hr after hCG injection. The ovaries were removed and follicles >5 mm in diameter were dissected out, opened with fine scissors, and the sticky cumulus-surrounded oocytes removed and held in phosphate-buffered saline (PBS) until transfer.

Recipient heifers were inseminated at the onset of estrus and again 24 to 30 hr later when oocytes were transferred to the oviducts. Some days later the fertilized eggs were recovered.

Oocytes without a dissociated cumulus had a relatively low potential for normal development (13%) compared with those having a dissociated cumulus (39%). After the retransfer of 16 blastocysts obtained in this manner to eight recipients, all became pregnant with a high fetal survival (13/16).

The efficacy of this method was confirmed by Shea,[7] who transferred oocytes to recipients for fertilization and after 5 days recovered 49% of which 32% were fertilized. Pregnancy rate after transfer approached 50% and the series included cows classed as "infertile".

Technically the dissection of follicles is not easy, and Trounson et al.[6] point out the variability in the proportion of normal embryos recovered from different recipients. Shea[7] used a surgical approach and an aspiration technique; 64% of aspirated follicles yielded oocytes.

Sperm penetration and cleavage were not observed when matured oocytes were transferred to rabbits inseminated with bull spermatozoa,[6,8] although similar transfers to the sheep[8] and pig[9] led to apparently normal cleavage.

#### *2. Maturation of Oocytes in Vitro*

The maturation of oocytes in vitro may be undertaken by either (1) culturing the oocyte within an explanted follicle or (2) removing the oocyte from the follicle and culturing it.

##### ***a. Follicle Culture***

The antral follicle represents a physiological entity capable of supporting nuclear and cytoplasmic maturation in vitro.[10] In several species, including hamster,[11] rabbit,[12] and sheep,[13] "follicle-matured" oocytes have been fertilized and have developed normally. In order to obtain resumption of meiosis within the follicle, steroids and gonadotrophins are added to the culture medium. In sheep Moor et al.[13] found that oocytes cultured within the follicle in hormone-free medium remained at the germinal vesicle stage. In the presence of FSH and LH 19% of oocytes in small (2 to 3 mm diameter) and 73% in larger (3 to 5 mm diameter) nonatretic follicles reached metaphase II. The addition of estradiol-17B to the culture medium increased the proportion of oocytes developing into blastocysts after transfer from less than 5% for follicles cultured without estrogen to 26% and 46% for oocytes from small and larger follicles, respectively. Of blastocysts derived from follicle-enclosed oocytes which had been matured in vitro and fertilized in vivo, 63% developed into lambs after transfer.

When 3- to 4-mm calf follicles were cultured by superfusion in a synthetic medium,[14] only one third of the oocytes reached metaphase II and cumulus cell dissociation seldom occurred. The lack of response when the oocyte and its cumulus were left inside such medium-sized follicles may be due to insufficient penetration of gonadotrophin or to the absence of LH receptors, so that the inhibitory control of granulosa cells in the oocyte and cumulus cells is not turned off.[10] With medium-sized follicles, a continuous-flow superfusion technique proved better than an organ culture grid method.[15] There was an increase of mitotic index and of follicle volume related to augmented amino acid consumption.

The culture of larger bovine follicles, even with superfusion techniques, is less likely to be successful due to the problems of nutrient and gaseous exchange within the follicle. Follicle dissection is in any event a painstaking procedure and unsuitable as a system for producing an inexpensive supply of fertilized eggs.

*b. Extrafollicular Oocyte Culture*

Oocytes from all the mammalian species studied are able to resume meiosis from the dictyate stage to metaphase II when cultured in a suitable medium.[10]

In laboratory animals oocytes matured in vitro cannot readily be fertilized. The ability of the zona to be penetrated by spermatozoa is acquired in vivo between 5 and 6 hr after hCG treatment of the hamster[16] and rat.[17] In vitro-matured oocytes from rabbits, pigs, cows, and women can be penetrated by spermatozoa, but the male nucleus does not dissociate and swell immediately after penetration of in vitro-matured oocytes as it does in freshly ovulated eggs[10] or in in vivo-matured oocytes.[6] The oocyte within the follicle in vivo develops the capacity for changing sperm nuclear structure 6 to 8 hr after hCG treatment in the rabbit[18] and hamster.[19]

Protein synthesis does not change during the resumption of meiosis in extrafollicular oocytes in vitro in the same way as it does within the follicle.[20] Maturation of extrafollicular oocytes in vitro has not been followed by normal embryonic development[8] except in the mouse[21,22] and cow.[23]

In the cow the maturation of ovarian oocytes in vitro[24,25] and fertilization after transfer to mated animals has been reported.[8,9,26,27]

Interpretation of claims of fertilization and cleavage must be cautious in the absence of information on normal development to term. In the experiments reported by Trounson et al.,[6] it was found that cow oocytes matured in vitro readily undergo parthenogenetic development to at least the eight-cell stage. Similar parthenogenetic development was noted by Newcomb et al.,[23] with apparent cleavage up to eight cells, although the number of nuclei did not exceed five. Parthenogenetic activation of calf oocytes cultured in follicular fluid in vitro has also been reported.[28]

The culture system which was used to mature extrafollicular oocytes in vitro and which, after in vivo fertilization and retransfer to a recipient resulted in the birth of two calves,[23] utilized Ham's F10 medium which was supplemented with 10 $\mu$g/m$\ell$ estradiol and 1 i.u./m$\ell$ hCG. When oocytes were cultured using this system in the absence of hCG, no advanced developmental stages (late morula or blastocyst) were obtained 7 days after transfer to a mated recipient.[29] This is noteworthy in view of recent findings in sheep which indicate that when high levels of LH are included in the culture medium, a high proportion of oocytes develop normally into blastocysts in a mated recipient.[30]

*3. In Vitro Fertilization of Matured Oocytes*

Only unsuccessful attempts[31] to fertilize in vitro cattle follicular oocytes matured in vitro had been reported until recently.[32,33] Brackett et al.[33] cultured oocytes in a modified Krebs-Ringer bicarbonate solution (KRB) for 20 to 24 hr and tested spermatozoa incubated in this solution, or in excised cow reproductive tracts or in the rabbit uterus in vivo. None of the 38 oocytes inseminated with spermatozoa incubated in KRB appeared to be fertilized or activated, but about 20% of those inseminated with spermatozoa incubated in reproductive tracts showed signs of penetration followed by activation. It was suggested that the rabbit uterus might be used for the study of the capacitation requirements of spermatozoa.

Yanagimachi[34,35] reported that the capacitation of hamster sperm can be induced by follicular fluid from both the hamster and the cow. With bovine spermatozoa, too, there is evidence that incubation with follicular fluid will induce an acrosome reaction similar to that which occurs during capacitation.[36] A close correlation has been found between the acrosome reaction and the ability of spermatozoa to fertilize. In the guinea pig only spermatozoa that have undergone a physiological acrosome reaction are capable of penetrating the zona pellucida and entering the ooplasm.[37]

The rabbit acrosome reaction inducing activity of bovine follicular fluid is a serum component and probably a high molecular weight protein,[38] which according to Lui et al.,[39] is serum albumin. Capacitation as measured in vitro by the loss of fluorescene-tetracycline HCl bound to bull spermatozoa can be induced with bovine serum albumin (BSA).[40] Hamster spermatozoa will undergo the acrosome reaction in the presence of BSA and the acrosome reaction-inducing ability of BSA can be increased severalfold by purification of the albumen.[39,41] The presence of calcium ions in the medium is apparently required for capacitation of spermatozoa in vitro.[37,40]

Once in vitro maturation and fertilization of bovine oocytes have been reliably achieved, a culture system will then be essential for the development of the eggs to stages suitable for transfer. Culture systems are discussed in Section III of this chapter.

**B. Fertilized Eggs**

*1. Recovery after Slaughter*

The first calves obtained by means of egg transfer were derived from eggs removed from the donor after slaughter and then transferred surgically to recipients.[42] Rowson et al.[1] compared the conception rates achieved after the surgical transfer of eggs recovered either after slaughter (33%) or surgically (91%). When the mean interval from slaughter of the donor to flushing was only 30 min,[43,44] a much improved rate of development was achieved either after transfer to the rabbit oviduct or to heifers, either directly or following a period in the rabbit oviduct. After the direct transfer of two eggs to heifers 67% of recipients became pregnant with 50% fetal survival and, after 3 or 4 days in the rabbit oviduct and retransfer to heifers, 86% of recipients became pregnant with 86% fetal survival.

These results indicate that with a good technique which ensures the minimum of delay between slaughter and flushing and where temperature changes are controlled, acceptable pregnancy rates may be achieved after the transfer of eggs removed from slaughtered cattle.

*2. Surgical Egg Recovery*

While the midventral approach has been the method of choice for the surgical recovery of bovine eggs, some workers have utilized a flank and others a transvaginal approach. Each of these methods will be reviewed.

***a. Midventral Laparotomy***

The animal is fasted for 12 to 48 hr before surgery. Usually no premedication is administered and anesthesia is simply induced with a short-acting intravenous barbi-

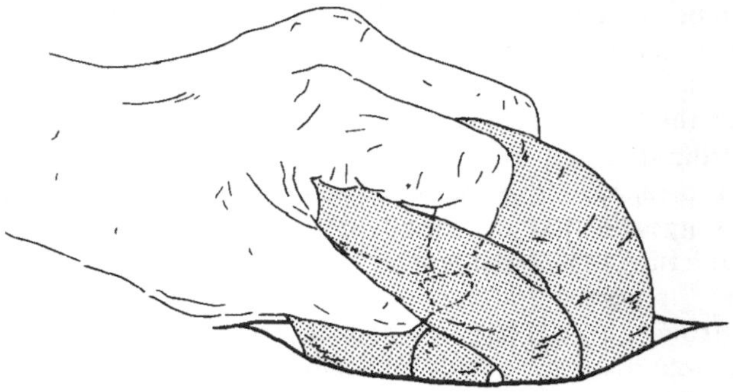

FIGURE 1. Exteriorization of the bovine uterus from a midventral laparotomy site.

turate (4 to 8 g thiopentone sodium in 40 to 50 m*l* water for injection). The recumbent animal is then intubated with an 18- to 25-mm diameter endotracheal tube and anesthesia is maintained with a halothane/nitrous oxide/oxygen gas mixture.

The animal is placed on its back on an operating table. The rear of the table is raised hydraulically in order to facilitate location of the uterus and minimize interference from the abdominal viscera. A site immediately in front of the udder is prepared and disinfected. It is usual for the surgeon to stand on the animal's right. An incision is made immediately in front of the udder along the mid line and sufficiently long (approximately 12 cm) to accommodate the operator's left hand.

The uterus is located either by its more muscular wall, which distinguishes it from the surrounding viscera, or by first locating an enlarged stimulated ovary and then moving the hand mesially until the uterus is located. The bovine uterus may offer considerable resistance to complete exteriorization and it is important, if trauma is to be reduced, to adopt a correct procedure. The base of the uterus (cervical extremity) is least movable, and it should not be grasped at this point but at its middle or ovarian third. Considerable damage may be done if a horn is excessively squeezed and pulled during exteriorization. It is useful to locate the external bifurcation of the two horns and introduce the first finger between the horns with the thumb and second finger around the right and left horns, respectively (Figure 1). The fingers may then be tightly pressed together through the broad ligament ensuring that the uterus itself is not bruised. Alternatively one uterine horn may be grasped in a similar manner.

The uterus may now be secured by tying a ligature around it and through the broad ligament to a stainless steel crossbar using umbilical tape (Figure 2). Alternatively, it may be held by an assistant; this approach is perhaps less traumatic but allows less freedom of movement at the operation site.

**Oviduct flushing** — This method has been described in outline by Rowson et al.[1] In order to cannulate the oviduct, it will in most cases also be necessary to exteriorize the enlarged ovary, but wherever possible, this is best avoided because young corpora lutea readily hemorrhage. The tip of the left uterine horn, i.e., the one furthest from the operator, is withdrawn and tension applied with the right hand. With the left index finger the fimbria is hooked out and is cannulated by an assistant (Figure 3). A small bent glass or nylon cannula (internal diameter 3 mm) with a rim at its end may be used. The cannula now located within the oviduct is grasped between the thumb and first two fingers and its free end directed into a glass collection dish. The operator pinches the uterus to occlude the lumen and, with a syringe attached to a smooth blunt 18-g needle, passes the needle into the lumen and forces fluid through the upper part of the uterus into the oviduct and thence through the cannula and into the collecting

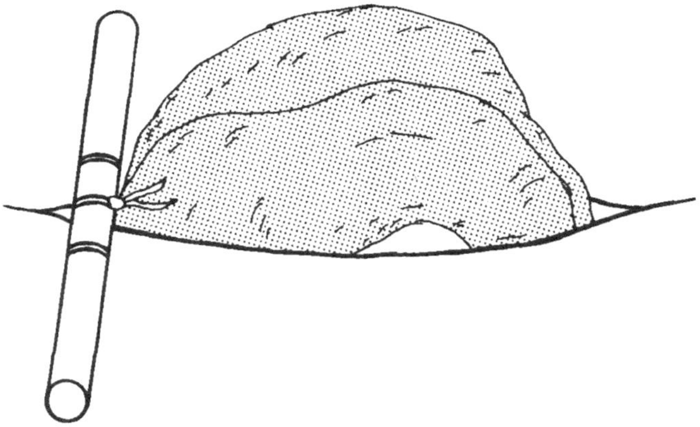

FIGURE 2. Technique for preventing the bovine uterus from returning to the abdomen with a suture of umbilical tape tied through the broad ligament and secured to a stainless steel crossbar.

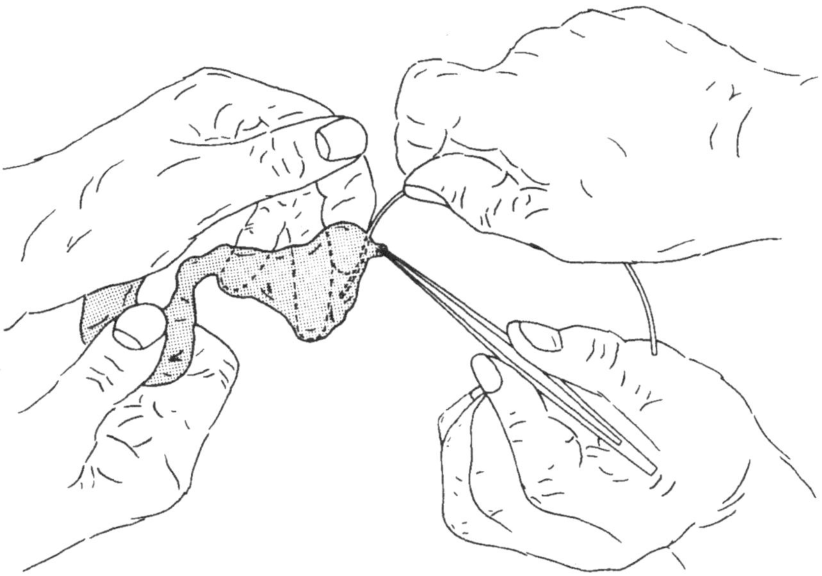

FIGURE 3. Technique of cannulating the oviduct via the fimbria in the cow.

cup (Figure 4). In most cases it will be necessary to pause during the flushing and force fluid through the utero-tubal junction. In the cow the pressures required to do this are such that the uterus may rupture and fluid then enter the broad ligament. This is much more likely to happen if a large portion of uterus is flushed than if only the tip of the uterus is flushed. Rupture elsewhere seldom occurs. To repeat the procedure in the opposite horn the operator and assistant change sides.

This technique is considered less suitable for the recovery of eggs from the uterus due to the possible risk of damaging the eggs, e.g., by rupture of the zona pellucida.

**Uterine flushing** — The uterus is exteriorized after which either one of two procedures may be followed:

1. The uterus is punctured near to its base with a pair of closed artery forceps so as

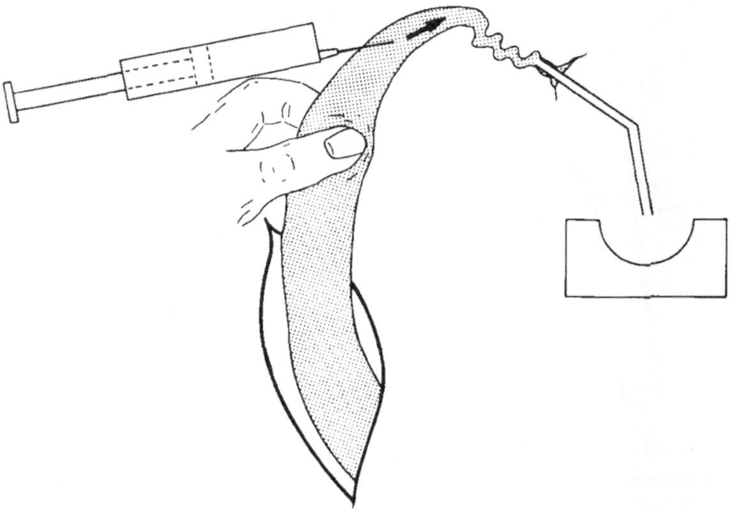

FIGURE 4. Technique for flushing eggs through the bovine oviduct. The lumen of the uterus is occluded and flushing medium directed into the oviduct.

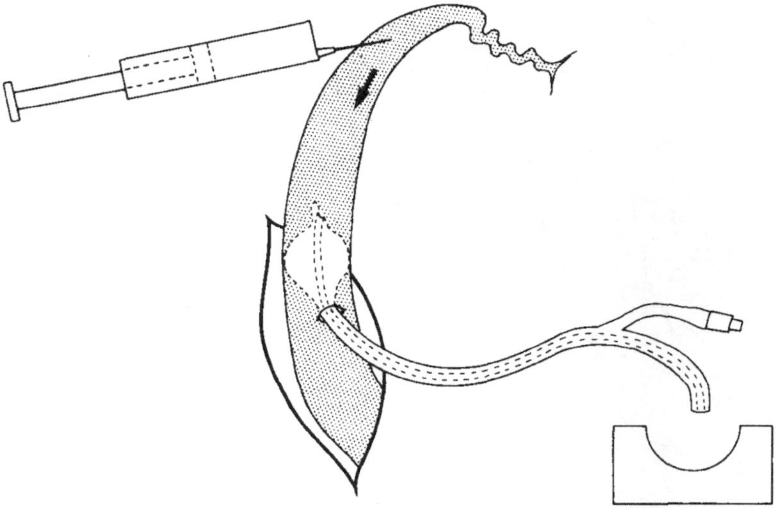

FIGURE 5. Method for recovering early uterine cow eggs. Eggs are dislodged by flushing medium into the tip of the uterus and collecting it via a urological balloon catheter inserted into the base of the uterus.

to permit the insertion of a small balloon catheter (14 Fg urological catheter) into the lumen, after which the balloon is inflated. With the free end of the catheter directed into a collecting cup, a blunt needle is inserted near to the tip of the uterus and flushing medium is passed from a syringe to the tip and recovered via the catheter at the base of the uterus (Figure 5). This method is easy to apply and in most situations is more suitable than method 2.

2. First the base of the uterus is firmly occluded using a ligature of umbilical tape. Eggs are flushed from the tip of the uterus and the uterus is pinched to retain the flushings in the base (Figure 6A). The uterus is then incised in its upper third

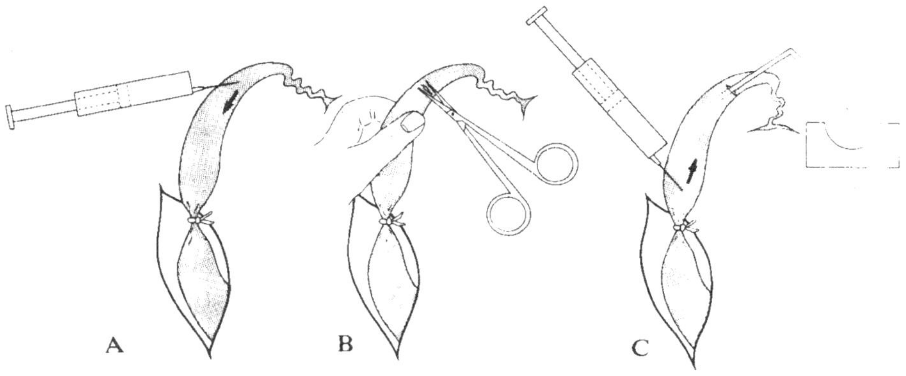


FIGURE 6. Method for recovering hatched and elongated blastocysts from the bovine uterus. (A) Eggs are dislodged from the tip region of the uterus. (B) Fluid is retained at the base of the uterus and an incision made into the uterine wall with scissors. (C) Fluid containing eggs is flushed from the base of the uterus through a glass cannula into a collecting vessel.

to allow a large diameter (1.5 cm) glass tube to be inserted (Figure 6B). This tube is held in place and fluid is then passed into the ligated base of the uterus and collected via the glass tube into special glass cups (Figure 6C). The uterine wound is repaired. This method is traumatic and most suited to investigative work requiring very late egg stages.

**Simultaneous flushing of oviduct and uterus** — This method[45] combines both uterine and oviduct flushing techniques and is most suitable for recovery eggs on days 4, 5, and 6, when it is expected that some of the eggs may be located in both the oviducts and the uterus.[46-48] Eggs are first flushed from the tip of the uterus and out through a balloon catheter inserted in the base of the uterus, then the very tip of the uterus is pinched and the remaining fluid is passed through the oviduct.

***b. Flank Recovery***

Some workers[49] have recovered eggs via a flank incision, but this approach is unsatisfactory due to the limitation on handling imposed by the broad ligament. It may be impossible to flush some cows and in others bilateral incisions may be necessary.

***c. Transvaginal Recovery***[50]

With this technique the peritoneal cavity is entered through an incision in the dorsal vaginal wall immediately above the cervix (Figure 7). An apparatus consisting of a metal teat cannula attached to a rubber balloon catheter is mounted on a ring, placed on the operator's second finger, and introduced through this incision. The cannula is inserted into the lumen of the uterine horn close to the cervix so that the balloon lies within the uterus; it is then inflated with 30 m$\ell$ air. The finger is removed from the ring and 20 to 80 m$\ell$ medium is introduced into the uterus through the catheter. The uterus is lifted and fluid is drained through the same catheter into a collecting vessel via a three-way valve.

The vaginal wound usually closes within 4 days, and few or no adhesions arise if both peritoneal and vaginal muscle layers are cut without separation of the peritoneum from the vagina. Several operations may be repeated on the same animal, but use of the technique is limited to parous cows.

***d. Intravaginal Oocyte Recovery***

Bedirian et al.[51] have described an intravaginal approach for oocyte recovery. The

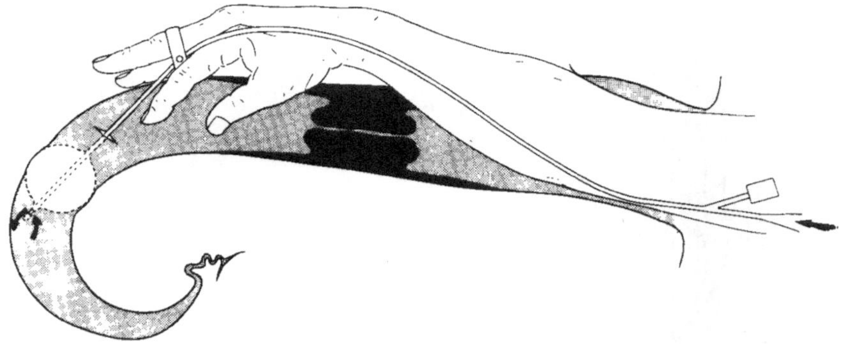

FIGURE 7. Transvaginal method for egg recovery. An incision is made in the roof of the vagina with a guarded knife. A modified 2-lm urological catheter attached to a teat cannula is mounted on a finger ring device and is passed through the incision. The teat cannula and balloon catheter are poked into the uterus and the cuff inflated. The hand may then be freed from the finger ring and used to manipulate the uterus while it is being flushed with medium in an ebb-and-flow manner. (Reproduced from Testart, J. and Godard-Siour, C., *Theriogenology*, 4, 157, 1975. With permission.)

method would seem to have little advantage over other surgical methods due to the difficulty in handling, its restriction to adult animals, and no reduction in subsequent adhesions. A similar method using a needle to transfix the vaginal wall and draw oocytes from the ovary has been described by Marcinkowski et al.[52]

*3. Nonsurgical Egg Recovery*

As early as 1949 two methods for the recovery of cow eggs had been developed. Rowson et al.[2] described the use of a long 3-lumen rubber catheter while a second method[3] was demonstrated by Dracy et al.[53] at the First National Egg Transfer Breeding Conference at San Antonio, Texas in 1949. More recently many authors have described methods for nonsurgical egg recovery; all can be classified into three basic systems, viz. fixed- and variable-distance circulation systems and to-and-fro-systems (Figures 8 and 9D). Circulation systems have a separate fluid inlet and outlet between which fluid circulates. The distance between inlet and outlet may be fixed or variable (Figures 9-4 and 8A). These devices are usually manipulated to the location of embryos at the ovarian extremity (tip) of the uterus. The "to-and-fro" systems rely on the ebb-and-flow movement of fluid to collect embryos and, hence, need only to be placed within the base of a uterine horn (Figure 8B).

An example of each of these three-basic systems will be described in detail and other similar methods noted.

***a. Fixed Distance Circulation System[54]***

The vulva and perineum are disinfected with 70% alcohol. A sterile metal speculum is then passed into the vagina to the cervix (Figure 9A). The central core of the speculum is removed and a metal introducer (Figure 9B) passed through the speculum as far as the cervix. The metal introducer consists of an outer cannula and central insert (much like an elongated blunt trochar and cannula) with its leading end formed into a smooth bullet shape which facilitates its passage through the cervix. For heifers the overall length of the introducer is 52 cm and its maximum external diameter is 7.2 mm; for cows the length is 60 cm. The introducer is then passed through the cervix, using the rectal technique of cervical fixation, and is carefully manipulated into the uterine horn which is to be flushed (Figure 9C). When the leading end of the introducer

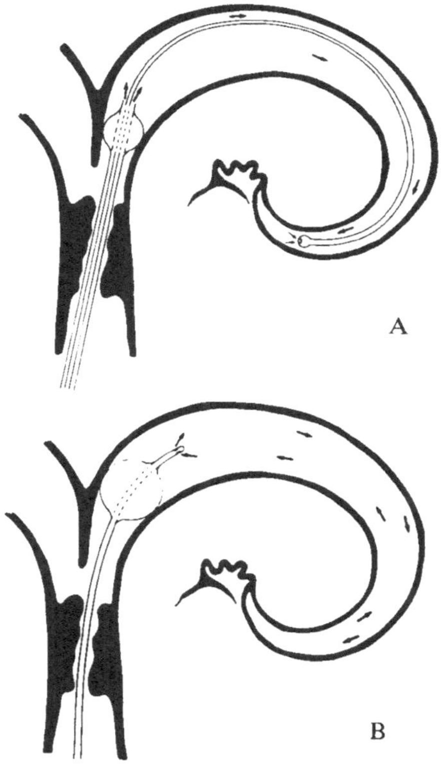


FIGURE 8. (A) Variable distance three-way method of flushing. (B) To-and-fro method of flushing using a 2-lm catheter. (Reproduced from Sugie, T., Soma, T., Fukumitsu, S., and Otsuki, K., *Bull. Natl. Inst. Anim. Prod. (Chiba)*, 25, 27, 1972. With permission.)

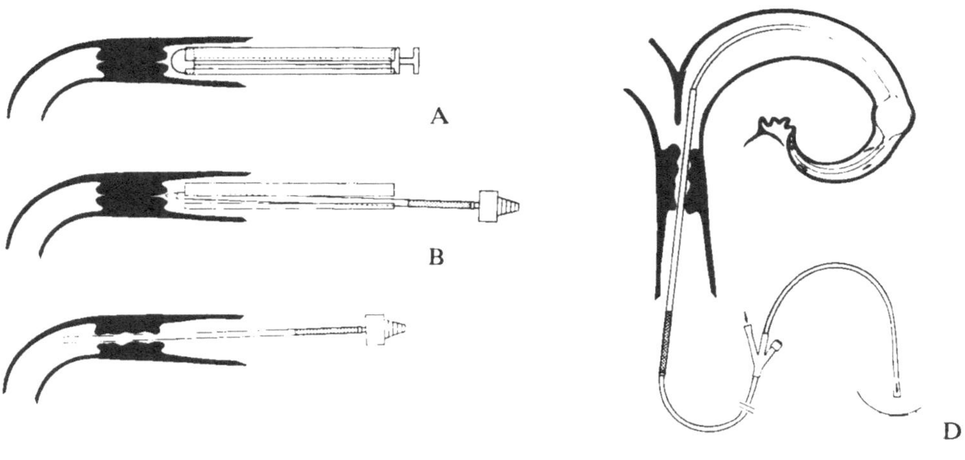


FIGURE 9. Three-way circulation system for the nonsurgical recovery of cow eggs with a fixed distance between fluid inlet and outlet. (A) Speculum passed into vagina to the cervix; (B) central core of speculum removed and introducer passed to cervix; (C) speculum removed and introducer passed through cervix into uterine horn; (D) central core of introducer removed and catheter passed to tip of uterine horn and balloon inflated — arrows indicate direction of flow.

is in or beyond the region of the intercornual ligament, the hand within the rectum is used to fix the cannula in position while an assistant removes the central core of the introducer and maintains asepsis. The catheter is passed through the cannula into the uterine horn by an assistant wearing sterile surgical gloves.

The catheter is constructed from a 14 Fg (Charrière gauge) PVC 3-lumen extrudate. Its length is 115 cm with a 42.5-cm extension to the fluid outlet. Fluid is passed into the uterus through an inlet hole at the leading extremity of the catheter and recovered through an outlet hole immediately anterior to a latex balloon of 20-mℓ capacity. The distance between the catheter tip and the fluid outlet hole is 14 cm. Generally the catheter passes freely to the ovarian extremity of the uterus, but when resistance is encountered the assistant must stop feeding the catheter into the cannula. The operator should only then release his grasp on the cannula and move his hand around the uterus until the tip of the catheter is located. After correcting the cause of the obstruction the hand is replaced around the cannula and the assistant instructed to continue passing the catheter. This procedure should be followed until the tip of the catheter is within 5 cm of the utero-tubal junction or until no further improvement in positioning can be obtained. The operator should then move his hand to the catheter tip and back to the vicinity of the catheter cuff. The assistant should slowly inflate the cuff with a small volume (2 mℓ) of air and sterile water (2 mℓ) until the operator is able to locate it. Then the assistant continues inflation with sterile water until the cuff is considered by the operator to have adequately occluded the uterus.

Flushing is then commenced using medium from 50-mℓ catheter syringes (Figure 9D). Initially, medium is flushed steadily into the uterus until a gentle flow of fluid is recovered, after which surges of fluid may be forced into the uterus which is thereby allowed to gently expand and empty. As little as 150 mℓ of medium may be used. During the flushing the operator may simply grasp the utero-tubal junction without manipulating it in any way (Figure 10A), or various gentle manipulative procedures may be carried out. The uterus may be pinched onto the catheter to direct fluid to the tip of the uterus (Figure 10B) and then finger pressure is released (Figure 10C), or fluid in the tip of the horn may be milked back (Figure 10D). When the flushing of the first uterine horn is completed, the catheter is withdrawn and the central core of the introducer replaced. The introducer is then manipulated into the opposite horn and the procedure repeated using another sterile catheter. The recovered medium may be collected into large test tubes and the eggs allowed to settle. After 10 to 20 min all but 10 mℓ of the supernatant fluid in each tube should be carefully withdrawn, the tube swirled, and its contents decanted into glass cups for microscopic examination.

#### *b. Variable Distance Circulation System*[55]

This device consists of an outer stainless steel tube (6-mm exterior diameter) at the leading end of which a latex balloon is fixed. Within this tube is a polyethylene tube at the leading end of which is a metal head with openings. This head locates within the leading extremity of the outer tube for introduction.

The instrument is inserted through the cervix and into the base of the uterine horn to be flushed. The latex balloon is inflated and the inner flexible polyethylene tube is gradually passed to the upper part of the uterus (Figure 8A).

Warm irrigation fluid is injected between the outer steel tube and inner polyethylene tube into the uterus from a reservoir, through a fall of 1 m. The fluid passes up the horn and into openings in the metal head at the tip of the inner polyethylene tube. The flushings are collected into 100-mℓ test tubes.

#### *c. "To-and-Fro" System*[56]

The vulval lips are parted and a cervical expander passed into the vagina to the

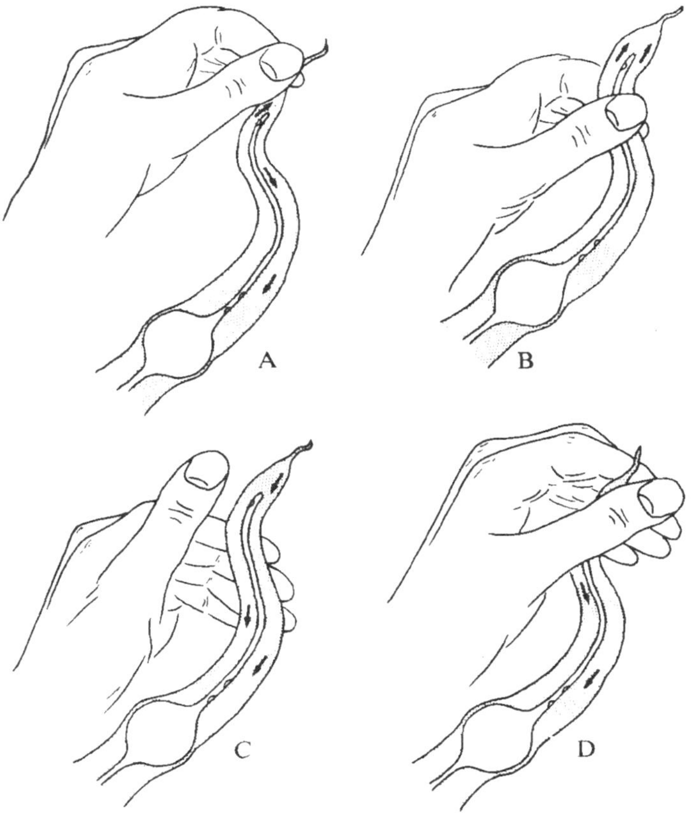


FIGURE 10. Manipulation procedures used during nonsurgical recovery of eggs with a fixed distance three-way catheter. (A) Utero-tubal junction occluded; (B) uterus pinched onto catheter directing fluid to tip; (C) uterus released, fluid drains to outlet; (D) fluid milked from tip of horn. (Reproduced from Newcomb, R., Christie, W. B., and Rowson, L. E. A., *Vet. Rec.*, 102, 414, 1978. With permission.)

external cervical os. The expander is manipulated through the lumen of the cervix until the tip is within the body of the uterus. The expander is then replaced by a sterile three-way Foley catheter fitted with a central stylet, and modified by enlarging the holes at the leading extremity and by lengthening inlet, outlet, and air tubes at its trailing end. The device is manipulated through the cervix and into one uterine horn just beyond the palpable bifurcation of the horns. The balloon is inflated carefully and the stylet removed.

About 20 m$\ell$ of flushing medium are run from a reservoir through the inlet tube of the Foley catheter and from the outlet to ensure that the system is patent. A drainage tube attached to the outlet of the Foley catheter is then clamped, and the uterine horn filled with medium while the utero-tubal junction is occluded by rectal manipulation. Then the clamp on the drainage tube is released and placed on the inlet tube. As the fluid from the horn flows into a collecting vessel the horn is gently massaged. The process is repeated until all fluid in the reservoir (800 m$\ell$) is used (Figure 11).

### *4. Factors Affecting Egg Recovery*

#### ***a. The Location of Eggs***

In untreated cattle, eggs enter the uterus about 4 days after estrus.[57] Using a tech-

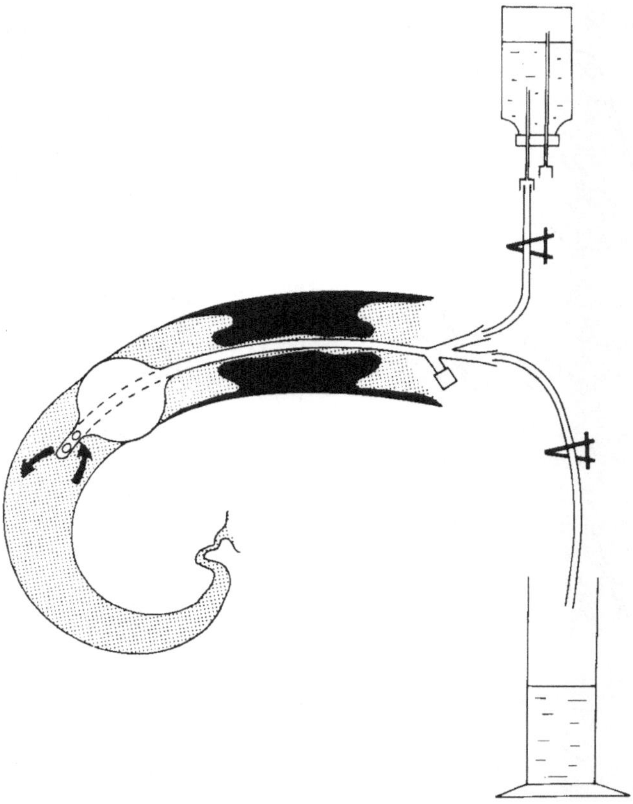

FIGURE 11. Method for to-and-fro flushing using a 3-lm catheter. Fluid is alternately run into the uterus from a reservoir and collected into a graduated cylinder. (Reproduced from Elsden, R. P., Hasler, J. F., and Seidel, G. E., Jr., *Theriogenology*, 6, 523, 1976. With permission.)

**Table 1**
**THE LOCATION, STAGE OF DEVELOPMENT, AND PROPORTION OF DEGENERATE EGGS RECOVERED ON DIFFERENT DAYS AFTER ESTRUS IN THE COW[47]**

| | Days after estrus | | | | |
|---|---|---|---|---|---|
| | 3 | 5 | 6 | 7 | 8 |
| No. of eggs | 146 | 354 | 281 | 307 | 163 |
| Proportion (%) of eggs recovered from the oviduct | 96 | 17 | 14 | 8 | 7 |
| Principal developmental stage recovered | 8-Cell | 16—32 cell | 32—64 cell | Blastocyst | Hatched blastocyst |
| Proportion (%) of degenerate eggs recovered (weighted means) | 13 | 16 | 23 | 24 | 55 |

nique to flush the oviduct and uterus separately Newcomb et al.[48] showed that in superovulated cattle there is a rapid reduction in the proportion of eggs recovered from the oviduct between 3 and 5 days after estrus (Table 1). However, even 7 and 8 days after estrus some eggs are retained within the oviduct. Similar findings were recorded by Moore.[58] These results confirm earlier findings on slaughtered cattle.[46]

### Table 2
### EFFECT OF POSITION OF TIP OF THREE-LUMEN CATHETERS ON THE PROPORTION OF EGGS RECOVERED FROM THE UTERINE HORN OF HEIFERS[54]

| Position (distance of catheter tip from utero-tubal junction) | 1 (0—2.5 cm) | 2 (2.5—5 cm) | 3 (5—10 cm) | 4 (10 cm) |
|---|---|---|---|---|
| Proportion (%) of eggs recovered | 56[a] | 61[a] | 38[b] | 23[b] |

*Note:* Values with dissimilar superscripts (a,b) are significantly different.

When a ligation technique was used to determine the location of eggs within the uterus 6, 7, and 8 days after estrus, it was found that 73% of all eggs were recovered from the tip (10 cm) of the uterine horn.[47,48]

*b. Method of Egg Recovery*

Commercially, it is unlikely that surgical methods of recovery will have further application except in very special situations. However, surgery has a definite place in a research setting where very early tubal or late developmental stages may be required.

Because eggs may be recovered from different sites at different times after estrus, the choice of method of recovery will inevitably influence success and will vary with the interval from estrus.

Surgical methods of egg recovery ultimately lead to the development of adhesions. These occur commonly between the greater omentum and the abdominal wall at the site of surgery, between the fimbria and ovary and between parts of the uterus. The development of adhesions may be minimized by careful handling, limiting exposure of the tract, ensuring that serosal surfaces are kept moistened with warm saline, and by using synthetic absorbable sutures (polyglycolic acid) in the peritoneal and muscular layers of the abdomen. Whatever techniques are used to minimize the development of adhesions, the tension required to exteriorize the bovine uterus is such that adhesions will be the ultimate sequel to repeated surgery. The development of adhesions can affect the future breeding potential of a donor and is, therefore, the overriding reason for using nonsurgical methods.

It is unlikely that nonsurgical methods can be as efficient as surgical methods of recovery, and a direct comparison suggests that although nonsurgical egg recovery techniques may approach quite closely surgical success rates, there is a disparity in favor of surgery.[54,59,60] In commercial practice recovery will be mostly performed nonsurgically, but the actual method used may well vary.

Using a fixed distance circulation system of recovery, the proximity of the catheter tip to the tip of the uterine horn had a highly significant effect ($p < 0.001$) on the efficiency of recovery[54,59,60] (Table 2) in heifers. Sreenan[61] has also shown that, using a circulation system of flushing with a three-way apparatus, the length of the catheter and thus the location within the uterine horn significantly ($p < 0.005$) affects results, favoring the use of a longer catheter (70 cm cf. 32 cm). Drost et al.[62] place the balloon of their three-way catheter in the body of the uterus and the tip of the device in the uterine horn being flushed and use a circulation system. This probably accounts for the poor recovery rate obtained from superovulated donors by these workers (2.1 ova per donor, 32% recovery rate). Similarly, because the three-way device of Baker et al.[63] was rigid, the curvature of the uterine horn would prevent its location being optimal for a circulation system. Hence, the poor recovery rate with this device could be explained (3.1 ova per donor, 35.7% recovery rate), although the device was not used conventionally for a circulation system. Devices placed in a similar location but which utilize a to-and-fro system rather than a circulation system for flushing have given acceptable results: 6.9 eggs per donor;[54] 6.3 eggs per donor, 54% recovery;[64] and 4.1 eggs per donor, 46% recovery.[65]

In adult cows the location of a 3-lumen circulation system catheter or a 2-lumen to-and-fro system catheter did not affect the success of recovery. A direct comparison between the two systems showed that 7 days after estrus the 3-lumen system provided a more efficient means of recovering eggs than the 2-lumen to-and-fro system, despite the fact that double the volume of fluid was used with the latter. However, there may be a change in favor of the two-way system on day 8, possibly associated with movement of eggs along the uterine horn.[60,66]

It is probable that a more efficient egg recovery will be obtained with a circulation system passed to the tip of the uterus than with a to-and-fro system when small volumes of fluid are used. With large volumes of flushing medium and the creation of turbulence, the to-and-fro systems have been shown to be efficient.[64,65] If many recoveries are to be made a circulation system is less time consuming since fluid enters and leaves the uterus simultaneously, unlike to-and-fro methods where fluid enters and leaves alternately and, therefore, takes longer to complete.

The age of the donor appears to influence egg recovery both with surgical[67] and with nonsurgical methods.[59,60] A greater proportion (approximately 10% more) of eggs is recovered from cows than from heifers. It may be because the adult tract is larger and the eggs are, therefore, more available to flushing procedures, or because heifers tend to respond with a greater number of ovulations than cows, but it may also be related to body size and a "dilution" of the effect of ovarian steroids on the uterus of the adult.

In heifers the cervix is narrow and the use of a narrow rigid or semirigid device is an advantage.[54,68,69] In cases where cervical size is greater, the diameter and rigidity of the instrument are less important. Flexible catheters with an internal stylet may be equally well passed into the uterus of parous cows.

### c. Interval from Estrus

When the differences in the efficiency of surgical egg recovery 3 to 8 days after estrus are compared, there is a significant ($p<0.001$) effect of "day" on the proportion of eggs recovered.[47] Significantly more eggs were recovered between days 1 and 4 (69.0 ± 2.8%) when compared with recovery between days 5 and 8 (49.8 ± 2.3%) ($t = 5.3$) or after day 9 (49.2 ± 6.4%) ($t = 2.8$). The proportion of eggs recovered on day 3 was 73.6 ± 3.1% compared with 58.2 ± 6.5% on day 4. Similar findings were recorded by Sreenan et al.[70] (day 3 — 78.3%, day 4 — 53.4%). Thus, when surgery is performed at a time after estrus when the eggs would normally be expected to have entered the uterus, the efficiency of recovery declines significantly. If maximum numbers of fertilized eggs are required then the recovery should be undertaken before the eggs enter the uterus.

The reasons for the lower recovery of eggs after day 4 might simply be explained by the larger volume and surface area of the uterine lumen making it physically more difficult to recover eggs. However, a more plausible explanation could be that a proportion of eggs is expelled shortly after entering the uterus. In superovulated cows estrogen levels remain at high levels until day 6.[71] It would be surprising, therefore, if transport was completely normal, although it has been suggested[72] that tubal transport of eggs in superovulated cattle is essentially similar to that recorded in untreated animals. Other authors have noted abnormal tubal transport and premature entry of eggs into the uterus.[46,73] Booth et al.[71] cite one particular cow in which estrogen rose to very high levels (280 pg/m$\ell$) and progesterone declined after day 4. Although this animal had 16 ovulations, no eggs were recovered on day 6 after estrus. Uterine activity usually disappears by day 4 or 5.[74]

### *d. hCG Treatment*

The intravenous administration of hCG at the beginning of estrus may affect subsequent egg recovery. The evidence is conflicting, for Moore[58] found that in cows, but not in heifers, hCG treatment appeared to reduce recovery and fertilization rates. On the other hand, Newcomb[66] found a significant improvement in egg recovery in cows after injecting 3000 i.u. hCG at the onset of estrus. These findings deserve reappraisal.

### *e. Ovarian Response*

Although the degree of ovarian response would be expected to influence steroid production,[71] neither the numbers of follicles and ovulations in the corresponding ovary nor their distribution between the ovaries appears to influence the surgery recovery of eggs following superovulation.[47,48] Scanlon et al.[72] found no differences in egg recovery rate when comparing animals with 1 to 12 ovulations (71% recovery) with those with more than 24 ovulations (70% recovery). Likewise neither Shea et al.[73] nor Beehan et al.[74] found any influence of ovarian response on surgical egg recovery. However, Hafez et al.[46] did find that the proportion of eggs recovered tended to fall with increasing numbers of ovulations, and Dowling[75] reported that rapid ovum transport is often associated with a high level of superovulation.

In contrast both Sreenan[76] and Newcomb[66] found that nonsurgical recovery rates decrease with increasing numbers of ovulations. If there really is a genuine difference between surgical and nonsurgical methods in this respect, its interpretation is a matter for conjecture.

### *f. Egg Viability*

The viability of the recovered eggs may be affected by several factors. Both semen quality and method of insemination have been shown to influence the proportion of unfertilized eggs after insemination with frozen semen.[60,66,76] When frozen semen is used to inseminate superovulated donors, "between bull" differences in the nonreturn rate — normally quite small — may be exaggerated, thereby producing significant differences in the proportion of unfertilized eggs recovered.[66,77] With frozen semen, the method of insemination has a most interesting effect. Thus, when insemination is restricted to the expected time of estrus the proportion of unfertilized eggs is higher than when it is repeated after the end of estrus. However, when insemination is performed after the end of estrus the proportion of eggs subsequently recovered is reduced,[66] possibly due to the later handling of the tract causing expulsion. It is, therefore, best if each donor is treated as an individual, and to ensure that none is inseminated late after the end of standing estrus, i.e., once ovulation has occurred. The normal development stages of eggs from superovulated donors at various times after estrus are indicated in Table 1. As the interval from estrus lengthens the proportion of degenerate eggs recovered increases (Table 1),[45,78] and between days 7 and 8 there is a significant increase in degenerate forms.[48] This appears to result from some adverse effect of the uterine environment following superovulation. When eggs were prematurely flushed from one oviduct into the uterus on day 3 and both utero-tubal junctions were ligated, then on day 7 or 8 few (18.5%) degenerate eggs were recovered from the opposite oviduct, whereas 2/3 of the eggs from the ipsilateral uterine horn were degenerate.[48] When cow eggs were removed from the oviduct on day 3 and transferred to the rabbit oviduct, a high proportion (85%) showed normal development on day 7.[79] Boland et al.[78] found that the proportion of abnormal eggs increases as the degree of ovarian stimulation increases. The level of PMSG (1500 or 2500 i.u.) affected, though not significantly, egg survival after transfer (70% and 56.7% pregnancy rates),[70] and there was no effect of the donor's ovulation rate on the pregnancy rate after transfer.[80] The quality of embryos from donors synchronized with progestagen-impregnated sponges was much poorer than from those synchronized with prostaglandin[81] (64% vs. 16% poor embryos, respectively).

#### *g. Nutritional Status of Donor*

This subject has been reviewed by Dunn,[82] who concluded that undernutrition does not have any marked effect upon either ovarian responsiveness or fertilization rate. However, he suggested that extremes of nutrition should be avoided from insemination until egg recovery.

Examination of the "tailhead condition" score in lactating Friesian cows has not revealed any significant relationship with ovarian response, egg recovery, or viability, although more viable embryos were recovered from animals in better condition.[66] The protein composition of the ration has been shown to influence basal levels and reserves of the gonadotrophin LH. With higher levels of protein in the ration, levels of LH and response to gonadotrophin releasing hormone were greater.[83]

The pool of follicles available for stimulation is probably determined several months previously[84] and, therefore, nutritional influences at the time of treatment would not necessarily be expected to produce any obvious relationships.

## III. MEDIA AND CONDITIONS FOR RECOVERY AND STORAGE OF EGGS

### A. Recovery, Storage, and Transfer Media

It is important that recovery and transfer media should not have any deleterious effect upon egg viability. The longer eggs can be maintained in vitro without adverse effect the greater the degree of flexibility in arranging for transfer and recipient use. There is little doubt that in the past the poor success rates were at least partly due to use of inadequate media.

Rowson et al.,[1] using TCM 199, were the first to obtain high rates of success after the surgical transfer of bovine eggs. However, with homologous serum no pregnancies were obtained. The first pregnancies to result from egg transfer in cattle were obtained with homologous serum,[4,42] and this presumably explains why bovine serum continued to be used despite the fact that Chang[85] had demonstrated that it was ovicidal.

Willett et al.,[42] who reported the birth of three calves from five transfers, used homologous serum as the medium for transfer. The medium was prepared by incubating the blood at 37°C for 30 to 60 min, centrifuging twice, and then storing the serum in a refrigerator for 1 or 2 days before the transfer. It was thought that the storage period would remove the ovicidal factor. The ovicidal factor is also destroyed by heat treatment (56°C for 30 min). Recovery and transfer media should have a macromolecular component such as albumin or heat-treated serum; without protein in the medium eggs adhere more readily to non-siliconized glassware.[59]

The use of TCM 199 with bicarbonate buffer and Hank's salts[1] represented a major breakthrough in bovine egg transfer. Subsequently, the suitability of TCM 199 was confirmed.[86-91] With this medium eggs transferred up to 2½ hr after recovery gave a satisfactory pregnancy rate.[1] This result was confirmed by Sreenan et al.[89] who, however, found that after 2 hr the duration of storage did have a significant effect ($p < 0.01$) on subsequent egg survival (Table 3). There is also an age effect on the "susceptibility" of eggs to storage in different media. Trounson et al.[79] compared the storage of day 3 and day 5 eggs in TCM 199 (with Hank's salts and bicarbonate buffer) with a phosphate buffered saline (PBS).[92] For example, with day-5 eggs there was no

Table 3
EFFECT OF STORAGE TIME IN MEDIUM TCM199 ON THE SURVIVAL OF COW EGGS FOLLOWING TRANSFER[89]

| | Storage time (min) | | | |
|---|---|---|---|---|
| | 40—80 | 81—120 | 121—180 | 181—480 |
| No. of eggs transferred | 30 | 32 | 28 | 20 |
| No. of eggs surviving (%) | 21 (70) | 24 (75) | 14 (50) | 6 (30) |

*Note:* Percentages in parentheses.

significant difference between these media as judged by the development of the eggs within the rabbit oviduct (PBS — 88% cf. TCM 199 — 71%). However, with day-3 eggs storage for 2 hr revealed differences (PBS — 71% of TCM 199 — 49%). Nevertheless, it does not necessarily follow that TCM 199 is a poor medium for egg recovery and transfer. The TCM 199 used by Trounson et al.[79] was unbuffered and, as $CO_2$ was lost, the pH of the medium changed (pH 6.9 to 7.5 after 1½ to 2 hr storage); moreover, whereas the PBS contained bovine serum albumin, the TCM 199 did not possess any macromolecular components.

Tervit[44] suggested that TCM 199 was superior to PBS as a recovery medium for cow eggs. There are, however, many reports which indicate that PBS will give consistent success when used as a recovery and transfer medium for cow eggs.[93-97] Other media which have been used for the transfer of cow eggs are Menezo's medium[98] (Schneider et al.[97]), Brinster's medium (Hahn et al.[96]), and Ham's F10 medium.

It is not necessary to maintain eggs at body temperature; in fact, it is probably preferable to maintain them in a suitable medium at room temperature (18° to 21°C).[99] Excessive cooling may be deleterious; although the critical temperature below which damage occurs has not been precisely defined, it has been shown that only 30 to 40% of day 5 to 6 morulae survive cooling to 0° and 7.5°C.[79] However, the early blastocyst is resistant to cooling.[100] Other workers have found that temperature control is important for the success of transfer,[101] although an earlier study[102] indicated that even four- to ten-cell eggs may be stored at 10°C in follicular fluid and still retain their viability up to 24 hr. Eggs have been kept for 24 to 48 hr at room temperature in PBS.[99]

**B. In Vitro Culture**

The requirements for culturing cow eggs in vitro have never been systematically examined. Early attempts to culture eggs used various natural media such as blood serum[103] and follicular fluid.[103-105] In general, 1 to 4 cell eggs seldom cleaved beyond 8 to 24 cells in culture. Tervit[44] cultured cow eggs under conditions similar to those used successfully for sheep eggs.[106] The medium was based on the composition of sheep oviduct fluid (SOF) described by Restall et al.,[107] and an atmosphere of 5% $CO_2$, 5% $O_2$, and 90% $N_2$, which had been found to be superior to 5% $CO_2$ in air,[40] was used. Development from eight-cell to morula and blastocyst stages was obtained in culture and transfer of some of these resulted in two pregnancies, but the success rate was low compared with that in sheep.

Kane[108] has reviewed the subject of bovine egg culture; he considers it likely that cow eggs will require relatively complex media. It is interesting to contrast the results of Wright and co-workers[109] in sheep and in cattle.[109,110] These workers cultured eggs in various media and gaseous atmospheres. In the ewe more cleavage divisions were obtained in the simpler defined media with SOF giving the best results, whereas in the cow most cleavage was obtained in the more complex media. Ham's F10 plus heat-treated fetal calf serum (HTFC) gave the best results, and none of the other media supported development beyond the early blastocyst stage (32 to 64 cells). Tervit's[44] observation that an atmosphere with a reduced oxygen concentration (5% $CO_2$, 5% $O_2$, and 90% $N_2$) gave the best results with sheep eggs was also true for cow eggs, giving significantly better results ($p < 0.01$) than an atmosphere of 5% $CO_2$ in air.[110] The stage at which eggs are first cultured is important to success, later stages being more easily cultured than earlier ones. The successful culture of one-cell eggs to the blastocyst stage has been reported, but later cleavage stages may more readily develop into blastocysts.[111]

Wright et al.[111] were unable to obtain the hatching of bovine blastocysts in culture without the addition of HTFC. Because bovine serum albumin (BSA) did not enable hatching to take place, it was concluded that some constituent present in HTFC was, therefore, essential. Renard et al.,[112] also using hatching as a criterion of successful culture, found that Menezo's medium[98] was superior to Brinster's medium, $BMOC_3$[113] for the culture of day-6 and -7 eggs to the day-10 stage. After the transfer of two eggs collected on day 6 and cultured for 4 days in Menezo's medium, embryos were present at day 20. After the culture of eggs for 24 hr similar pregnancy rates have been obtained to those from transfers of fresh eggs,[114] but culture for longer periods may reduce egg viability.[100,114] Embryonic wastage is increased after transfer of cultured embryos.[115]

### C. In Vivo Culture

There are several reports on the successful development of cow eggs in the rabbit oviduct.[105,116,117] Apparently, the endocrine status of the rabbit, whether in the follicular or luteal phase, is not important for success.[116] Eight-cell eggs transferred to the rabbit oviduct developed to late morulae or blastocysts after 2 or 4 days, respectively, and showed normal viability when retransferred to recipient heifers.[118] There is evidence that a higher conception rate may be obtained with those eggs which survive a period of culture in the rabbit oviduct than with those transferred directly to recipient heifers,[44,119] and fewer are degenerate[79] in comparison with others recovered at the same stage from the cow uterus.[48] The retention of cow eggs within the bovine oviduct by ligation of the utero-tubal junction 3 days after estrus resulted in a high proportion of morphologically normal eggs 4 days later.[48] Cow eggs retained within the sheep oviduct will also develop normally after retransfer to heifers.[120]

### D. Storage at Low Temperatures

For more prolonged storage of cow eggs attempts have been made to preserve them at low temperatures.

The first successful deep-freezing of cow eggs and subsequent birth of a calf was described by Wilmut et al.[121] using blastocyst stages (day-12). The possibility of using early morulae was also examined, but with little success, due to their inability to survive cooling even to 0°C. Wilmut postulated[122] that because later stage blastocysts were unaffected by cooling to 0°C, the egg would become resistant to cooling at some stage between eight-cell (day-3) and blastocyst.

Trounson et al.[100] showed that this stage was reached after day-6 by the blastocyst, although Wilmut et al.[123] had found satisfactory development after the culture of previously cooled day-6 morulae. Cooling of day-5 morulae had a deleterious effect, but a high proportion of day-7 early blastocyst stages remained viable.[124] After storage of blastocysts at 0°C for 48 hr before transfer to ten recipients, six pregnancies were obtained and 8 out of 20 transferred embryos survived.[100] It does not seem advisable to store eggs at 0°C for periods longer than 48 hr.[100] After very slow cooling Sreenan et al.[102] stored earlier eggs (days 2 to 4) at 10°C and then transferred them to rabbit oviducts to assess viability. All eggs so treated showed some further development following transfer, but no eggs stored for 46 hr survived.

The identification of an egg stage which survived cooling to 0°C led to immediate improvement in the success of deep freezing of bovine eggs.[125] Later Willadsen devised a two-step procedure for the freezing of bovine eggs which resulted in a high survival rate after thawing and transfer.[126] The validity of this method has been confirmed by field trial[127,128] and in commercial use,[127,129] but the results have not been as good as those of Willadsen et al.[126] Willadsen[130] also devised a rapid freezing technique for sheep eggs, using a two-step procedure, which gave a high success rate. Similar methods applied to the cow also gave good results after thawing and culture. However, the use of this rapid freezing technique by other workers has given little success,[131,132] until a similar method was used by Bilton.[133]

The ultimate length of time that bovine eggs will remain viable in liquid nitrogen is unknown, but it is likely to be considerable.

## IV. TRANSFER OF EGGS

### A. Procedures

#### *1. Surgical Methods of Egg Transfer*

##### ***a. Midventral Laparotomy***

The uterus is exteriorized in the same manner as for the surgical recovery of eggs, except that because manipulation is minimal, only a very small incision is necessary. The ovary containing the corpus luteum should be located and the tip of the adjacent uterine horn exteriorized. An inexperienced operator may find it necessary to exteriorize both horns and examine both ovaries before transferring an egg.

Various methods of transfer may be used. Excellent results have been obtained using a fire polished glass pipette attached to either a 1- or 2-m*ℓ* disposable syringe.[1,88] To facilitate the transfer, the uterine wall is first pierced with a blunt-ended 18-g needle; the pipette containing the egg in 0.02-m*ℓ* medium is then passed into the uterine lumen and its contents evacuated. Alternatively, a flexible plastic cannula (2 to 3 fg) may be used to effect the transfer,[134] either being introduced through a puncture wound as described above or through a metal cannula made from a needle,[54] thereby reducing the chance of blood contaminating the cannula or the egg.

##### ***b. Flank Transfer***

With this method it is essential to perform a rectal examination to establish the presence and location of a corpus luteum which determines which flank should be prepared. A paravertebral block (L. 1, 2, and 3) may be performed and supported when necessary by local infiltration, or local infiltration anesthesia only may be used. When using this technique it is an advantage if animals are used to restraint. The animal is also rendered more tractable if a low dose of a tranquilizer (10 mg acepromazine IV) is administered. It is also advantageous to give a low volume epidural anesthetic (4 to 5 m*ℓ* 2% lignocaine) after other anesthetic preparations are completed, if only to immobilize the tail. Some operators squeeze the animal in a crate in order to immobilize it for this procedure. An incision 11.5 cm long is made downwards and forwards below the tuber coxae. The muscle is separated by blunt dissection and the peritoneum punctured with a poking action of the fingers.

It is not always possible to examine the ovary visually so as to confirm the rectal diagnosis of the presence of a corpus luteum, but if the incision has been made sufficiently far back, it is usually possible to exteriorize the tip of the uterine horn and do the transfer, using the methods described for the midventral approach. However, the recipient is never completely immobile and it is, therefore, preferable to use a flexible cannula (rather than a glass pipette) which is passed through a hole made into the lumen with a blunt needle. Some operators do not exteriorize the uterus but perform the transfer to the uterine horn within the abdomen.

### *2. Transvaginal Egg Transfer*

#### ***a. Testart's Method***[135]

This method employs a similar finger ring technique to that described by Testart et al.[50] for the recovery of eggs. The operator's hand is passed through an incision in the dorsal vaginal wall close to the cervix and a cannula containing an egg is introduced into the uterine horn (Figure 12). When the cannula can be moved freely in the uterine lumen the egg is injected in 0.2-m*ℓ* medium.

#### ***b. Sugie's Method***[136]

The apparatus used with this method consists of three concentric tubes. The outer tube has a closed rounded leading end with a small aperture at its center. It is passed to the anterior fornix of the vagina and a second tube with a long pointed needle attached to its leading end is passed within the outer tube. The needle is passed through the aperture in the outer tube and through the wall of the dorsal vaginal fornix. The needle is guided per rectum into the lumen of the uterus (Figure 13-1). This difficult procedure is facilitated by passing carbon dioxide gas ($CO_2$) through the needle tip from a supply line attached to the trailing end of the tube. The pressure of the $CO_2$ is controlled so that it will not dilate the muscular wall of the uterus, but when within the uterine lumen will dilate it. The egg is delivered via a long tubular pipette with a blunt needle at its leading end which projects beyond the sharp point of the needle attached to the intermediate tube (Figure 13-2). With this apparatus 17 of 68 recipients (25%) became pregnant after the transfer of 83 eggs.[53,55]

### *3. Transcervical Egg Transfer*

In order to transfer eggs via the cervix without surgery, existing artificial insemination equipment has usually been used, including glass insemination pipettes[137] and Cassou insemination guns.[138] The use of special instruments and modified insemination apparatus has also been described.[93,139,140]

### *4. Factors Affecting the Success of Egg Transfer*

#### ***a. Evaluation of Egg Quality***

Of primary importance to the success of transfer is the ability of the operator to distinguish between unfertilized and fertilized eggs and between viable and nonviable eggs.

When early cleavage stages are recovered then the difference between one-cell unfertilized eggs and fertilized eggs is usually quite obvious. However, the difference may be much less obvious when examining compacted morulae and blastocyst stages. Some unfertilized eggs can cause confusion even to experienced workers, and it may be only after an egg is stained and a single nucleus is observed that the diagnosis is confirmed. Distinguishing viable from nonviable eggs is again not always straightforward. Even-sized blastomeres with clearly delineated cell walls and clear cytoplasm are useful criteria for judging the viability of early-stage eggs. Conversely, uneven blastomeres and cloudy cytoplasm with indistinct cell outlines are suspect. Once the compacted morula stage is reached assessment may become more difficult. It is not uncommon to find individual blastomeres which remain isolated together with a smaller than usual compacted group. The first impression of such an egg may be that it is retarded, since it may resemble quite closely a two- or four-cell stage. The egg, however, may be quite viable, although because some cells have not compacted, it has less cytoplasmic mate-

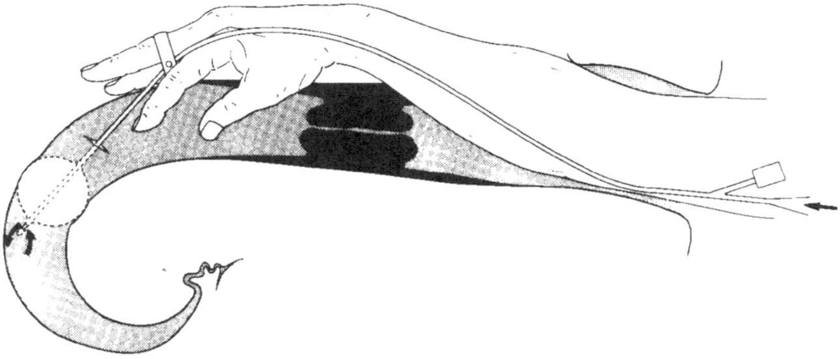

FIGURE 12. Transvaginal method for transferring eggs. The hand is passed through a dorsal vaginal incision. Using a finger ring a cannula containing the egg is inserted into the uterus and the egg expelled into the lumen. (Reproduced from Testart, J., Godard-Siour, C., and du Mesnil du Buisson, F., *Theriogenology,* 4, 163, 1975. With permission.)

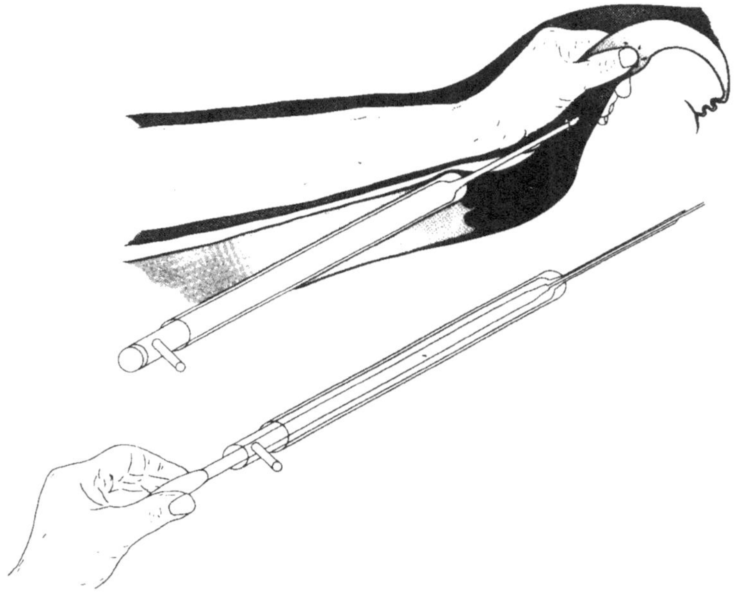

FIGURE 13. Transvaginal method for transferring eggs. (1) The uterus is grasped per rectum and a needle passed through an opening in the end of a stainless steel tube. The needle transfixes the dorsal vaginal and uterine walls. A carefully controlled flow of $CO_2$ gas through the needle expands the uterine lumen once correctly positioned. (2) An egg is expelled through a blunt needle, passed through the transfixing needle, attached to a long pipette. (Reproduced from Sugie, T., *J. Reprod. Fert.*, 10, 197, 1965. With permission.)

rial and should not be given the same assessment as a completely normal egg. Shea et al.[73] scored eggs on the basis of their appearance, using a 1-to-5 scale, with 3 being average, 1 very poor, and 5 excellent. In general the transfer of grade-5 eggs resulted in the highest pregnancy rate.

A similar system of grading eggs as excellent, good, fair, or poor has been used by Elsden et al.[141] with pregnancy rates following transfer of 63%, 58%, 31%, and 12%, respectively ($p < 0.05$). A significant ($p < 0.01$) effect of egg quality on success after transfer has also been found by Tervit et al.[142] ("good" vs. "poor" eggs — 64% vs. 27% pregnancy rate) and by Schneider et al.[97] ($p < 0.05$) ("good" vs. "fair" — 71% and 55%, respectively).

Less subjective ways of assessing egg quality have also been investigated. For example, Renard et al.[143] examined glucose uptake by culturing for 20 hr blastocysts, obtained from cows on day 10 or 11. Most of the blastocysts (81.3%) grew in culture and about 50% took up glucose. There was no detectable morphological difference between blastocysts which did or did not take up glucose, but development in vivo was superior among those taking up glucose (69% vs. 14%).

Mohr et al.[144] examined the use of fluorescein diacetate to assess the viability of mouse eggs, showing that nonviable eggs were not able to accumulate intracellular fluorescein. Preliminary work with cow eggs[145] has shown that normally developed eggs could accumulate fluorescein, but eggs which were retarded or those cultured in suboptimal media accumulated little or no intracellular fluorescein. The development of such techniques may in the future prove to be most useful.

#### *b. Transfer Medium*

The type of medium used to recover, store, and transfer eggs can have a considerable effect on the success of transfer; that aspect is discussed in the section on the short-term storage of eggs.

#### *c. Synchronization of Estrus in Donor and Recipient*

To obtain optimal pregnancy rates it is essential that the onset of estrus in donor and recipient should be closely synchronized.[1,87-90] Significantly poorer results are obtained when estrus in the recipient is ±1 day out of phase with that in the donor[90] (Figure 14). Information has now accrued with larger numbers of cattle under commercial situations,[97,146] but the results are not always consistent.

It is probable that the degree of synchronization required varies for different ages of egg, and that various mechanisms may be involved. For example, with early uterine transfers both age of egg and ejection may complicate the synchronization picture. The difficulty of evaluating some egg stages, particularly late morulae, may also play a part in the apparent requirements of synchronization. Recent evidence[140,147] would suggest that most uterine egg stages considered viable at transfer will persist at least until the beginning of luteolysis (day 16). Embryos must be present in the uterus by this time if pregnancy is to ensue.[148]

In view of findings in the sheep that the luteotrophic/luteolytic signal is specific to a certain age of embryo,[149] synchronization requirements at the time of egg transfer could be explained by different stages of development of embryo and uterus around the time of luteolysis. Our current knowledge of the mechanisms involved in synchronization is sparse.

#### *d. Age of Egg*

The age of egg transferred to the uterus has a significant effect on pregnancy rate.[47,90] Thus, the transfer of day-3 eggs is significantly less successful than the transfer of day-4, -5, -6, and -7 eggs, and day 4 transfers less successful than those of day-6 eggs (Figure 15). The poor conception rate achieved with uterine transfers of day-3 cow eggs is in marked contrast to results with sheep eggs of the same age, despite their morphological similarity.[150] In the cow, Hamilton et al.[57] could find no eggs in the uterus before 96 hr after the end of estrus, whereas in sheep most eggs are present in the uterus 66 hr after ovulation.[151]

Ejection of eggs from the uterus may be a factor in the poor conception rate after

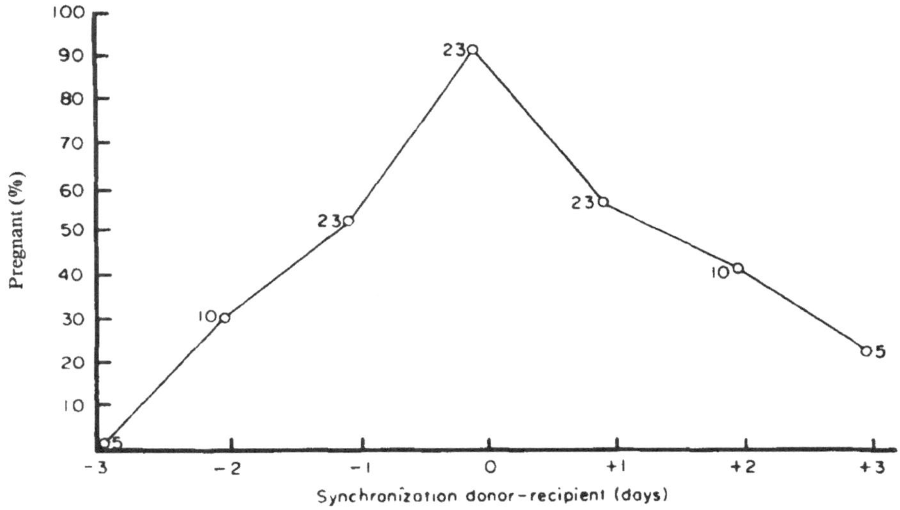


FIGURE 14. Conception rate after the transfer of cow eggs to synchronous or asynchronous recipients. (Reproduced from Rowson, L. E. A., Lawson, R. A. S., Moor, R. M., and Baker, A. A., *J. Reprod. Fert.*, 28, 427, 1972. With permission.)

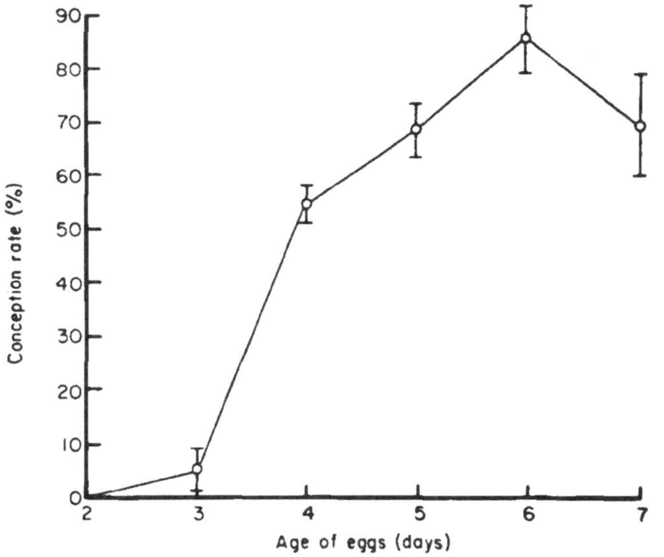


FIGURE 15. Conception rate after the transfer of cow eggs of different ages to uteri that were synchronous or out of synchrony by +1 or −1 day. Vertical bars represent ±SEM. (Reproduced from Newcomb, R. and Rowson, L. E. A., *J. Reprod. Fert.*, 43, 539, 1975. With permission.)

the uterine transfer of day-3 eggs. However, that the effect is principally one of egg age is evident from the finding that the transfer of day-4 eggs to a day-3 uterus was significantly more successful ($p < 0.01$) than that of day-3 eggs.[47,90] This suggests that day-3 cow eggs are unable to withstand the uterine environment. This conclusion is further supported by the observation that when day-3 eggs are prematurely flushed into the uterus there is an increase in the proportion of degenerate eggs compared with the population retained in the oviduct.[47,48] The highest pregnancy rate after surgical transfer to the uterus was obtained with day-5, -6, and -7 eggs.[47,90] The transfer of early morulae resulted in a lower pregnancy rate (61%; $p < 0.05$) than that of late morulae (67%), early blastocyst (67%), or late blastocyst (71%).[97] With nonsurgical transfers there is a suggestion that pregnancy rates are higher when using more advanced embryos.[95,152]

However, not all authors have observed the same effect. Newcomb[93] found no difference in success between transfers of day-7, -8, or -10 eggs. Possibly, the discrepancy in success rate found between day-6, -7, and later-stage eggs may be attributable to the greater difficulty in assessing the viability of younger stages.[93,153]

#### *e. Egg Expulsion*

Early attempts to transfer eggs nonsurgically using artificial insemination techniques were unsuccessful,[75,154-159] and it was not until 1964 that Mutter et al.[160] reported a pregnancy. Expulsion of eggs was considered to be responsible for the failure of nonsurgical transfers.[161-163] As a model to demonstrate the expulsion of eggs, radioactive microspheres have been used.[162,163] Expulsion of microspheres from the uterus was rapid (1½ hr after insertion) following their nonsurgical transfer through the cervix.

It had been previously shown that cervical stimulation induced the release of oxytocin and that oxytocin injections caused uterine contractions.[164-167] Harper et al.[162] considered that the expulsion of microspheres and eggs from the uterus was caused by uterine contractions stimulated by cervical manipulation during transfer. However, when drugs were administered to counteract the effects of oxytocin, the expulsion of microspheres was not prevented. This led Rowson et al.[163] to conclude that it was unlikely that oxytocin release after cervical stimulation was the cause of the ejection phenomenon.

Lawson et al.,[137] in an attempt to produce relaxation of the myometrium, transferred eggs nonsurgically to recipient heifers under halothane anesthesia. Despite its relaxant effect on the uterus, this method had no advantage over a control procedure.

One device was specifically designed to by-pass the cervix and deposit an egg into the uterus through a needle via the anterior fornix of the vagina, resulting in a second successful "nonsurgical" bovine egg transfer.[136] The successful egg was a 32-cell morula. In the unsuccessful attempts reported by Rowson,[154] none of the nonsurgically transferred eggs had more than 16 cells, and the age of eggs may have influenced the results.

Tervit[44] further examined the ejection of radioactive microspheres and found that it occurred only in the early luteal phase and that at later stages retention was good. These findings are supported by the nonsurgical transfer studies of Lawson et al.[137]

The electrical activity of the bovine myometrium has been recorded during the estrous cycle.[139] Maximum activity occurred at estrus and disappeared on day 4 or 5 of the cycle. Sham nonsurgical transfers on day 6 or day 8 did not induce electrical activity of the myometrium.

During the luteal phase of the estrous cycle the bovine uterus is relatively unresponsive to oxytocin treatment.[168] Treatment with oxytocin during the early luteal phase (days 1 to 8) results in an elevation of prostaglandin $F\alpha$ (PGF-$\alpha$) in posterior vena cava blood.[47,169] There were significant ($p < 0.02$) differences in the release of PGF between days, with a suggestion that the peak release occurs on day 3. Sham tran scervical transfer on day 3 resulted in what appeared to be a small release of PGF-$\alpha$ in uterine venous and posterior vena caval plasma, but on day 6 no release was observed.[47,170]

Taken together these observations indicate that, until day 4 after estrus, myometrial activity may be a substantial cause of failure of nonsurgical (or surgical) egg transfer. After that time, it is unlikely to influence results.

### *f. Infection*

The luteal phase bovine uterus is more susceptible to infection than that at estrus,[171,172] and the failure of transcervical transfer has been attributed to this susceptibility.[171,173]

Workers examining the deleterious effects of intrauterine insemination during pregnancy showed that the majority of pregnancies were terminated by this procedure, but the addition of antibiotics to the semen before insemination mostly prevented the effect.[174,175]

Dziuk et al.[158] tested the effect of including antibiotics in a study of sham egg transfer to previously inseminated cows and concluded that cervical "intervention" was not sufficiently harmful to prevent the normal development of the fertilized egg in all cases.

In a similar experiment, a concentric cannula device designed for egg transfer[176] was passed through the cervix of heifers which had been inseminated 6 days previously, and into the uterine horn ipsilateral to the corpus luteum. A strictly aseptic technique was used and no antibiotic was given. Twelve of the 25 treated heifers remained pregnant compared with 14 of 26 untreated control heifers.[177]

Brand et al.[176] performed similar sham transfers to the contralateral uterine horn of ten cows, five of which became pregnant. The frequency of bacteriological contamination of the device after sham transfers was also examined. The device consisted of two concentric stainless steel cannulae. It was introduced into the cervix and then the inner cannula was forced through a paper seal in the outer cannula and guided into the uterine horn. A ureteric catheter was then passed through this inner cannula as far as possible into the uterine horn and the embryo or "sham transfer" delivered in 0.5-mℓ medium. During sham transfer contamination occurred 100%, 66%, and 50% of times for the outer and middle cannulae and inner catheter, respectively. Of 12 recipients slaughtered 48 to 72 hr after sham transfer, bacteria were recovered from the uteri of two. The possibility exists, therefore, that bacteria or trauma inflicted upon the endometrium may alter the uterine environment and interfere with pregnancy without necessarily producing a clinically detectable endometritis. Other workers have found that bacteriological contamination of the transfer instrument is unrelated to the success or failure to establish pregnancy.[152]

Thus, although bacterial contamination cannot be completely excluded as a cause of failure of nonsurgical transfer, if a good aseptic technique is used and if antibiotics are included in the transfer medium, infection is unlikely to exert a major influence upon success.

### *g. Site of Transfer*

It is important that the egg should be placed within the uterine horn ipsilateral to the corpus luteum if a high pregnancy rate is to be achieved, because surgical transfer to the contralateral horn has resulted in poor survival.[178,179]

The actual position of an egg within the ipsilateral uterine horn also influences success. With a rigid device nonsurgical transfer cannot be effected deep into the uterus due to its curvature. In consequence, it is not possible to deposit eggs nonsurgically much beyond the intercornual ligament. The site of nonsurgical transfer is therefore quite different from that in surgical transfer, which is usually made to the tip of the uterus.

Anatomically the disposition of the uterine blood vessels in cattle is consistent with the hypothesis that the local pathway for uterine-induced luteolysis is a veno-arterial pathway. Furthermore, the uterine branch of the ovarian artery and the utero-ovarian arterial anastomosis are significantly more prominent on the side with a corpus luteum.[180] Anatomical considerations indicate that the direction of flow in the anasto-

Table 4
LOCATION OF THE CONCEPTUS WHEN UNILATERAL PREGNANCY IS ESTABLISHED AFTER BILATERAL EGG TRANSFER

| | Ipsilateral to corpus luteum | | Contralateral to corpus luteum | |
|---|---|---|---|---|
| | 1[a] | 2[b] | 1[a] | 2[b] |
| Number of eggs transferred | 13 | 13 | 13 | 13 |
| Number of embryos surviving (%) | 9 (69) | 10 (77) | 4 (31) | 3 (23) |

*Note:* Percentages in parentheses.
[a] From Sreenan, Beehan, and Mulvehill.[89]
[b] From Newcomb.[47]

mosis on the side of the corpus luteum should be principally towards the uterus. Thus, the role of the utero-ovarian arterial anastomosis may be to contribute to the preparation of the ipsilateral tube and uterine horn for pregnancy.

Possibly, an egg within the tip of the bovine uterine horn exerts a trophic effect upon the corpus luteum and in return the disposition of the blood vessels to the tip of the uterus ensures a locally increased supply of ovarian hormones which may influence egg development. Wintenberger-Torres[181] has shown that an increase in the level of progesterone increases the rate of embryonic development. Such an interrelationship could be important at the blastocyst stage and might help to explain the nonsurgical transfer results of Trounson et al.[152] In an experiment in which one egg was transferred to each uterine tip or to each uterine base, no influence of transfer site on success was revealed.[70] However, recent reports suggest that position is important.[93,140,182,183]

Transfer of a single egg to either the tip or base of the ipsilateral uterine horn has indicated the superiority ($p < 0.025$) of the tip as a site for transfer.[140] Similarly, bilateral transfer to different sites showed that fetal survival was higher in the case of the ipsilateral tip.[179] Overall fetal survival was greater when the egg was placed in the tip of the ipsilateral horn than in the base.

After transfers to each uterine tip and when only one fetus survives, significantly more single pregnancies establish within the ipsilateral uterine horn[47,179] (Table 4). With contralateral transfers, Del Campo et al.[184] showed that the luteotrophic/antiluteolytic signal is transferred from the contralateral to the ipsilateral horn via the uterine lumen. After contralateral transfer performed surgically the corpus luteum was maintained and the conceptus continued to develop in 8 out of 12 recipients until slaughter on day 24. The latter findings have been confirmed by Christie et al.,[147] who also showed that there was a substantial reduction in pregnancy rate between day 24 and 42. Previous studies indicating a low survival rate in the contralateral uterine horn[178,179] were, therefore, also substantiated. The reduction in pregnancy rate between day 24 and 42 could be prevented by the administration of hCG daily from day 13 to 35.[147]

The poor results after surgical contralateral transfer may reflect in exaggerated form the physiological problem associated with the suboptimal location of nonsurgically transferred eggs in the ipsilateral uterine horn. Christie et al.[140] showed that after nonsurgical transfer a high proportion of embryos survived both to day 16 (78.6%) and to day 24 to 26 (66.7%). Further loss to day 42 was prevented by the administration of hCG as above (pregnancy diagnosis at day 42, control 50%; hCG treated 65%).

Available evidence[185] suggests that the tip of the ipsilateral uterine horn is a favored site for egg development. Although a replicate comparison of surgical and nonsurgical transfers has not revealed differences in success rate,[93] the generally lower pregnancy rates achieved after nonsurgical compared with surgical transfers[97] — 66% and 44%, respectively — may be attributed in part to a suboptimal location and either an effect of site per se or its effect on the transmission of an embryonic signal.

### *h. Nonsurgical Technique and Operator Differences*

A "within animal" comparison of nonsurgical and surgical transfer[94] showed that despite an extremely high fetal survival after surgical transfer to the ipsilateral uterine horn (85%), the fetal survival rate following nonsurgical transfer to the contralateral horn was poor (35%). This could not be explained by an effect of site alone,[183] for similar transfers performed surgically resulted in a higher fetal survival in the base of the contralateral uterine horn than after nonsurgical transfer to this site, and a similar fetal survival in the ipsilateral tip.

Some component of the nonsurgical transfer technique itself may, therefore, be involved in addition to the effect of "site". For example, Boland's work[183] indicates that the time taken to perform a nonsurgical transfer influences the success rate. Thus, transfers taking more than 3 min, i.e., the more difficult ones, resulted in a lower pregnancy rate than easier transfers.[186]

Tervit et al.[142] have also found the highest rate of success was achieved with the fastest transfers. The presence of cervical mucus or trauma and serum exudation which would be expected to be greater during extended manipulation might be implicated. Comparison of nonsurgical transfers with a Cassou gun and transfer to the tip of the uterus using a catheter showed that more recipients became pregnant with the gun technique ($p < 0.05$),[142] and it was presumed that stress and possible endometrial injury caused the lower pregnancy rates after transfers by catheter. Similar results have been found by other workers.[187]

Further evidence that the ease with which the cervix is traversed may be important was provided by the work of Newcomb,[93] who found a higher pregnancy rate after the nonsurgical transfer of an egg to cows (80%) than to heifers (49%).

Significant differences in success after nonsurgical transfer have been found between operators.[97,188] So far the nonsurgical transfer of deep-frozen eggs has resulted in little success.[189]

### *i. Nutritional Status of Recipient*

It is probable that all those nutritional influences known to affect fertility after insemination will also affect success after egg transfer. In recipients in better condition, as judged by the tailhead score, the pregnancy rate was higher than in those in poor condition.[93] Gombe et al.[190] consider that ovarian hypofunction under conditions of restricted energy intake is due to a reduced ability of ovarian tissue to respond to LH. A submaintenance diet has been found to influence progesterone output within 5 days of the beginning of the period of undernourishment.[191]

Since the administration of progesterone to ewes can influence synchronization requirements,[192] changes in progesterone level may affect fertility after egg transfer. In practice, however, no significant difference in the serum progesterone levels was found between recipients which remained pregnant and those which did not.[193] Dunn[82] considers that the nutrition of the recipient must be controlled from the onset of estrus until about 40 days after estrus and that sudden changes, particularly those lasting a week or more, should be avoided. In order to provide adequate nutrition for fetal growth and body maintenance of the recipient, he suggests that they should be fed to gain 0.6 to 0.7 kg/day during the last trimester of gestation.

## B. Induction of Twin Pregnancy Using Egg Transfer

The highest rates of twin pregnancy have been obtained by surgical egg transfer rather than by injecting gonadotrophins to cause multiple ovulations.[93,178,194,196] Gonadotrophin treatment produces a very variable ovarian response which is not conducive to twinning. However, a major factor influencing success was noted by Gordon et al.,[197] who found that if two ovulations occurred unilaterally, the twin pregnancy rate was low, whereas when one ovulation occurred in each ovary it was much higher.

This observation was confirmed by Rowson et al.[86] using surgical egg transfer. The transfer of one egg to each uterine horn or of two eggs to the uterine horn ipsilateral to the corpus luteum gave similar pregnancy rates (72% and 73%), but the twin pregnancy rate after bilateral transfer was superior (72% and 45% of pregnant animals with twins, respectively). The difference between these groups has been attributed to competition between embryos. When both eggs are transferred to the same uterine horn the number of cotyledonary attachments is fewer than after bilateral transfer.[86] That a higher twin pregnancy rate may be achieved by surgical egg transfer has been confirmed by others.[89,183,198,199] However, field attempts to produce twin pregnancies using nonsurgical transfer have been much less successful.

There are various ways in which twin pregnancies might be induced with nonsurgical transfer. Either two eggs may be transferred to an unmated recipient[76,95,101,138,199-201] or one egg to a previously mated recipient.[76,101,197,198,201] The latter system obviously requires fewer eggs and has been the one most commonly examined. The same method had also been employed using a transvaginal surgical approach,[135] but it suffers from the disadvantage that the additional egg is usually transferred to the contralateral horn. Transfers only to the contralateral uterine horn result in a very low pregnancy rate.[170] Continuation of a pregnancy after nonsurgical transfer to the contralateral horn of a mated recipient is therefore largely dependent upon the survival ipsilaterally of the native egg. Boland et al.[200] did not find any recipient in their series pregnant to the transferred egg alone.

Although comparisons between single egg transfers or single egg transfer to mated recipients and artificial insemination have not revealed differences in success,[77,202,203] pregnancy rates have not usually approached those obtained by transferring two eggs surgically. Transfers to the contralateral horn of mated recipients have resulted in a similar pregnancy rate to that after artificial insemination, but the fetal survival rates of the nonsurgically transferred eggs has been poor (approximately 25 to 40%).[76,101] After the nonsurgical transfer of a single blastocyst to each uterine horn, Renard et al.[95] obtained a pregnancy rate of 60%, and of all recipients pregnant 50% carried twins.

In a "within-animal" comparison of surgical and nonsurgical transfers in which one egg was transferred surgically to the tip of the ipsilateral horn and one nonsurgically to the contralateral horn, thereby mimicking egg transfer to the mated recipient, there was a low survival of the nonsurgically transferred egg (35%) despite an extremely high fetal survival rate in the ipsilateral horn (85%).[94] In a subsequent experiment in which bilateral surgical transfers were made to each uterine tip, or to the ipsilateral tip and contralateral base, or to the ipsilateral base and contralateral tip or each base, it was shown that the only favored site for embryo survival appeared to be the tip of the ipsilateral uterine horn.[183] As survival rates after transfer to the other sites were similar, this experiment does not explain the poor survival after nonsurgical transfer in the earlier "within-animal" comparison of surgical and nonsurgical transfer.

Transfer of two eggs to the ipsilateral uterine horn[183] resulted in higher fetal survival rate than that reported by Rowson et al.[86] When one egg was transferred surgically to the tip of the ipsilateral horn and one to its base, one embryo migrated to the contralateral horn in a high proportion of resulting twin pregnancies (55.5%).

There is little doubt that the success achieved with bilateral surgical transfer is unlikely to be realized using nonsurgical transfer to establish twin pregnancy.

### C. Disease Control

The preservation of eggs of farm species by deep freezing offers considerable scope for an international trade in eggs. The first successful international transport of frozen bovine eggs — from New Zealand to Australia — was recorded by Bilton et al.[204]

Deep freezing brings with it the problem of ensuring that diseases exotic to the importing country are not introduced to that country, and thus necessitates the imposition of certain safeguards.

It is probable that the type of regulations governing the distribution of frozen semen will be imposed on the distribution of frozen eggs. There are, however, certain distinct differences in the manner of collection of eggs and semen which will influence what restrictions may be imposed if international trade in eggs is to be economically viable.

For example, at present it would seem that the macromolecular component in the medium used to freeze eggs should be provided by serum albumin or serum. In the rabbit Tsunoda et al.[205] have demonstrated the advantage of including rabbit serum in various media used to freeze rabbit eggs (49% live young after transfer cf. 14%). However, from a veterinary viewpoint such animal products might be considered to be undesirable, despite the fact that heat treatment (56°C for 30 min) of bovine fetal serum will destroy viral contaminants.[206] It is probable that restrictions on the source of such products to particular countries may be imposed by the veterinary authorities of some importing countries, although such restrictions could present considerable problems, for example, during any shortfall in supply.

Unless the urethra of the bull is cannulated during the collection of semen, it is unlikely that semen recovered from it will be uncontaminated. However, the surgical collection of eggs should produce a sterile product. The degree of contamination from vaginal secretions or cervical mucus during nonsurgical collection will depend on the operator and the technique which is being used.

For some countries and for certain diseases a sample of the collected semen from each animal is tested. For eggs this approach is ruled out because such restriction would render any commercial exercise nonviable, since the average number of viable eggs produced per donor per collection is only four.[65,66,207]

An alternative approach is to examine the flushing medium used, but this may be quite illogical in view of the manner in which eggs are handled. Firstly, the volumes of recovery medium used are large (100 to 1000 mℓ per uterine horn). Eggs are transferred from this initial "pool" in approximately 0.02-mℓ medium into a volume of 25-mℓ medium for bench storage. If there is any contamination with blood cells during this process the eggs would again be transferred to fresh medium. The present method used to freeze eggs commercially is that described by Willadsen et al.[99] The eggs are transferred to 0.5, 1.0, and 1.5 *M* solution of DMSO in PBS[92] and are then transferred to fresh 1.5 *M* DMSO in PBS in a glass vial for freezing.

During thawing the eggs are placed progressively in 1.5, 1.25, 1.0, 0.75, 0.5, and 0.25 *M* DMSO in PBS and then into PBS. Since it has been shown that after five serial dilutions of eggs which were initially placed in a viral contaminated solution the medium is sterile,[208] it is most unlikely that there would be any transmission of disease in the medium used to freeze eggs from any well-run laboratory. The remaining risks would appear to be from intrazonal pathogens. Although the zona pellucida acts as an effective barrier to most viruses, it has been shown that certain viruses can infect eggs of experimental animals.[209]

With a proper donor testing program the exportation of frozen eggs could be a safe and economically viable method of introducing new strains or breeds of livestock. Within a country the transfer of freshly collected eggs on farm to recipients in the same ownership would seem to impose little risk of disease transmission. However, if eggs are to be collected and frozen for distribution into other herds, certain restrictions must be imposed on the conduct of such work and on the disease status of donors.

From an animal health point of view the harvesting and distribution of eggs from slaughtered animals imposes considerable risks. It will be necessary, therefore, to ensure that donor animals undergo some screening process for transmissible diseases. It may also be necessary for these animals to come from approved premises where all resident animals undergo such screening. If the technology for transferring frozen eggs from slaughtered animals nonsurgically for twin pregnancy production is developed, whether or not it will ever be adopted commercially may be largely determined by the animal health restrictions applied and their consequent economic effect.

## REFERENCES


1. **Rowson, L. E. A., Moor, R. M., and Lawson, R. A. S.,** Fertility following egg transfer in the cow; effect of method, medium and synchronization of oestrus, *J. Reprod. Fert.*, 18, 517, 1969.
2. **Rowson, L. E. A. and Dowling, D. F.,** An apparatus for the extraction of fertilized eggs from the living cow, *Vet. Rec.*, 61, 191, 1949.
3. **Dracy, A. E. and Petersen, W. E.,** Isolation of ova from the living bovine, *J. Dairy Sci.*, 33, 797, 1949.
4. **Willett, E. L., Black, W. G., Casida, L. E., Stone, W. H., and Buckner, P. J.,** Successful transplantation of a fertilized bovine ovum, *Science (New York)*, 113, 247, 1951.
5. **Hervey, C.,** Thirty calves a year from your best cow, *Farm. J.*, 73, 46, 1949.
6. **Trounson, A. O., Willadsen, S. M., and Rowson, L. E. A.,** Fertilization and development capability of bovine follicular oocytes matured in vitro and in vivo and transferred to the oviducts of rabbits and cows, *J. Reprod. Fert.*, 51, 321, 1977.
7. **Shea, B. F.,** Recovery of bovine follicular oocytes and their fertilization in a recipient animal, *Theriogenology*, 9, 101, 1978.
8. **Sreenan, J.,** In vitro maturation and attempted fertilization of cattle follicular oocytes, *J. Agric. Sci. (Camb.)*, 75, 393, 1970.
9. **Shea, B. F., Bedirian, K. N., and Baker, R. D.,** Sperm penetration of bovine follicular oocytes following in vitro and in vivo maturation, *Biol. Reprod.*, 9 (Abstr. 59), 84, 1973.
10. **Thibault, C.,** Are follicular maturation and oocyte maturation independent processes?, *J. Reprod. Fert.*, 51, 1, 1977.
11. **Mandelbaum, J., Plachot, M., and Thibault, C.,** Role of the follicle in final zone pellucida maturation of hamster oocyte, *Ann. Biol. Anim. Biochem. Biophys.*, 17, 389, 1977.
12. **Thibault, C. and Gérard, M.,** Facteur cytoplasmique nécessaire à la formation du pronucleus mâle dans l'ovocyte de lapine, *C. R. Hebd. Seanc. Acad. Sci. (Paris)*, 270, 2025, 1970.
13. **Moor, R. M. and Trounson, A. O.,** Hormonal and follicular factors affecting maturation of sheep oocytes in vitro and their subsequent developmental capacity, *J. Reprod. Fert.*, 49, 101, 1977.
14. **Menezo, Y., Gérard, M., and Thibault, C.,** Culture du follicule de Graaf de bovin dans un système à courant liquide et gazeux continu, *C. R. Hebd. Séanc. Acad. Sci. (Paris)*, 283, 1309, 1976.
15. **Menezo, Y., Gérard, M., Szöllösi, D., and Thibault, C.,** In vitro exchange between the follicle and its culture medium, in *Control of Reproduction in the Cow*, Sreenan, J. M., Ed., Martinius Nyhoff, The Hague, 1978, 418.
16. **Plachot, M. and Mandelbaum, J.,** Non-fécondabilité de l'ovocyte de hamster maturé in vitro, *C. R. Hebd. Séanc. Acad. Sci. (Paris)*, 284, 953, 1977.
17. **Niwa, K. and Chang, M. C.,** Fertilization of rat eggs in vitro at various times before and after ovulation with special reference to fertilization of ovarian oocytes matured in culture, *J. Reprod. Fert.*, 43, 435, 1975.
18. **Thibault, C. and Gérard, M.,** Cytoplasmic and nuclear maturation of rabbit oocytes in vitro, *Annl. Biol. Anim. Biochim. Biophys.*, 13, 145, 1973.
19. **Usui, N. and Yanagimachi, R.,** Behaviour of hamster sperm nuclei incorporated into eggs at various stages of maturation, fertilization and early development, *J. Ultrastruct. Res.*, 57, 276, 1976.
20. **Warnes, G. M., Moor, R. M., and Johnson, M. H.,** Changes in protein synthesis during maturation of sheep oocytes in vivo and in vitro, *J. Reprod. Fert.*, 49, 331, 1977.
21. **Cross, P. C. and Brinster, R. L.,** In vitro development of mouse oocytes, *Biol. Reprod.*, 3, 298, 1970.
22. **Mukhertee, A. B.,** Normal progeny from fertilization in vitro of mouse oocytes matured in culture and spermatozoa capacitated in vitro, *Nature (London)*, 237, 397, 1972.

23. **Newcomb, R., Christie, W. B., and Rowson, L. E. A.**, Birth of calves after in vivo fertilization of oocytes removed from follicles and matured in vitro, *Vet. Rec.*, 102, 461, 1978.
24. **Pope, C. E. and Stephens, D. F.**, In vitro maturation of bovine follicular oocytes, *J. Anim. Sci.*, 38 (Abstr.), 216, 1974.
25. **Pope, C. E. and Turman, E. J.**, Maturation of bovine oocytes in vitro and in vivo, *J. Anim. Sci.*, 32 (Abstr.), 221, 1974.
26. **Hunter, R. H. F., Lawson, R. A. S., and Rowson, L. E. A.**, Maturation, transplantation and fertilization of ovarian oocytes in cattle, *J. Reprod. Fert.*, 30, 325, 1972.
27. **Shea, B. F., Latour, J. P. A., Bedirian, K. N., and Baker, R. D.**, Maturation in vitro and subsequent penetrability of bovine follicular oocytes, *J. Anim. Sci.*, 43, 809, 1976.
28. **Thibault, C.**, In vitro maturation and fertilization of rabbit and cattle oocytes, in *Regulation of Mammalian Reproduction*, Crozier, S. K., Corfman, P., and Condliffe, P. K., Eds., Charles C Thomas, Springfield, Ill., 1973, 231.
29. **Newcomb, R., Christie, W. B., and Rowson, L. E. A.**, unpublished data, 1978.
30. **Moor, R. M., Cahill, L. P., and Stewart, F.**, Ovarian stimulation or egg production as a limiting factor of egg transfer, *Proc. 9th Int. Congr. Anim. Reprod. A. I. (Madrid)*, 1, 43, 1980.
31. **Foote, W. D. and Thibault, C.**, Recherches expérimentales sur la maturation in vitro des ovocytes de truie et de veau, *Ann. Biol. Anim. Biochim. Biophys.*, 9, 329, 1969.
32. **Iritani, A. and Niwa, K.**, Capacitation of bull spermatozoa and fertilization in vitro of cattle follicular oocytes matured in culture, *J. Reprod. Fert.*, 50, 119, 1977.
33. **Brackett, B. G., Oh, Y. K., Evans, J. F., and Donawick, W. J.**, In vitro fertilization of cow ova, *Theriogenology*, 9, 89, 1978.
34. **Yanagimachi, R.**, In vitro capacitation of hamster spermatozoa by follicular fluid, *J. Reprod. Fert.*, 18, 275, 1969.
35. **Yanagimachi, R.**, In vitro acrosome reaction and capacitation of golden hamster spermatozoa by bovine follicular fluid and its fractions, *J. Exp. Zool.*, 170, 269, 1969.
36. **Breuer, D. J. and Wells, M. E.**, Effects of in vitro incubation of bovine spermatozoa in bovine follicular fluid, *J. Anim. Sci.*, 44, 262, 1977.
37. **Yanagimachi, R.**, In vitro fertilization of guinea-pig ova, *Anat. Rec.*, 172 (Abstr.), 430, 1972.
38. **Oliphant, G., Cabot, C. L., and Singhas, C. A.**, Nature of the rabbit acrosome reaction-inducing activity of follicular fluid, *J. Reprod. Fert.*, 50, 245, 1977.
39. **Lui, W., Cornett, E., and Meizel, S.**, Identification of the bovine follicular fluid protein involved in the in vitro induction of the hamster sperm acrosome reaction, *Biol. Reprod.*, 17, 34, 1977.
40. **Byrd, E. W., Goodeaux, L. L., Pool, S. H., and Godke, R. A.**, Effect of bovine serum albumin and $Ca^{++}$ on capacitation of bovine spermatozoa, *Theriogenology*, 11, 93, 1979.
41. **Lui, C. W. and Meizel, S.**, Biochemical studies of the in vitro acrosome reaction-inducing activity of bovine serum albumin, *Differentiation*, 9, 59, 1977.
42. **Willett, E. L., Buckner, P. J., and Larson, G. L.**, Three successful transplantations of fertilized bovine eggs, *J. Dairy Sci.*, 36, 520, 1953.
43. **Tervit, H. R. and Rowson, L. E. A.**, Viability of ova from slaughtered cattle, *7th Int. Congr. Anim. Reprod. A. I. (Munich)*, 1, 489, 1972.
44. **Tervit, H. R.**, Culture and Transfer of Sheep and Cattle Ova, Ph.D. thesis, University of Cambridge, England, 1973.
45. **Newcomb, R. and Rowson, L. E. A.**, A technique for the simultaneous flushing of ova from the bovine oviduct and uterus, *Vet. Rec.*, 96, 468, 1975.
46. **Hafez, E. S. E., Sugie, T., and Gordon, I.**, Superovulation and related phenomena in the beef cow. I. Superovulatory responses following PMS and hCG injections, *J. Reprod. Fert.*, 5, 359, 1963.
47. **Newcomb, R.**, Investigation of Factors Affecting Superovulation, Egg Recovery, and Transfer in Cattle, M.Sc. thesis, University of Cambridge, England, 1976.
48. **Newcomb, R., Rowson, L. E. A., and Trounson, A. O.**, The entry of superovulated eggs into the uterus, in *Egg Transfer in Cattle*, Rowson, L. E. A., Ed., Commission of the European Communities, Luxembourg, 1976, 1.
49. **Bell, L. and Bee, B.**, personal communication, 1974.
50. **Testart, J. and Godard-Siour, C.**, Transvaginal recovery of uterine eggs in the cow, *Theriogenology*, 4, 157, 1975.
51. **Bedirian, K. N. and Baker, R. D.**, An intravaginal method for the recovery and transfer of bovine eggs, *Can. J. Anim. Sci.*, 53, 67, 1973.
52. **Marcinkowski, D. P., Ludwick, T. M., and Ghafouri, M. T. A.**, A non-surgical technique for the recovery of follicular fluid from cattle, with possible applications to ova and embryo culture, *Theriogenology*, 13, 104, 1980.
53. **Dracy, A. E. and Petersen, W. E.**, Technique for isolating fertilized bovine ova by flushing the uterus with physiological solutions, in *Proc. Natl. Egg Transfer Breed. Conf.*, Foundation of Appl. Res., San Antonio, Tex., 1949.

54. **Newcomb, R., Christie, W. B., and Rowson, L. E. A.,** The non-surgical recovery and transfer of bovine embryos, in *Control of Reproduction in the Cow,* Sreenan, J. M., Ed., Martinius Nyhoff, The Hague, 1978, 292.
55. **Sugie, T., Soma, T., Fukumitsu, S., and Otsuki, K.,** Studies on ovum transfer in cattle with special reference to collection of ova by means of non-surgical techniques, *Bull. Natl. Inst. Anim. Prod. (Chiba),* 25, 27, 1972.
56. **Elsden, R. P., Hasler, J. F., and Seidel, G. E., Jr.,** Non-surgical recovery of bovine eggs, *Theriogenology,* 6, 523, 1976.
57. **Hamilton, W. J. and Laing, J. A.,** Development of the egg of the cow up to the stage of blastocyst formation, *J. Anat. (London),* 80, 194, 1946.
58. **Moore, N. W.,** The control of time of oestrus and ovulation and the induction of superovulation in cattle, *Aust. J. Agric. Res.,* 26, 295, 1975.
59. **Newcomb, R., Christie, W. B., and Rowson, L. E. A.,** Non-surgical recovery of bovine embryos, *Vet. Rec.,* 102, 414, 1978.
60. **Newcomb, R.,** Non-Surgical Recovery and Transfer of Bovine Ova, Fellowship thesis, Royal College of Veterinary Surgeons, London, 1979.
61. **Sreenan, J. M.,** Non-surgical egg recovery and transfer in the cow, *Vet. Rec.,* 102, 58, 1978.
62. **Drost, M., Brand, A., and Aarts, M. H.,** A device for non-surgical recovery of bovine embryos, *Theriogenology,* 6, 503, 1976.
63. **Baker, A. A. and Jillella, D.,** Technique of surgical and non-surgical ova collection of superovulated cows, *Vet. Rec.,* 103, 558, 1978.
64. **Rowe, R. F., Del Campo, M. R., Eilts, C. L., French, L. R., Winch, R. P., and Ginther, O. J.,** A single cannula technique for non-surgical collection of ova from cattle, *Theriogenology,* 6, 471, 1976.
65. **Greve, T., Lehn-Jensen, H., and Rasbech, N. O.,** Non-surgical recovery of bovine embryos, *Theriogenology,* 7, 239, 1977.
66. **Newcomb, R.,** Investigation of factors affecting superovulation and non-surgical embryo recovery from lactating British Friesian cows, *Vet. Rec.,* 106, 48, 1980.
67. **Newcomb, R., Rowson, L. E. A., and Trounson, A.,** unpublished data, 1976.
68. **Alexander, A. M., Marcus, A. N., and Hooton, J. N.,** Non-surgical bovine embryo recovery, *Vet. Rec.,* 99, 221, 1976.
69. **Rasbech, N. O.,** Instruments for non-surgical collection and transfer of bovine embryos, *Br. Vet. J.,* 135, 185, 1979.
70. **Sreenan, J. M. and Beehan, D.,** Methods of induction of superovulation in the cow and transfer results, in *Egg Transfer in Cattle,* Rowson, L. E. A., Ed., Commission of the European Communities, Luxembourg, 1976, 19.
71. **Booth, W. D., Newcomb, R., Strange, H., Rowson, L. E. A., and Sacher, H. B.,** Plasma oestrogen and progesterone in relation to superovulation and egg recovery in the cow, *Vet. Rec.,* 97, 366, 1975.
72. **Scanlon, P., Sreenan, J., and Gordon, I.,** Hormonal induction of superovulation in cattle, *J. Agric. Sci. (Camb.),* 70, 179, 1968.
73. **Shea, B. F., Hines, D. J., Lightfoot, D. E., Ollis, G. W., and Olson, S. M.,** The transfer of bovine embryos, in *Egg Transfer in Cattle,* Rowson, L. E. A., Ed., Commission of the European Communities, Luxembourg, 1976, 145.
74. **Beehan, D. and Sreenan, J. M.,** unpublished observations, 1978; quoted by Sreenan, J. M., 1959.
75. **Dowling, D. F.,** Problems of the transplantation of fertilized ova, *J. Agric. Sci. (Camb.),* 39, 374, 1949.
76. **Sreenan, J. M.,** Non-surgical egg recovery and transfer in the cow, *Vet. Rec.,* 102, 58, 1978.
77. **Newcomb, R., Rowson, L. E. A., and Trounson, A. O.,** The Sacrewell project. An on farm demonstration of the potential of egg transfer, *Vet. Rec.,* 103, 415, 1978.
78. **Boland, M. P., Crosby, T. F., and Gordon, I.,** Morphological normality of cattle embryos following superovulation using PMSG, *Theriogenology,* 10, 175, 1978.
79. **Trounson, A. O., Willadsen, S. M., Rowson, L. E. A., and Newcomb, R.,** The storage of cow eggs at room temperature and at low temperatures, *J. Reprod. Fert.,* 46, 173, 1976.
80. **Shea, B. F., Hines, D. J., Lightfoot, D. E., Ollis, G. W., and Olson, S. M.,** The transfer of bovine embryos, in *Egg Transfer in Cattle,* Rowson, L. E. A., Ed., Commission of the European Communities, Luxembourg, 1976, 145.
81. **Tervit, H. R., Cooper, M. W., Goold, P. G., and Haszard, G. M.,** Non-surgical embryo transfer in cattle, *Theriogenology,* 13, 63, 1980.
82. **Dunn, T. G.,** Relationship of nutrition to successful embryo transplantation, *Theriogenology,* 13, 27, 1980.
83. **Jordan, E. R. and Swanson, L. V.,** Serum progesterone and luteinizing hormone in dairy cattle fed varying levels of crude protein, *J. Anim. Sci.,* 48, 1154, 1979.
84. **Moor, R. M., Cahill, L. P., and Stewart, F.,** Ovarian stimulation of egg production as a limiting factor of egg transfer, *Proc. 9th Int. Congr. Anim. Reprod. A. I. (Madrid),* 1, 1980, 43.

85. **Chang, M. C.**, Effects of heterologous sera on fertilized rabbit ova, *J. Gen. Physiol.*, 32, 291, 1949.
86. **Rowson, L. E. A., Lawson, R. A. S., and Moor, R. M.**, Production of twins in cattle by egg transfer, *J. Reprod. Fert.*, 25, 261, 1971.
87. **Rowson, L. E. A., Lawson, R. A. S., Moor, R. M., and Baker, A. A.**, Egg transfer in the cow: synchronization requirements, *J. Reprod. Fert.*, 28, 427, 1972.
88. **Sreenan, J. M. and Beehan, D.**, Egg transfer in the cow: pregnancy rate and egg survival, *J. Reprod. Fert.*, 41, 497, 1974.
89. **Sreenan, J. M., Beehan, D., and Mulvehill, P.**, Egg transfer in the cow: factors affecting pregnancy and twinning rates following bilateral transfers, *J. Reprod. Fert.*, 44, 77, 1975.
90. **Newcomb, R. and Rowson, L. E. A.**, Conception rate after uterine transfer of cow eggs in relation to synchronization of oestrus and age of eggs, *J. Reprod. Fert.*, 43, 539, 1975.
91. **Betteridge, K. J. and Mitchell, D.**, Embryo transfer in cattle. Experience of twenty-four completed cases, *Theriogenology*, 1, 69, 1974.
92. **Whittingham, D. G.**, Survival of mouse embryos after freezing and thawing, *Nature (London)*, 233, 125, 1971.
93. **Newcomb, R.**, Surgical and non-surgical transfer of bovine embryos, *Vet. Rec.*, 105, 432, 1979.
94. **Newcomb, R., Christie, W. B., and Rowson, L. E. A.**, Comparison of the fetal survival rate in heifers after the transfer of an embryo surgically to one uterine horn and non-surgically to the other, *J. Reprod. Fert.*, 52, 395, 1978.
95. **Renard, J.-P., Heyman, Y., and du Mesnil du Buisson, F.**, Unilateral and bilateral cervical transfer of bovine embryos at the blastocyst stage, *Theriogenology*, 7, 189, 1977.
96. **Hahn, J., Moustafa, L. A., Schneider, U., Hahn, R., Romanowski, W., and Roselius, R.**, Survival of cultured and transplanted bovine embryos following surgical and non-surgical transfers, in *Control of Reproduction in the Cow*, Sreenan, J. M., Ed., Martinius Nyhoff, The Hague, 1978, 356.
97. **Schneider, H. J., Jr., Castleberry, R. S., and Griffin, J. L.**, Commercial aspects of bovine embryo transfer, *Theriogenology*, 13, 73, 1980.
98. **Menezo, M. Y.**, Milieu synthetique pour la survie et la maturation des gametes et pour la culture de l'oeuf fécondé, *C. R. Acad. Sci. (Paris)*, 282, 1967, 1976.
99. **Willadsen, S. M., Polge, C., and Rowson, L. E. A.**, In vitro storage of cattle embryos, in *Control of Reproduction in the Cow*, Sreenan, J. M., Ed., Martinius Nyhoff, The Hague, 1978, 427.
100. **Trounson, A. O., Willadsen, S. M., and Rowson, L. E. A.**, The influence of in vitro culture and cooling on the survival and development of cow embryos, *J. Reprod. Fert.*, 47, 367, 1976.
101. **Gordon, I. and Boland, M. P.**, Cattle twinning by non-surgical egg transfer, in *Control of Reproduction in the Cow*, Sreenan, J. M., Ed., Martinius Nyhoff, The Hague, 1978, 336.
102. **Sreenan, J., Scanlon, P., and Gordon, I.**, Storage of fertilized cattle ova in vitro, *J. Agric. Sci. (Camb.)*, 74, 593, 1970.
103. **Brock, H. and Rowson, L. E. A.**, The production of viable bovine ova, *J. Agric. Sci. (Camb.)*, 42, 479, 1952.
104. **Thibault, C.**, La culture in vitro de l'oeuf de vache, *Ann. Biol. Anim. Biochim. Biophys.*, 6, 159, 1966.
105. **Sreenan, J., Scanlon, P., and Gordon, I.**, Culture of fertilized cattle eggs, *J. Agric. Sci. (Camb.)*, 70, 183, 1968.
106. **Tervit, H. R., Whittingham, D. G., and Rowson, L. E. A.**, Successful culture in vitro of sheep and cattle ova, *J. Reprod. Fert.*, 30, 493, 1972.
107. **Restall, B. J. and Wales, R. G.**, The fallopian tube of the sheep. III. The chemical composition of the fluid from the fallopian tube, *Aust. J. Biol. Sci.*, 19, 687, 1966.
108. **Kane, M. T.**, Culture of mammalian ova, in *Control of Reproduction in the Cow*, Sreenan, J. M., Ed., Martinius Nyhoff, The Hague, 1978, 383.
109. **Wright, R. W., Anderson, G. B., Jr., Cupps, P. T., Drost, M., and Bradford, G. E.**, In vitro culture of embryos from adult and prepuberal ewes, *J. Anim. Sci.*, 42, 912, 1976.
110. **Wright, R. W., Jr., Anderson, G. B., Cupps, P. T., and Drost, M.**, Successful culture in vitro of bovine embryos to the blastocyst stage, *Biol. Reprod.*, 14, 157, 1976.
111. **Wright, R. W., Jr., Anderson, G. B., Cupps, P. T., and Drost, M.**, Blastocyst expansion and hatching of bovine ova cultured in vitro, *J. Anim. Sci.*, 43, 170, 1976.
112. **Renard, J.-P., du Mesnil du Buisson, F., Wintenberger-Torres, S., and Menezo, Y.**, In vitro culture of cow embryos from day 6 and day 7, in *Egg Transfer in Cattle*, Rowson, L. E. A., Ed., Commission of the European Communities, Luxembourg, 1976, 159.
113. **Brinster, R. L.**, In vitro culture of the embryo, in *Pathways to Conception*, Charles C Thomas, Springfield, Ill., 1971.
114. **Peters, D. F., Anderson, G. B., Bon Durant, R., Cupps, P. T., and Drost, M.**, Transfer of cultured bovine embryos, *Theriogenology*, 10, 337, 1978.
115. **Renard, J. -P., Heyman, Y., and Ozil, J.-P.**, Importance of gestation losses after non-surgical transfer of cultured and non-cultured bovine blastocysts, *Theriogenology*, 13, 109, 1980.

116. **Adams, C. E., Moor, R. M., and Rowson, L. E. A.**, Survival of cow and sheep eggs the in rabbit oviduct, Proc. *6th Int. Congr. Anim. Reprod. A. I. (Paris),* 1, 1968, 573.
117. **Sreenan, J. M. and Scanlon, P.**, Continued cleavage of fertilized bovine ova in the rabbit, *Nature (London),* 217, 867, 1968.
118. **Lawson, R. A. S., Rowson, L. E. A., and Adams, C. E.**, The development of cow eggs in the rabbit oviduct and their viability after re-transfer to heifers, *J. Reprod. Fert.,* 28, 313, 1972.
119. **Boland, M. P., Crosby, T. F., and Gordon, I.**, unpublished data; cited by **Gordon, I.**, *Egg Transfer in Cattle,* Rowson, L. E. A., Ed., Commission of the European Communities, Luxembourg, 1976, 305.
120. **Willadsen, S.**, personal communication, 1980.
121. **Wilmut, I. and Rowson, L. E. A.**, Experiments on the low-temperature preservation of cow embryos, *Vet. Rec.,* 92, 686, 1973.
122. **Wilmut, I.**, personal communication, 1973.
123. **Wilmut, I., Polge, C., and Rowson, L. E. A.**, The effect on cow embryos of cooling to 20, 0 and −196°C, *J. Reprod. Fert.,* 45, 409, 1975.
124. **Willadsen, S. M., Trounson, A. O., Rowson, L. E. A., Polge, C., and Newcomb, R.**, Preservation of cow embryos in vitro, *Proc. 8th Int. Cong. Anim. Reprod. A. I. (Krakow),* 3, 329, 1976.
125. **Willadsen, S. M., Trounson, A. O., Polge, C., Rowson, L. E. A., and Newcomb, R.**, Low temperature preservation of cow eggs, in *Egg Transfer in Cattle,* Rowson, L. E. A., Ed., Commission of the European Communities, Luxembourg, 1976, 117.
126. **Willadsen, S. M., Polge, C., and Rowson, L. E. A.**, The viability of deep-frozen cow embryos, *J. Reprod. Fert.,* 52, 391, 1978.
127. **Lehn-Jensen, H.**, Deep freezing of bovine embryos — a field trial, in *Control of Reproduction in the Cow,* Sreenan, J. M., Ed., Martinius Nyhoff, The Hague, 1978, 437.
128. **Lehn-Jensen, H. and Greve, T.**, Low temperature preservation of cattle blastocysts, *Theriogenology,* 9, 313, 1978.
129. **Coulthard, H.**, personal communication, 1978.
130. **Willadsen, S. M.**, Factors affecting the survival of sheep embryos during deep-freezing and thawing, in *The Freezing of Mammalian Embryos,* C.I.B.A. Foundation Symp. 52 (new series), Elsevier, Amsterdam, 1977, 175.
131. **Massip, A., Van der Zwalmen, P., Ectors, F., De Coster, R., D'Ieteren, G., and Hanzen, C.**, Deep freezing of cattle embryos in glass ampules or French straws, *Theriogenology,* 12, 79, 1979.
132. **Newcomb, R.**, unpublished data, 1979.
133. **Bilton, R. J.**, Preservation of embryos of the large domestic species, *Proc. 9th Int. Congr. Anim. Reprod. A. I. (Madrid),* 2, 1980, 245.
134. **Alexander, A. M. and Marcus, A. N.**, A modified technique for the transplantation of embryos into recipient cows, *Vet. Rec.,* 100, 73, 1977.
135. **Testart, J., Godard-Siour, C., and du Mesnil du Buisson, F.**, Transvaginal transplantation of an extra egg to obtain twinning in cattle, *Theriogenology,* 4, 163, 1975.
136. **Sugie, T.**, Successful transfer of a fertilized bovine egg by non-surgical techniques, *J. Reprod. Fert.,* 10, 197, 1965.
137. **Lawson, R. A. S., Rowson, L. E. A., Moor, R. M., and Tervit, H. R.**, Experiments on egg transfer in the cow and ewe: dependence of conception rate on the transfer procedure and stage of the oestrous cycle, *J. Reprod. Fert.,* 45, 101, 1975.
138. **Sreenan, J. M.**, Successful non-surgical transfer of fertilized cow eggs, *Vet. Rec.,* 96, 490, 1975.
139. **Brand, A., Drost, M., Aarts, M. H., and Gunnink, J. W.**, A device for non-surgical transfer of bovine embryos and its effect on uterine contamination, *Theriogenology,* 6, 509, 1976.
140. **Christie, W. B., Newcomb, R., and Rowson, L. E. A.**, Non-surgical transfer of bovine eggs: investigation of some factors affecting embryo survival, *Vet. Rec.,* 106, 190, 1980.
141. **Elsden, R. P., Nelson, L. D., and Seidel, G. E., Jr.**, Superovulating cows with follicle stimulating hormone and pregnant mare's serum gonadotrophin, *Theriogenology,* 9, 17, 1978.
142. **Tervit, H. R., Cooper, M. W., Goold, P. G., and Haszard, G. M.**, Non-surgical embryo transfer in cattle, *Theriogenology,* 13, 63, 1980.
143. **Renard, J.-P., Philippon, Y., and Menezo, Y.**, In vitro uptake of glucose by bovine blastocysts, *J. Reprod. Fert.,* 58, 161, 1980.
144. **Mohr, L. R. and Trounson, A. O.**, The use of fluorescein diacetate to assess embryo viability in the mouse, *J. Reprod. Fert.,* 58, 189, 1980.
145. **Church, R. B. and Raines, K.**, Biological assay of embryos utilizing fluorescein diacetate, *Theriogenology,* 13, 91, 1980.
146. **Kunkel, R. N. and Stricklin, W. R.**, Donor-recipient asynchrony, stage of embryo development and the post-transfer survival of bovine embryos, *Theriogenology,* 9, 96, 1978.
147. **Christie, W. B., Newcomb, R., and Rowson, L. E. A.**, Embryo survival in heifers after transfer of an egg to the uterine horn contralateral to the corpus luteum and the effect of treatments with progesterone or hCG on pregnancy rates, *J. Reprod. Fert.,* 56, 701, 1979.

148. **Betteridge, K. J., Mitchell, D., Eaglesome, M. D., and Randall, G. C. B.,** Embryo transfer in cattle 10 to 17 days after oestrus, *Proc. 8th Int. Congr. Anim. Reprod. A. I. (Krakow),* 3, 237, 1976.
149. **Rowson, L. E. A. and Moor, R. M.,** The influence of embryonic tissue homogenate infused into the uterus, on the life span of the corpus luteum in the sheep, *J. Reprod. Fert.,* 13, 511, 1967.
150. **Averill, R. L. W. and Rowson, L. E. A.,** Ovum transfer in the sheep, *J. Endocrinol.,* 16, 326, 1958.
151. **Holst, P. J.,** The time of entry of ova into the uterus of the cow, *J. Reprod. Fert.,* 36, 427, 1974.
152. **Trounson, A. O., Rowson, L. E. A., and Willadsen, S. M.,** Non-surgical transfer of bovine embryos, *Vet. Rec.,* 102, 74, 1978.
153. **Heyman, Y., Renard, J.-P., Ozil, J.-P., and du Mesnil du Buisson, F.,** Cervical embryo transfer at different stages in cattle, in *Control of Reproduction in the Cow,* Sreenan, J. M., Ed., Martinius Nyhoff, The Hague, 1978, 330.
154. **Rowson, L. E. A.,** Methods of inducing multiple ovulations in cattle, *J. Endocrinol.,* 7, 260, 1951.
155. **Willett, E. L.,** Egg transfer and superovulation in farm animals, *Iowa State Coll. Sci.,* 28, 83, 1953.
156. **Dracy, A. E.,** The future of ova transfer, *Iowa State Coll. Sci.,* 28, 101, 1953.
157. **Dziuk, P. J. and Petersen, W. E.,** Attempts at non-surgical transfer of bovine ova, *J. Anim. Sci.,* 13, 1019, 1954.
158. **Dziuk, P. J., Donker, J. D., Nichols, J. R., and Petersen, W. E.,** Problems associated with the transfer of ova between cattle, Minn. Agric. Exp. Stn. Tech. Bull. No. 222, 1958.
159. **Umbaugh, R. E.,** Superovulation and ovum transfer in cattle, *Am. J. Vet. Res.,* 10, 295, 1949.
160. **Mutter, L. R., Graden, A. P., and Olds, S.,** Successful non-surgical bovine embryo transfer, *A. I. Dig.,* 12, 3, 1964.
161. **Bennett, J. P. and Rowson, L. E. A.,** The use of radioactive artificial eggs in studies of egg transfer and transport in the female reproductive tract, *Proc. 4th Int. Congr. Anim. Reprod. A. I. (The Hague),* 2, 360, 1961.
162. **Harper, M. J. K., Bennett, J. P., and Rowson, L. E. A.,** A possible explanation for the failure of non-surgical ovum transfer in the cow, *Nature (London),* 190, 789, 1961.
163. **Rowson, L. E. A., Bennett, J. P., and Harper, M. J. K.,** The problem of non-surgical egg transfer to the cow uterus, *Vet. Rec.,* 76, 21, 1964.
164. **Vandemark, N. L. and Hays, R. L.,** The effect of oxytocin, adrenalin, breeding techniques and milking on uterine motility in the cow, *J. Anim. Sci.,* 10, 1083, 1951.
165. **Vandemark, N. L. and Hays, R. L.,** Uterine motility responses to mating, *Am. J. Physiol.,* 170, 518, 1952.
166. **Hays, R. L. and Vandemark, N. L.,** Effects of oxytocin and epinephrine on uterine motility in the bovine, *Am. J. Physiol.,* 172, 557, 1953.
167. **Hays, R. L. and Vandemark, N. L.,** Effect of stimulation of the reproductive organs of the cow on the release of an oxytocin-like substance, *Endocrinology,* 52, 634, 1953.
168. **Rowson, L. E. A.,** The movement of radio-opaque material in the bovine uterine tract, *Br. Vet. J.,* 111, 334, 1955.
169. **Newcomb, R., Booth, W. D., and Rowson, L. E. A.,** The effect of oxytocin treatment on the levels of prostaglandin F in the blood of heifers, *J. Reprod. Fert.,* 49, 17, 1977.
170. **Newcomb, R. and Rowson, L. E. A.,** Aspects of the non-surgical transfer of bovine eggs, *Proc. 8th Int. Congr. Anim. Reprod. A. I. (Krakow),* 3, 262, 1976.
171. **Rowson, L. E. A., Lamming, G. E., and Fry, R. M.,** Influence of ovarian hormones on uterine infection, *Nature (London),* 171, 749, 1953.
172. **Hawk, H. W., Brinsfield, T. H., Turner, G. D., Whitmore, G. W., and Norcross, M. A.,** Effect of ovarian status on induced acute inflammatory responses in cattle uteri, *Am. J. Vet. Res.,* 25, 362, 1964.
173. **Lamming, G. E. and Rowson, L. E. A.,** Ovarian hormones and uterine infection in cattle, *Proc. R. Soc. Med.,* 46, 387, 1953.
174. **Vandemark, N. L., Salisbury, G. W., and Boley, L. E.,** Pregnancy interruption and breeding techniques in the artificial insemination of cows, *J. Dairy Sci.,* 35, 219, 1952.
175. **Tanabe, T. Y., Heist, C. E., and Almquist, J. O.,** Factors affecting pregnancy interruption in artificial insemination of dairy cattle, *J. Dairy Sci.,* 38, 601, 1955.
176. **Brand, A., Drost, M., Aarts, M. H., and Gunnink, J. W.,** A device for non-surgical transfer of bovine embryos and its effect on uterine contamination, *Theriogenology,* 6, 509, 1976.
177. **Seidel, G. E., Jr., Bowen, J. M., Homan, N. R., and Okun, N. E.,** Fertility of heifers with sham embryo transfer through the cervix, *Vet. Rec.,* 97, 307, 1975.
178. **Newcomb, R. and Rowson, L. E. A.,** Multiple ovulation, egg transplantation, towards twinning, in *Principles of Cattle Production,* Swan, H. and Broster, W. H., Eds., Butterworths, London, 1976, 59.
179. **Sreenan, J. M.,** Egg transfer in the cow: effect of site of transfer, *Proc. 8th Int. Congr. Anim. Reprod. A. I. (Krakow),* 3, 269, 1976.
180. **Ginther, O. J. and Del Campo, C. H.,** Vascular anatomy of the uterus and ovaries and the unilateral luteolytic effect of the uterus: cattle, *Am. J. Vet. Res.,* 35, 193, 1974.

181. Wintenberger-Torres, S., Modification du milieu uterin chez les brebis superovulies et dévéloppment des blastocysts, *Proc. 6th Int. Congr. Anim. Reprod. A. I. (Paris)*, 1, 581, 1968.
182. Boland, M. P., Crosby, T. F., and Gordon, I., unpublished data cited by Gordon, I., 1976.
183. Newcomb, R., Christie, W. B., and Rowson, L. E. A., Fetal survival rate after the surgical transfer of two bovine embryos, *J. Reprod. Fert.*, 59, 31, 1980.
184. Del Campo, M. R., Rowe, R. F., French, L. R., and Ginther, O. J., Unilateral relationship of embryos and the corpus luteum in cattle, *Biol. Reprod.*, 16, 580, 1977.
185. Newcomb, R. and Rowson, L. E. A., Investigation of physiological factors affecting non-surgical transfer, *Theriogenology*, 13, 41, 1980.
186. Gordon, I., Cattle twinning by the egg transfer approach, in *Egg Transfer in Cattle*, Rowson, L. E. A., Ed., Commission of the European Communities, Luxembourg, 1976, 305.
187. Newcomb, R., Christie, W. B., and Rowson, L. E. A., unpublished data, 1978.
188. Halley, S. M., Rhodes, R. C., III, McKellar, L. D., and Randel, R. D., Successful superovulation, non-surgical collection and transfer of embryos from Brahman cows, *Theriogenology*, 12, 97, 1979.
189. Trounson, A. O., Brand, A., and Aarts, M. H., Non-surgical transfer of deep-frozen bovine embryos, *Theriogenology*, 10, 111, 1978.
190. Gombe, S. and Hansel, W., Plasma luteinizing hormone (LH) and progesterone levels in heifers on restricted energy intakes, *J. Anim. Sci.*, 37, 728, 1973.
191. Hill, J. R., Jr., Lamond, D. R., Henricks, D. M., Dickey, J. F., and Niswender, G. D., The effects of undernutrition on ovarian function and fertility in beef heifers, *Biol. Reprod.*, 2, 78, 1970.
192. Lawson, R. A. S. and Cahill, L. P., The survival of embryos transferred asynchronously to ewes treated with progesterone, *J. Reprod. Fert.*, 43, 385, 1975.
193. Hasler, J. F., Bowen, R. A., Nelson, L. D., and Seidel, G. E., Jr., Serum progesterone concentrations in cows receiving embryo transfers, *J. Reprod. Fert.*, 58, 71, 1980.
194. Anderson, G. B., Methods for producing twins in cattle, *Theriogenology*, 9, 3, 1978.
195. Gordon, I. and Boland, M. P., Towards cattle twins by egg transfer, *World Rev. Anim. Prod.*, 14, 5, 1978.
196. Newcomb, R., Towards twinning, *Rhod. Vet. J.*, 9, 2, 1978.
197. Gordon, I., Williams, C., and Edwards, J., The use of serum gonadotrophin (PMS) in the induction of twin-pregnancy in the cow, *J. Agric. Sci. (Camb.)*, 59, 143, 1962.
198. Sreenan, J. M. and Beehan, D., Effect of site of transfer on pregnancy and twinning rates following bilateral egg transfer in the cow, *J. Reprod. Fert.*, 48, 223, 1976.
199. Anderson, G. B., Cupps, P. T., Drost, M., Horton, M. B., and Wright, R. W., Induction of twinning in beef heifers by bilateral embryo transfer, *J. Anim. Sci.*, 46, 449, 1978.
200. Boland, M. P., Crosby, T. F., and Gordon, I., Twin pregnancy in cattle established by non-surgical egg transfer, *Br. Vet. J.*, 131, 738, 1975.
201. Boland, M. P., Crosby, T. F., and Gordon, I., Birth of twin calves following a simple transcervical non-surgical egg transfer technique, *Vet. Rec.*, 99, 274, 1976.
202. Boland, M. P. and Gordon, I., Twinning in lactating Friesian cows by non-surgical egg transfer, *Vet. Rec.*, 103, 241, 1978.
203. Sreenan, J. M. and McDonagh, T., Comparison of the embryo survival rate in heifers following artificial insemination, non-surgical blastocyst transfer or both, *J. Reprod. Fert.*, 56, 281, 1979.
204. Bilton, R. J. and Moore, N. W., Successful transport of frozen cattle embryos from New Zealand to Australia, *J. Reprod. Fert.*, 50, 363, 1977.
205. Tsunoda, Y. and Sugie, T., Effect of the freezing medium on the survival of rabbit eggs after deep freezing, *J. Reprod. Fert.*, 50, 123, 1977.
206. King, A. A. and Harkness, J. W., Viral contamination of foetal bovine serum, *Vet. Rec.*, 97, 16, 1975.
207. Brand, A., Trounson, A. O., Aarts, M. H., Drost, M., and Zaayer, D., Superovulation and non-surgical embryo recovery in the lactating dairy cow, *Anim. Prod.*, 26, 55, 1978.
208. Bolin, S. R., Runnels, L. J., and Sawyer, C. A., Exposure of fertilized porcine ova to pseudorabies virus, *Theriogenology*, 11, 92, 1979.
209. Bowen, R. A., Viral infections of mammalian preimplantation embryos, *Theriogenology*, 11, 5, 1979.

Chapter 7

# EGG TRANSFER IN THE SHEEP AND GOAT

N. W. Moore

## TABLE OF CONTENTS

## I. INTRODUCTION

Although egg transfer has been used more frequently in sheep than in goats, the procedures presently in use are essentially similar.

The early transfer works of Warwick et al.[1] and Lopyrin et al.[2,3] were only marginally successful, but they did show that sheep and goat embryos could survive transfer. In studies on hybridization between domestic sheep and goats, Warwick et al.[1] were not able to achieve the birth of hybrids, though they did report the birth of a kid to a doe following the collection and return to her of one four-cell embryo, while one of four ewes which received embryos from donor ewes produced a lamb at term. In an initial experiment Lopyrin et al.[2] obtained 8 lambs from 90 recipient ewes, one of which was born following the transfer of an unfertilized egg to a ewe which had been previously mated to a fertile ram. In a subsequent study by Lopyrin et al.[3] 9 lambs were born to 46 recipient ewes. Unfortunately, none of the early workers provided details of their transfer procedures.

Substantial improvement in the survival rate of transferred embryos has been achieved, and procedures developed and adopted for use in sheep[4-6] are basically those presently in use in both sheep and goats. Improvement in the survival of transferred embryos has been to a large degree due to strict attention to synchronization of the reproductive status of respective donors and recipients and to the collection and storage of embryos prior to transfer in media capable of supporting development of embryos in vitro.

## II. PREPARATION OF DONORS

### A. Superovulation

Serum gonadotrophins, e.g., PMSG and pituitary extracts of equine[7,8] and ovine[9] origin, will readily induce superovulation when given during the latter stage of the luteal phase of the estrous cycle. In the cyclic ewe PMSG is administered as a single intramuscular or subcutaneous injection on the 11th to 13th day of the cycle, whereas pituitary extracts need to be given as multiple subcutaneous or intramuscular injections commencing around the 12th day. The need for multiple treatment with pituitary extracts is presumably due to their greater rate of inactivation. In the cyclic doe PMSG should be given between days 16 and 18 and treatment with extracts should commence around day 17.

A crude extract (HAP) prepared from horse anterior pituitaries[8] has been extensively used in sheep.[10-13] Division of the total dose of HAP into three equal doses given on consecutive days commencing on day 12 is much more effective than when treatment is over 2 days (Table 1), but no advantage is achieved by further division into twice-daily treatments. In comparison with PMSG, pituitary extracts have the disadvantages of not being readily available, and the necessity for multiple injections. However, HAP appears to have somewhat different dose-response relationships, as can be seen from Tables 1 and 2. Frequently, a large proportion of the follicles which develop in response to high doses of PMSG fail to ovulate, and when high doses of PMSG are given ovulation rate can decrease due to failure of many follicles to ovulate. This does not appear to be a characteristic of HAP. The number of unovulated follicles remains reasonably constant as dose increases. Hence, when large numbers of ovulations are required, it may be more appropriate to use HAP rather than PMSG. Similar comparative data are not available for the doe, but observations on the Angora doe[15,16] suggest that high doses of HAP are not associated with a marked increase in numbers of unovulated follicles.

hCG, LH, LHRH, and their analogues have been used in varied combinations with

**Table 1**
**OVARIAN RESPONSE OF THE MERINO EWE TO A HORSE ANTERIOR PITUITARY EXTRACT (HAP)**

| Duration of treatment (days) | Total dose HAP (mg) | Mean number[a] | |
|---|---|---|---|
| | | Corpora lutea | Unovulated follicles |
| 2 | 60 | 2.1 | 2.8 |
| | 90 | 4.1 | 3.5 |
| | 135 | 9.4 | 3.8 |
| 3 | 60 | 6.4 | 2.6 |
| | 90 | 9.1 | 2.0 |
| | 135 | 13.3 | 3.8 |

[a] Twelve ewes per group.

Modified from Moore, N. W. and Shelton, J. N., *J. Reprod. Fert.*, 7, 79, 1964.

**Table 2**
**OVARIAN RESPONSE OF BORDER LEICESTER X MERINO EWES TO GRADED DOSES OF PMSG AND HAP**

| Preparation | Dose | Mean number[a] | |
|---|---|---|---|
| | | Corpora lutea | Unovulated follicles |
| PMSG[b] | 3 mℓ | 2.9 | 0.3 |
| | 6 mℓ | 4.8 | 3.5 |
| | 12 mℓ | 4.1 | 9.0 |
| HAP | 30 mg | 2.6 | 0.1 |
| | 60 mg | 4.5 | 1.6 |
| | 120 mg | 5.8 | 1.6 |

[a] Nine ewes per group.
[b] Raw PMSG.

Modified from Shelton, J. N. and Moore, N. W., *J. Reprod. Fert.*, 14, 175, 1967.

PMSG, but in general they have proved of little value. In ewes given a relatively low dose of PMSG (750 i.u.), hCG (400 to 800 i.u. intravenously) given within 3 hr after the onset of estrus did increase the proportion of follicles which ruptured,[17] but there is little evidence to suggest any beneficial effect of hCG following larger doses of PMSG. In the doe, Nishikawa et al.[18] found that 500 to 1000 i.u. hCG given intravenously on the 1st or 2nd day of estrus increased the proportion of follicles which ovulated. However, the does had been treated with multiple doses of PMSG prior to hCG. In Angora does Moore and Eppleston[16] found that a single injection of PMSG, or HAP given over 3 consecutive days, produced few unovulated follicles, providing evidence that little or no benefit is to be gained from the use of hCG.

Differences in ovulation rate between breeds and between strains within breeds have

Table 3
**OVULATORY RESPONSE TO PMSG OF MERINO EWES SELECTED FOR AND AGAINST MULTIPLE BIRTHS**

| Selection | Dose PMSG (i.u.) | Number of ewes | Mean number of ovulations |
|---|---|---|---|
| Against multiples | 0 | 9 | 1.0 |
| | 375 | 10 | 1.1 |
| | 750 | 10 | 1.5 |
| | 1500 | 12 | 3.7 |
| For multiples | 0 | 11 | 1.5 |
| | 375 | 15 | 2.5 |
| | 750 | 15 | 3.8 |
| | 1500 | 15 | 7.1 |

Modified from Bindon, B. M., Ch'ang, T. S., and Turner, H. N., *Aust. J. Agric. Res.*, 22, 809, 1971.

been well documented, and there is good evidence suggesting that ovulatory response to exogenous gonadotrophin reflects these differences. Merinos selected for multiple births show an increase in ovulation rate and their ovulatory response to graded doses of PMSG is almost twice that of ewes selected against multiples[19,20] (Table 3). Bindon et al.[21] and Cahill et al.[22] were unable to relate circulatory levels of gonadotrophins to differences in ovulation rate between Romanov and Ile-de-France ewes, and they concluded that differences in ovulation rate were primarily due to differences in ovarian sensitivity to gonadotrophins. Thus, it seems likely that animals characterized by high ovulation rates will produce larger numbers of eggs in response to exogenous gonadotrophin than those which normally shed few ova. Hence, superovulation and transfer could prove a useful aid in preferential selection for the more fecund animals within populations.

A major and as yet unresolved problem of transfer is the marked individual variation in ovulatory response to standard doses of gonadotrophin. Breed certainly contributes to the variation; advancing age as has been shown in the cow[23] depresses ovulatory response to PMSG and HAP and nutritional status may play a role. In the ewe treated with HAP during the latter part of the luteal phase of the cycle, ovulatory response has been shown to be related to the time elapsing between the end of treatment and onset of estrus.[8] Peak ovulatory responses were shown by ewes first detected in estrus 24 to 48 hr after treatment, whereas ovulatory response was depressed in those which were in estrus within 24 or after 48 hr of the end of treatment with HAP. However, none of the above, either singly or collectively, accounts for all of the individual variation in ovulatory response to gonadotrophins.

**B. Control of Time of Ovulation in Donors**

The use of gonadotrophins during the latter stages of the estrous cycle restricts transfer to the breeding season and does not allow precise prediction of dates of transfer.

In the cyclic ewe and doe progestagens and prostaglandin $F_{2\alpha}$ (PG) and its analogues effectively control the time of ovulation, and when used with gonadotrophins, marked precision of the time of superovulation can be achieved (see Table 4). In the ewe and doe progesterone is administered daily by intramuscular injection (10 to 12 mg/day), while progestagens can be effectively administered by intravaginal pessary (Synchromate® containing 30 mg cronolone — G. D. Searle; Repromap® containing 60 mg M.A.P. — Upjohn). Treatment should be continued for 2 to 3 days beyond the ex-

### Table 4
### TREATMENT REGIMES FOR THE CONTROL OF TIME OF SUPEROVULATION IN THE CYCLIC EWE AND DOE

| Regime | Time of estrus[a] | Ref. Ewe | Ref. Doe |
|---|---|---|---|
| **Progestagen/PMSG** | | | |
| Progesterone (10—12 mg/day SC or IM); PMSG day of last injection | 48—60 | 20 | 16 |
| Progestagen by intravaginal pessary; PMSG 1—1½ days before removal of pessary | 24—36 | 44 | 44 |
| **Progestagen/Pituitary Extract (HAP)** | | | |
| Progesterone (10—12 mg/day SC or IM); HAP on 3 consecutive days commencing day of penultimate injection | 48—60 | 24,25 | 44 |
| Progestagen by intravaginal pessary; HAP on 3 consecutive days commencing 2 days before removal of pessary | 24—36 | 44 | 44 |
| **Prostaglandins (PG)/PMSG** | | | |
| PMSG given 1—3 days before PG | 24—48 | 30 | 44 |

[a] Hours after last injection of progesterone, removal of pessary or injection of PG.

pected date of estrus, or when the expected date of estrus is not known treatment by injection or by pessary should be for 12 to 14 days in the ewe and 16 to 18 in the doe. PMSG is given on the day before the last injection of progesterone or 1 to 1½ days before removal of pessaries, while pituitary extracts (e.g., HAP) should be administered on 3 consecutive days commencing on the day of the penultimate injection of progesterone or 2 days before removal of pessaries. The majority of animals show estrus 48 to 60 hr after the final injection of progesterone, or 24 to 36 hr after removal of pessaries.

PGs are luteolytic in the ewe[26-28] and doe,[16,28,29] but they have not been widely used with gonadotrophins to induce superovulation at predetermined times. Trounson et al.[30] found 85% of ewes in estrus within 36 hr of a single intramuscular injection of the PG analogue Cloprostenol (Estrumate®) given 24 to 72 hr after PMSG. There is no published information available on the use of PGs and gonadotrophins for the induction of superovulation at predetermined times in the doe, but limited data from this laboratory suggest that the doe might respond in a similar manner to the ewe.

Progestagen/gonadotrophin treatment is effective for the induction of superovulation in the doe and ewe during periods of nonbreeding activity, but it is extremely unlikely that procedures involving PGs would be effective in animals devoid of corpora lutea. In the Merino ewe there appears to be little difference in response to progestagen/PMSG or HAP between animals showing either no or only limited breeding activity,[31] but the same may or may not be true of those breeds of sheep showing a well-defined period of anestrum.

The Angora doe, which does show a well-defined period of nonbreeding activity during the spring and early summer, will readily respond to progestagen/PMSG or HAP regimes during anestrum, but in some 30% or more of does, the induced corpora lutea regress prematurely within 4 to 5 days of ovulation with subsequent loss of viability of embryos.[32] This may not be so of milch breeds, as Corteel[33] has reliably induced estrus and obtained acceptable conception rates following the induction of estrus in anestrous does. However, the dose of PMSG used was low and not designed to induce superovulation.

Ovulation can be induced in the prepubertal lamb as early as 8 weeks of age by treatment with progesterone and PMSG,[34] and fertilization can be achieved by intrauterine insemination.[35] Trounson et al.[36] reported superovulation in 10 to 16-week-old Welsh mountain lambs given 750 to 1000 i.u. PMSG after progesterone treatment (10 mg/day for 3 to 18 days). Ovulatory response increased from a mean of 0.5 after 3 days pretreatment with progesterone to 9.8 following 18 days pretreatment. Some 80% of ova recovered after uterine insemination were fertilized. Four of six embryos transferred to ligated rabbit oviducts showed apparently normal development and one of three transferred to mature ewes developed into a normal lamb.

Limited work on the doe[37] indicates that similar procedures will induce superovulation with subsequent fertilization by natural service in prepubertal does of 5 to 6 months of age.

### C. Dose of Gonadotrophin

Dose/ovulatory response relationships have been established for PMSG[38,39] and for HAP,[8] and ideally the dose of gonadotrophin should be adjusted according to size and breed of donor to maximize yield of embryos. However, dose/response relationships exist only over a narrow range of doses. High doses of gonadotrophin can depress ovulatory response due to failure of a high proportion of follicles to ovulate, and in practice it would seem appropriate to administer a dose a little below that estimated to give a maximum ovulatory response in order to minimize the chances of excessive follicular response in individual animals.

PMSG and HAP have been used extensively in this laboratory and Table 5 summarizes our results. The doses shown have been found to induce high ovulation rates without overstimulation, as evidenced by excessive follicular responses.

### D. Response to Repeated Treatment

There is little published information for the ewe and none for the doe on the ovulatory response to repeated treatment with gonadotrophins.

In an early study involving only one ewe given 900 i.u. PMSG followed 16 days later by a further treatment with 1000 i.u. PMSG, 5 and 14 corpora lutea, respectively, resulted from the first and second treatment.[40] Subsequently, Hulet and Foote[41] suggested that the ewe does become refractory to PMSG, but remains refractory for only a brief period. They found that the first of two injections, each of 1000 i.u. PMSG spaced 33 days apart, depressed the ovulatory response to the second injection, whereas when 340 days elapsed between injections the ovulatory response to the second injection did not differ from that in ewes which had not been previously treated with PMSG. Similarly, Moore and Shelton[10] found no effect of an initial treatment with HAP on the ovarian response to a second treatment given 1 year later. Hence, it would seem that the ewe can become refractory to exogenous gonadotrophins, but it takes some little time for the refractory state to develop, and once developed it persists for a period of time less than 1 year.

The doe appears to become rapidly refractory to PMSG and HAP, and remains refractory. Eppleston[37] found that the mean ovulatory response of does given two injections each of 700 i.u. PMSG 38 days apart declined from 8.1 to 2.5. When 1 or more years elapsed between successive treatments with HAP or PMSG Moore and Eppleston[42] found that in no case, irrespective of dose, did Angora does produce more than two ovulations to the second treatment.

Table 5
EXPECTED OVULATORY RESPONSE TO GONADOTROPHIN

| Species | Breed | Body weight (kg) | Dose: PMSG (i.u.) | Dose: HAP[a] (mg) | Mean number of ovulations |
|---|---|---|---|---|---|
| Sheep | Merino | 40—50 | 1200—1300 | 80 | 10—12 |
| | Border Leicester | 60—70 | 1400—1600 | 100 | 12—15 |
| Goat | Angora | 25—30 | 1000—1100 | 45 | 10—12 |
| | Saanen | 50—60 | 1500 | 60 | 12—15 |

[a] Horse anterior pituitary extract prepared and administered as described by Moore and Shelton.[8]

## E. Mating of Donors

The method of mating is critical if high rates of fertilization are to be achieved. Complete and partial failure of fertilization to both artificial insemination (AI) and repeated natural service occur particularly if large doses of gonadotrophin are used. Failure seems to be due to faulty transport of spermatozoa through the cervix rather than to lack of fertilizability of ova shed by superovulated females. Failure of fertilization occurs more frequently in the ewe than in the doe and may well be associated with differences in patency of the cervix. At estrus the cervix of a high proportion of does can be penetrated with an insemination pipette of some 8-mm diameter,[29] whereas Fukui and Roberts[43] were able to penetrate the cervix of only 14 of 33 estrous ewes with a much smaller cannula made from a 17-gauge hypodermic needle. In the ewe the problem of failure of fertilization can be overcome by the direct deposition of about 0.02 mℓ semen into the tip of each uterine horn.[25] The uterus is exposed by midventral laparotomy preferably under local anesthesia. General anesthesia can be used but there is the chance of anesthetics, particularly barbiturates, interfering with ovulation. The uterus should be handled carefully and the ovaries and oviducts not exposed. Fertilization rates in excess of 90% can be achieved by surgical insemination.[13,25] Insemination should be carried out around the time of onset of estrus, as surgical insemination late in estrus, particularly near the time of ovulation, can depress the rate of recovery of embryos due to rapid transport of embryos through the reproductive tract.[44,45]

The benefits to be gained from surgical insemination are fully realized when precise control of the time of estrus and ovulation is achieved (see Table 4). All ewes can be inseminated at a predetermined time corresponding to the anticipated time of onset of estrus as shown in Table 6 for progestagen-treated animals. Fixed-time surgical insemination has a further advantage. Some 10 to 15% of ewes may not show estrus after treatment,[25] and if mating is by natural service or AI on detection of estrus they would be lost. The ovulatory response of ewes which do not show estrus does not differ from that of those which do exhibit estrus, and further their ova are readily fertilized. Hence, ewes which do not show estrus can be recovered if surgically inseminated at a fixed time after treatment. When large numbers of embryos of known age are required, as is almost invariably the case with transfer, control of time of superovulation coupled with surgical insemination should be adopted as a routine procedure.

Failure of fertilization in the doe does not appear to be the problem that it is in the ewe. In extensive studies on superovulation in the Angora doe fertilization rates of 65%, 79%, and 88%, respectively, were recorded following surgical insemination,

Table 6
TIME OF SURGICAL INSEMINATION IN THE EWE FOLLOWING TREATMENT WITH PROGESTAGENS AND GONADOTROPHINS

| Treatment[a] | Time of insemination[b] (hr) |
|---|---|
| Progesterone by injection/PMSG | 48—54 |
| Progesterone by injection/HAP | 48—54 |
| Progestagen pessary/PMSG | 24—30 |
| Progestagen pessary/HAP | 24—30 |

[a] Treatments as shown in Table 4.
[b] Hours after final injection of progesterone, or removal of pessary.

hand mating at 12 hourly intervals, and repeated natural mating.[16] Does were surgically inseminated within 12 hr of the onset of estrus, and in the light of recent experience[46] it is necessary to delay insemination to about 24 hr after the onset of estrus in order to achieve fertilization rates comparable to those obtained by repeated natural mating. When bucks producing high quality semen are available, then natural mating should be used. When frozen semen is to be used or individual bucks are to be mated to a number of does, then surgical insemination may be warranted. In view of the relative ease of penetration of the cervix in the estrous doe, insemination directly into the uterine horns via the cervix might be considered.

Insemination directly into the uterine horns using laparoscopic techniques has not been reported for either the ewe or doe, but it could well be a procedure worthy of investigation.

## III. PREPARATION OF RECIPIENTS

Potential recipients should be proven breeders, capable of carrying embryos to term and rearing young, and it is advisable that they be of similar size or larger than their respective donors. Recipients need not be of the same breed as their donors, there being no evidence in sheep of any difference between intra- and inter-breed transfers in the proportion of recipients supporting embryos.[4,6,47-50] Under some circumstances, interbreed transfers may be preferred. Recipients of highly fecund breeds will produce more multiple births than those of lesser fecundity,[47,49] while lambs of small breeds born to recipients of a larger breed tend to be heavier at birth than when born to mothers of their own breed.[47,48,50] Transfers between and within breeds of differing lengths of gestation have clearly shown the role played by the fetus in determining the length of gestation (Table 7).

### A. Synchronization between Donors and Recipients

Survival of embryos is dependent upon close synchrony between the time of onset of estrus in donors and recipients. In the ewe, maximum survival rates are achieved when donors and recipients are less than 24 hr out of synchrony.[51,52] Recipients will support embryos 2 days "older" or "younger" than themselves, but when donors and recipients are 3 or more days out of synchrony very few embryos survive transfer. Synchrony in the doe appears to be less critical than in the ewe. Recipients that are out of synchrony by as much as 48 hr with their donors will support embryos.[16]

The number of recipients required for each donor will depend upon expected ovulation rate in donors and number of embryos to be transferred to individual recipients,

Table 7
EFFECT OF BREED OF EMBRYO TRANSFERRED ON THE LENGTH OF GESTATION IN MERINO AND BORDER LEICESTER RECIPIENTS

| Breed | | Mean length of gestation (days) | |
|---|---|---|---|
| Embryo | Recipient | Single births | Multiple births |
| Merino | Merino | 150.1 | 150.4 |
| Merino | Border Leicester | 152.0 | 150.8 |
| Border Leicester | Border Leicester | 145.5 | 146.0 |
| Border Leicester | Merino | 145.4 | 146.8 |

Modified from Moore, N. W., *Aust. J. Agric. Res.*, 19, 295, 1968.

but on average ten potential recipients/donor should be adequate. Control of the time of estrus in recipients may be necessary when insufficient numbers are naturally synchronized with the donors. Progesterone by injection (10 to 12 mg/day), progestagen-impregnated pessaries, and PGs effectively control the time of estrus in the cyclic ewe and doe, and treatment with progesterone or pessaries does not affect the capacity of recipients to support embryos.[16,53] Some care may be needed in the use of PGs. Effective synchronization of estrus in groups of cyclic ewes and does in which dates of estrus are not known is achieved by two spaced PG injections.[29,54] However, the time elapsing between injections seems to be important. In the doe, Moore and Eppleston[16] found that the proportion of recipients which kidded increased from 39% to 67% when the interval between injections of the PG analogue Cloprostenol (Estrumate®) was increased from 10 to 14 days; suggestive evidence that the length of the luteal phase prior to the transfer estrus may be critical. A similar effect of interval between injections on fertility to AI in sheep has been observed.[55]

The majority of ewes and does will show estrus 3 to 5 days after the final injection of progesterone, 2 to 3 days after removal of pessaries, or 2 to 4 days after PG, and when similar procedures to control the time of estrus are to be used in donors and recipients the termination of progestagen treatment (final injection or removal of pessary) or injection of PG in recipients should be 1 day earlier than in donors.

## IV. EMBRYO RECOVERY

Essentially similar procedures are used for collection of embryos from the ewe and doe, and they differ little from those originally described by Hunter et al.[4]

The ovaries, oviducts, and uterine horns are exteriorized by midventral laparotomy under general anesthesia. A cannula of about 2 mm external diameter is inserted 2 to 3 cm into the oviduct via the fimbria and held in place with the fingers or soft-nosed artery forceps. Warmed (25 to 30°C) flushing medium (4 to 6 m$\ell$) is injected into the lumen of the uterine horn and carefully expressed through the oviduct and cannula into a round-bottomed collecting vessel. Using these procedures about 85% of ova shed, as estimated by counting corpora lutea, should be collected. Collections are usually carried out 3 to 7 days after mating. In the ewe and doe embryos enter the uterus around the 4th day after the onset of estrus, after which most of the uterus and oviducts should be flushed. Flushing can be carried out on day 8 and later, but by this time embryos will have developed to hatched blastocysts which frequently collapse during collection, making their identification in the flushing medium difficult. Embryos in the flushings are readily identified under a stereoscopic microscope (× 20 magnification). They should be removed from the flushings, transferred to fresh medium, and held at 25 to 30°C prior to transfer. Nonsurgical collection procedures, as used in the cow, have not been described for either the ewe or doe, and it would seem unlikely, because of the narrow and tortuous nature of the cervical canal of both species, that effective nonsurgical collection procedures will be developed.

The stage of development of sheep[8,56] and goat[16] embryos collected at various times after the onset of estrus has been described. Goat embryos tend to be some 12 to 24 hr less advanced than sheep embryos in their development. The difference is probably due to later ovulation (in relation to onset of estrus) in the doe rather than to any slower rate of cleavage of goat embryos.

Initially, homologous blood serum was used as flushing medium,[4-6] but it has now been replaced by balanced salt solutions enriched with 10 to 20% blood serum, 3 to 5 mg/mℓ bovine serum albumin, or 5% fetal calf serum. The most commonly used salt solution is Dulbecco's phosphate buffered saline. Antibiotic, e.g., 100 i.u. potassium penicillin + 50 i.u. streptomycin sulphate/mℓ, should be added and the medium filtered, e.g., Millipore® cellulose acetate filters of 0.3-μm pore size. After preparation media can be stored frozen for several months or held refrigerated for 3 to 4 days prior to use.

### A. Storage of Embryos Prior to Transfer

In most instances embryos are transferred within a few hours of collection with little evidence of any effect on viability of time elapsing between collection and transfer. Embryos should be stored in fresh flushing medium at 25 to 30°C, preferably in round-bottomed containers to aid identification. In early work on transfer, sheep embryos were held within dialysis membranes immersed in flushing medium.[5,6] It was thought that the dialysis membrane allowed dilution of metabolic products which might be harmful to embryos. However, survival rates achieved in subsequent studies in which dialysis chambers were not used indicate no need for this method of storage.

The duration of storage can be extended by in vitro culture, by cooling and holding embryos at 5 to 10°C, or by freezing in liquid nitrogen. Most collection media will support development in culture for 2 to 3 days when the appropriate culture conditions are maintained.[57-60] Development in culture should proceed at the same rate as in vivo and after culture embryos should be transferred to recipients synchronous with donors of the cultured embryos. Sheep and goat embryos have been successfully stored at 5 to 10°C, but the duration of storage without marked loss of viability is limited to about 4 days.[60-63] Little or no development of embryos takes place at 5 to 10°C, and it would seem advisable to transfer embryos after storage to recipients that are less advanced than donors.

Sheep embryos will survive and show continued development for up to 5 days in the oviduct of the rabbit,[64] and this novel method of storage has been used to transport sheep embryos between countries.[65]

## V. TRANSFER OF EMBRYOS

Transfer procedures are similar for sheep and goats and they are basically those developed and illustrated by Hunter et al.[4] Surgery is involved and it is likely, because of the nature of the cervix in both species, that nonsurgical transfer procedures similar to those used in the cow (see Chapter 6) will be extremely difficult or too time-consuming to be of value. Nonsurgical transfer has been attempted in the goat using very fine cannulae and cervical forceps.[66] Transfers were unsuccessful in all but one recipient which had been injected with hyoscine-*N*-butylbromide prior to transfer. Hyoscine-*N*-butylbromide blocks neural transmission across parasympathetic ganglia, and it was

Table 8
EFFECT OF SITE OF TRANSFER ON THE SURVIVAL OF SHEEP EMBRYOS

| Age of embryo[a] (days) | Stage of development | Site of transfer (O = oviduct; U = uterus) | Proportion of transfers successful[b] | Ref. |
|---|---|---|---|---|
| 2—4 | 2 to 6 cells | O | 25/40 — 63% | 6 |
| | 8 cells and >8 cells | U | 32/40 — 80% | 6 |
| 2—3½ | <8 cells | O | 41/69 — 59% | 11 |
| | 8 cells and >8 cells | U | 136/193 — 70% | 11 |
| 2—2½ | 2 to 6 cells | O | 13/27 — 48% | 51 |
| 2½—3 | 4 to 8 cells | O | 15/27 — 56% | 51 |
| 3—3½ | 6 to >8 cells | O | 20/27 — 74% | 51 |
| 2—2½ | 2 to 6 cells | U | 10/27 — 37% | 51 |
| 2½—3 | 4 to 8 cells | U | 10/27 — 37% | 51 |
| 3—3½ | 6 to >8 cells | U | 15/27 — 56% | 51 |
| 2 | ? | O | 12/20 — 60% | 68 |
| 3 | ? | O | 20/23 — 87% | 68 |
| 4 | ? | O | 14/21 — 67% | 68 |
| 2 | ? | U | 15/21 — 71% | 68 |
| 3 | ? | U | 11/20 — 55% | 68 |
| 4 | ? | U | 13/20 — 65% | 68 |
| 5 | Morulae | U | 12/16 — 75% | 52 |
| 7 | Blastocysts | U | 17/24 — 71% | 52 |
| 9 | Blastocysts | U | 15/24 — 63% | 52 |

[a] Days after onset of estrus in donors.

[b] Proportion of recipients which lambed or were pregnant at autopsy; only those recipients in which asynchrony between donors and recipients was 2 days or less.

suggested that it neutralized the cervical stimulation involved in transfer procedures. Transfer using laparoscopic procedures has not been attempted, but may well be worthy of investigation.

### A. Procedures

The reproductive tract is exposed by midventral laparotomy under general or local (e.g., Xylocaine®) anesthesia. Local anesthesia has the advantage of lowered cost and rapidity, but recipients should be restrained in a cradle similar to that described by Lamond and Urquhart.[67]

The most common method of transfer is by Pasteur pipette attached by flexible tubing to a 2-mℓ syringe. A smooth-ended pipette inserted through the fimbria should be used for transfer to the oviduct, while uterine transfers are carried out using a sharp-ended pipette inserted into the lumen some 3 to 5 cm from the utero-tubal junction.

### B. Site of Transfer

The section of the reproductive tract to which embryos are to be transferred is determined by their age. In the ewe and doe embryos enter the uterus on the 3rd to 4th day after the onset of estrus, and extensive studies in the ewe (see Table 8) show that embryos collected before day 4 (day 0 = day of estrus) should be transferred to the oviducts rather than the uterus. Those collected on day 4 can be transferred to either site, while embryos older than 4 days should be transferred to the uterus. When embryos are transferred to the appropriate site survival increases from approximately 50% for day 2 embryos to 70% or more for day 3 and older embryos. Similar detailed information is not available for the doe, but there are suggestions that the uterus is an inappropriate site for embryos of less than 4 days of age.[16]

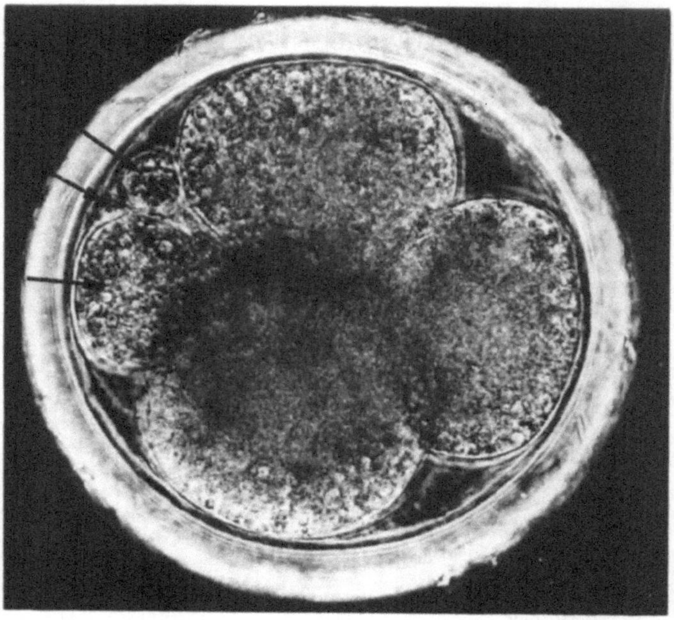

FIGURE 1. "Atypical" sheep embryo recovered 2 days after the onset of estrus. Note three anucleate particles (arrowed). Phase contrast. (Magnification × 250.) (From I. D. Killeen, Ph.D. Thesis, University of Sydney.)

In neither the ewe nor the doe is information available on:

1. Survival of embryos according to whether they are transferred ipsilateral or contralateral to the ovary containing the corpus luteum (or corpora lutea)
2. Frequency of transuterine migration of embryos, and any effect which spacing between embryos might have on their survival

Until these points are clarified then it would seem advisable to transfer embryos to the ipsilateral oviduct or uterine horn, and when recipients are to receive more than one embryo, then they should be transferred to both oviducts or uterine horns.

### C. Evaluation of Embryos

Embryos collected from individual donors are generally all at a uniform stage of development, but occasionally one or more may be retarded, e.g., two-cell embryos collected with eight-cell embryos. Information on the survival of retarded embryos transferred to recipients is lacking, but work in this laboratory shows that they rarely show continued development in culture indicating that they are not viable.

In the ewe it is not uncommon to see embryos containing one or more anucleate particles in association with normal-sized nucleated blastomeres, while in the doe embryos containing large vacuoles are common (see Figures 1 and 2). However, the presence of anucleate particles or large vacuoles appears to have no effect upon viability, and these embryos should be classed as "atypical" rather than abnormal.[44,45]

### D. Number of Embryos Transferred

The number of embryos transferred has little or no effect upon the proportion of recipients which subsequently establish pregnancy,[6,11,47] suggesting that the success or

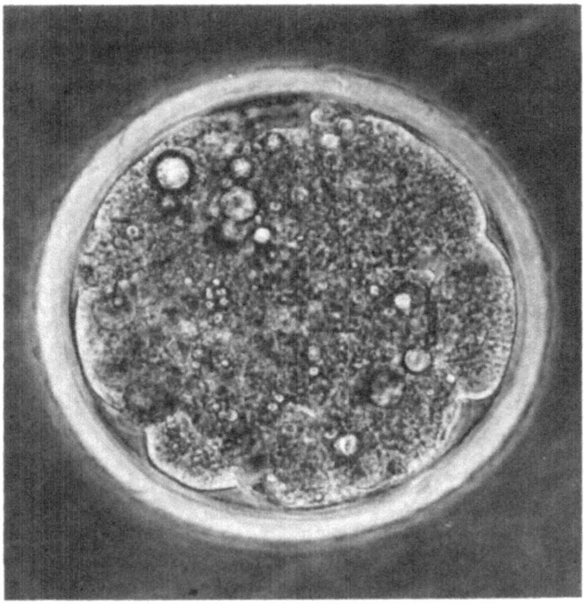

FIGURE 2. "Atypical" goat embryo recovered 5 days after the onset of estrus. Note large vacuoles. Phase contrast. (Magnification × 200.)

failure of transfer is dependent more upon the capacity of the reproductive tract of recipients to support embryos than on the intrinsic developmental capacity of embryos. However, the proportion of embryos which survive is dependent upon the number transferred to individual recipients. Transfer studies in which more than one embryo was transferred to individual recipients have clearly shown that the high incidence of multiple births in highly fecund breeds is due not only to their high ovulation rate, but also to the capacity of their uteri to support more than one embryo.[6,47,49] Hence, the incidence of multiple births within potential recipients should be considered before deciding on numbers of embryos to be transferred. Those which rarely have twins should receive only one embryo, while two or even three can be transferred to those which have a high incidence of multiple births.

## REFERENCES


1. **Warwick, B. L., Berry R. O., and Horlacher, W. R.**, Results of mating rams to Angora female goats, in *Proc. Am. Soc. Anim. Prod.*, 1934, 225.
2. **Lopyrin, A. I., Loginova, N. V., and Karpov, P. L.**, Experiments in interbreed transference of ova in sheep, *Sovetsk. Zooteh.*, 8, 50, 1950; *A.B.A.*, 1449 (Abstr.), 18, 1950.
3. **Lopyrin, A. I., Loginova, N. V., and Karpov, P. L.**, The effect of changed conditions during embryogenesis on growth and development of lambs, *Sovetsk. Zooteh.*, 11, 83, 1951; *A.B.A.*, 729 (Abstr.), 20, 1952.
4. **Hunter, G. L., Adams, C. E., and Rowson, L. E.**, Inter-breed ovum transfer in sheep, *J. Agric. Sci. (Camb.)*, 46, 143, 1955.
5. **Averill, R. L. W. and Rowson, L. E. A.**, Ovum transfer in sheep, *J. Endocrinol.*, 16, 326, 1958.
6. **Moore, N. W. and Rowson, L. E. A.**, Egg transfer in sheep. Factors affecting the survival and development of transferred eggs, *J. Reprod. Fert.*, 1, 332, 1960.

7. **Hammond, J., Jr., Hammond, J., and Parkes, A. S.**, Hormonal augmentation of fertility in sheep. I. Induction of ovulation, superovulation and heat in sheep, *J. Agric. Sci. (Camb.)*, 32, 308, 1942.
8. **Moore, N. W. and Shelton, J. N.**, Response of the ewe to a horse anterior pituitary extract, *J. Reprod. Fert.*, 7, 79, 1964.
9. **Casida, L. E., Warwick, E. J., and Meyer, R. K.**, Survival of multiple pregnancies in the ewe following treatment with pituitary gonadotrophins, *J. Anim. Sci.*, 3, 22, 1944.
10. **Moore, N. W. and Shelton, J. N.**, Oestrous and ovarian response of the ewe to a horse anterior pituitary extract, *Nature (London)*, 194, 1283, 1962.
11. **Moore, N. W. and Shelton, J. N.**, The application of the technique of egg transfer to sheep breeding, *Aust. J. Agric. Res.*, 13, 718, 1962.
12. **Van Rensburg, S. J.**, Ovum production: action of various gonadotrophins in sheep and goats, *Onderstepoort J. Vet. Res.*, 31, 97, 1964.
13. **Boland, M. P. and Gordon, I.**, Recovery and fertilization of eggs following natural service and uterine insemination in the Galway ewe, *Ir. Vet. J.*, 32, 123, 1978.
14. **Shelton, J. N. and Moore, N. W.**, The response of the ewe to pregnant mare serum and to horse anterior pituitary extract, *J. Reprod. Fert.*, 14, 175, 1967.
15. **Moore, N. W.**, Multiple ovulation and ovum transfer in the goat, *Proc. Aust. Soc. Anim. Prod.*, Vol. 10, Lloyd Davies, H., Ed., 1974, 246.
16. **Moore, N. W. and Eppleston, J.**, Embryo transfer in the Angora goat, *Aust. J. Agric. Res.*, 30, 973, 1979.
17. **Killeen, I. D. and Moore, N. W.**, The effect of pregnant mare serum gonadotrophin and human chorionic gonadotrophin on ovulation and on fertility in the ewe, *Aust. J. Agric. Res.*, 21, 807, 1970.
18. **Nishikawa, Y., Horie, T., and Onuma, H.**, Studies on transplantation of ova (artificial pregnancy) in goats. II. Experiments on a method of inducing ovulation of artificially developed follicles and estrus in treated animals, in *Proc. Jpn. Acad.*, Vol. 39, Japan Academy, Tokyo, 1963, 610.
19. **Bindon, B. M., Ch'ang, T. S., and Turner, H. N.**, Ovarian response to gonadotrophin by Merino ewes selected for fecundity, *Aust. J. Agric. Res.*, 22, 809, 1971.
20. **Trounson, A. O. and Moore, N. W.**, Ovulation rate and survival of fertilized ova in Merino ewes selected for and against multiple births, *Aust. J. Agric. Res.*, 23, 851, 1972.
21. **Bindon, B. M., Blanc, M. R., Pelletier, J., Terqui, M., and Thimonier, J.**, Periovulatory gonadotrophin and ovarian steroid patterns in sheep of breeds of differing fecundity, *J. Reprod. Fert.*, 55, 15, 1979.
22. **Cahill, L. P., Mariana, J. C., and Mauleon, P.**, Total follicular populations in ewes of high and low ovulation rate, *J. Reprod. Fert.*, 55, 27, 1979.
23. **Moore, N. W.**, The control of time of oestrus and the induction of superovulation in cattle, *Aust. J. Agric. Res.*, 26, 295, 1975.
24. **Moore, N. W.**, Fertilization in ewes treated with progesterone and equine anterior pituitary extract, *Endocrinology*, 46, 121, 1970.
25. **Trounson, A. O. and Moore, N. W.**, Fertilization in the ewe following multiple ovulation and uterine insemination, *Aust. J. Biol. Sci.*, 27, 301, 1974.
26. **McCracken, J. A., Glew, M. E., and Scaramuzzi, R. J.**, Corpus luteum regression induced by prostaglandin $F_{2\alpha}$, *J. Clin. Endocrinol. Metab.*, 30, 544, 1970.
27. **Douglas, R. H. and Ginther, O. J.**, Luteolysis following a single injection of prostaglandin $F_{2\alpha}$ in sheep, *J. Anim. Sci.*, 37, 990, 1973.
28. **Hearnshaw, H., Restall, B. J., Nancarrow, C. D., and Mattner, P. E.**, Synchronisation of oestrus in cattle, sheep and goats using a prostaglandin analogue, in *Proc. Aust. Soc. Anim. Prod.*, Vol. 10, Lloyd Davies, H., Ed., 1974, 242.
29. **Moore, N. W. and Eppleston, J.**, Control of oestrus, ovulation and fertility in relation to artificial insemination in the Angora goat, *Aust. J. Agric. Res.*, 30, 965, 1979.
30. **Trounson, A. O., Willadsen, S. M., and Moor, R. M.**, Effect of prostaglandin analogue Cloprostenol on oestrus, ovulation and embryonic viability in sheep, *J. Agric. Sci. (Camb.)*, 86, 609, 1976.
31. **Moore, N. W.**, unpublished data, 1977.
32. **Moore, N. W.**, unpublished data, 1976.
33. **Corteel, J. M.**, The use of progestagens to control the oestrous cycle of the dairy goat, *Ann. Biol. Anim. Biochem. Biophys.*, 15, 253, 1975.
34. **Mansour, A. M.**, The hormonal control of ovulation in the immature lamb, *J. Agric. Sci. (Camb.)*, 52, 87, 1959.
35. **Land, R. B. and McGovern, P. K.**, Ovulation and fertilization in the lamb, *J. Reprod. Fert.*, 15, 325, 1968.
36. **Trounson, A. O., Willadsen, S. M., and Moor, R. M.**, Reproductive function in prepubertal lambs: ovulation, embryo development and ovarian steroidogenesis, *J. Reprod. Fert.*, 49, 69, 1977.
37. **Eppleston, J.**, personal communication, 1980.

38. **Robinson, T. J.**, The control of fertility in sheep. II. The augmentation of fertility by gonadotrophin treatment of the ewe in the normal breeding season, *J. Agric. Sci. (Camb.)*, 41, 6, 1951.
39. **Averill, R. L. W.**, The production of living sheep eggs, *J. Agric. Sci. (Camb.)*, 50, 17, 1958.
40. **Rowson, L. E. and Adams, C. E.**, An egg transfer experiment in sheep, *Vet. Rec.*, 69, 849, 1957.
41. **Hulet, C. V. and Foote, W. C.**, Ovulatory response of the ewe to repeated injections of PMS, *J. Anim. Sci.*, 29, 457, 1969.
42. **Moore, N. W. and Eppleston, J.**, unpublished data, 1978.
43. **Fukui, Y. and Roberts, E. M.**, Sperm transport after non-surgical intrauterine insemination with frozen semen in ewes treated with prostaglandin $F_{2\alpha}$, *J. Reprod. Fert.*, 51, 141, 1977.
44. **Killeen, I. D. and Moore, N. W.**, Fertilization and survival of fertilized eggs in the ewe following surgical insemination at various times after the onset of oestrus, *Aust. J. Biol. Sci.*, 23, 1279, 1970.
45. **Killeen, I. D. and Moore, N. W.**, The morphological appearance and development of sheep ova fertilized by surgical insemination, *J. Reprod. Fert.*, 24, 63, 1971.
46. **Moore, N. W.**, unpublished data, 1980.
47. **Moore, N. W.**, The survival and development of fertilized eggs transferred between Border Leicester and Merino ewes, *Aust. J. Agric. Res.*, 19, 295, 1968.
48. **Dickinson, A. G., Hancock, J. L., Hovel, G. J. R., Taylor, St. C. S., and Wiener, G.**, The size of lambs at birth — a study involving egg transfer, *Anim. Prod.*, 4, 64, 1962.
49. **Land, R. B. and Wilmut, I.**, The survival of embryos transferred in large groups to sheep of breeds with different ovulation rates, *Anim. Prod.*, 24, 183, 1977.
50. **Hinkelman, L. A., Bradford, G. E., Pollak, E. J., Anderson, G. B., and Cupps, P. T.**, An embryo transfer study of birth weight in Finnish Landrace and Targee sheep, *J. Anim. Sci.*, 48, 298, 1979.
51. **Moore, N. W. and Shelton, J. N.**, Egg transfer in sheep. Effect of degree of synchronisation between donor and recipient, age of egg, and site of transfer on the survival of transferred eggs, *J. Reprod. Fert.*, 7, 145, 1964.
52. **Rowson, L. E. A. and Moor, R. M.**, Embryo transfer in sheep: the significance of synchronizing oestrus in the donor and recipient animal, *J. Reprod. Fert.*, 11, 207, 1966.
53. **Shelton, J. N. and Moore, N. W.**, Survival of fertilized eggs transferred to ewes after progesterone treatment, *J. Reprod. Fert.*, 11, 149, 1966.
54. **Fairnie, I. J., Cumming, I. A., and Martin, E. R.**, Use of the prostaglandin analogue, ICI 80996, to synchronise ovulation in sheep in an artificial insemination programme, *Proc. Aust. Soc. Anim. Prod.*, Vol. 11, Egan, A. R., Ed., 1976, 133.
55. **Fukui, Y. and Roberts, E. M.**, Fertility of ewes treated with prostaglandin $F_{2\alpha}$ and artificially inseminated at predetermined intervals thereafter, *Aust. J. Agric. Res.*, 28, 891, 1977.
56. **Rowson, L. E. A. and Moor, R. M.**, Development of the sheep conceptus during the first fourteen days, *J. Anat.*, 100, 777, 1966.
57. **Tervit, H. R. and Rowson, L. E. A.**, Birth of lambs after culture of sheep ova in vitro for up to 6 days, *J. Reprod. Fert.*, 38, 177, 1974.
58. **Trounson, A. O. and Moore, N. W.**, Attempts to produce identical offspring in the sheep by mechanical division of the ovum, *Aust. J. Biol. Sci.*, 27, 505, 1974.
59. **Trounson, A. O. and Moore, N. W.**, The survival and development of sheep eggs following complete or partial removal of the zona pellucida, *J. Reprod. Fert.*, 41, 97, 1974.
60. **Bilton, R. J. and Moore, N. W.**, In vitro culture, storage and transfer of goat embryos, *Aust. J. Biol. Sci.*, 29, 125, 1976.
61. **Averill, R. L. W. and Rowson, L. E. A.**, Attempts at storage of sheep ova at low temperatures, *J. Agric. Sci. (Camb.)*, 52, 392, 1959.
62. **Kardymowicz, M., Kardymowicz, O., Kuhl, W., and Lada, A.**, Storage of sheep ova at low temperature, *Acta Biol. Zool. Cracoviensia*, 6, 31, 1963.
63. **Moore, N. W. and Bilton, R. J.**, The storage of fertilized sheep ova at 5°C, *Aust. J. Biol. Sci.*, 26, 1421, 1973.
64. **Averill, R. L. W., Adams, C. E., and Rowson, L. E. A.**, Transfer of mammalian ova between species, *Nature (London)*, 176, 167, 1955.
65. **Hunter, G. L., Bishop, G. P., Adams, C. E., and Rowson, L. E.**, Successful long-distance aerial transport of fertilized sheep ova, *J. Reprod. Fert.*, 3, 33, 1962.
66. **Otsuki, K. and Soma, T.**, Transfer of fertilized ova through the cervix in goats, *Bull. Natl. Inst. Anim. Ind. (Aoba, Chiba, Japan)*, 6, 27, 1964.
67. **Lamond, D. R. and Urquhart, E. J.**, Sheep laparotomy cradle, *Aust. Vet. J.*, 37, 430, 1961.
68. **Killeen, I. D.**, The effects of age of egg and site of transfer on survival of transferred eggs in the ewe, *Theriogenology*, 6, 637, 1976.

Chapter 8

# EMBRYO TRANSFER IN THE HORSE

W. R. Allen

## TABLE OF CONTENTS

## I. INTRODUCTION

Up to now the horse has remained the "Cinderella" among the common farm and laboratory animals with regard to the development and use of the technique of embryo transfer. Only in recent years has any attention been paid to this aspect of equine reproduction, and, as will become apparent to the reader, we still have much to learn about many basic parameters which are likely to influence the success of the method when applied to the horse.

In many ways it is surprising that embryo transfer should have been so slow to develop in equids since the mare would seem an ideal subject for commercial exploitation of the technique. Individual mare-stallion mating combinations can have enormous potential monetary value due to the pedigrees of the parents and/or the proven performance of previous offspring of either or both of the pair. Embryo transfer could obviously increase the rate of production of progeny from any such desirable pair. In addition, the retention in the breeding population of aged mares which have special genetic value leads to a variety of infertility states which are related, not to a failure of conception, but more to the inability of the individual mare to carry a pregnancy to term and give birth to a viable foal. Embryo transfer could be a practical solution to the problem and so provide a means of continuing to obtain significant numbers of the desirable progeny from such cases.

Two major barriers exist to limit the use of embryo transfer in horses. First, and perhaps foremost, the registration authorities for most pure breeds of horses throughout the world will not permit the entry into their respective stud books of progeny which have resulted from embryo transfer. Secondly, and as will be discussed in more detail later, there does not yet exist an efficient method for inducing superovulation in the mare. Despite these problems, however, there has been a number of significant advances in both the methodology and results of equine embryo transfers during the past few years, and the present chapter attempts to review this progress.

## II. THE ESTROUS CYCLE OF THE MARE

The mare is a seasonally polyestrous animal with a distinct breeding season during spring, summer, and early autumn. The normal estrous cycle is 21 to 23 days, consisting of 16 to 17 days of diestrus and 4 to 7 days of estrus. The duration of estrus differs markedly between individual mares and is also influenced by season, being prolonged at the start of the breeding season in early spring, shortest at the summer solstice, and lengthening again in the autumn.[1,2]

Some 8 to 12 follicles $>$ 1 cm in diameter are usually present in the ovaries on the 1st day of estrus, but only one of these, occasionally two in some breeds of horses, progresses to ovulation. The ovulatory follicle reaches a diameter of 3 to 5 cm and is thus readily palpable through the wall of the rectum. Ovulation occurs only through the small circular region on the ventro-medial surface of the ovary which is not covered by tunica albuginea, the ovulation fossa,[3] and it takes place 12 to 36 hr before cessation of the behavioral signs of estrus.[4] Luteinization of the corpus hemorrhagicum occurs quickly so that peripheral plasma progesterone concentrations have generally risen to 1 to 2 ng/m$\ell$ by 24 hr, and frequently reach as high as 3 to 5 ng/m$\ell$ by 48 hr after ovulation (Figure 1). This rapid rise permits the use of daily plasma progesterone estimations during estrus as an accurate method for determining the day of ovulation for the purpose of embryo transfer.

Pituitary gonadotrophin release patterns in the cycling mare differ markedly from those shown by the other large domestic species. Serum FSH concentrations show a biphasic pattern with a pronounced mid-diestrous surge and a second, generally

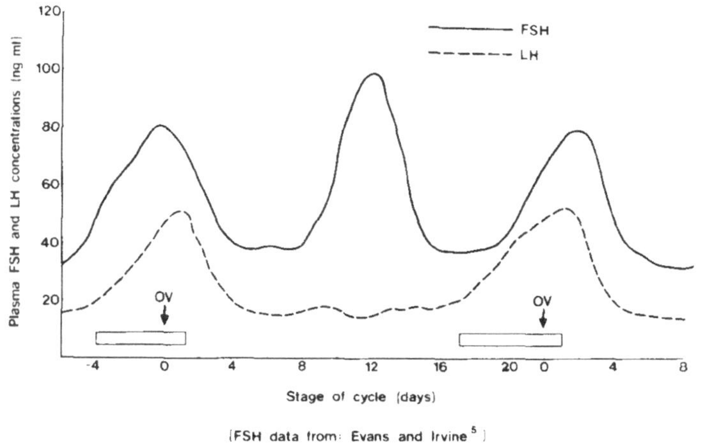


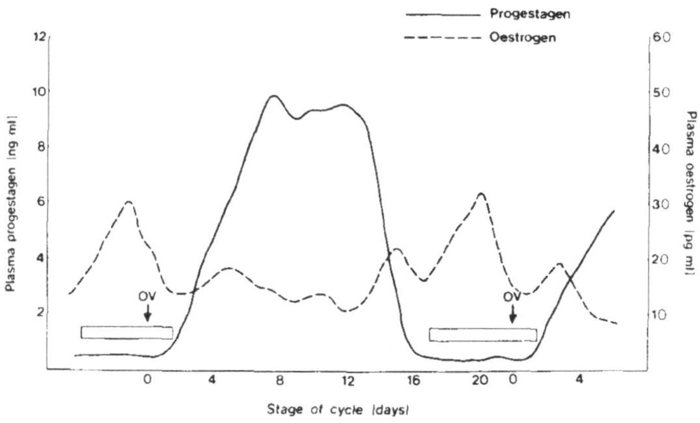


FIGURE 1.. Diagrammatic representation of the changes in concentrations of gonadotrophins (FSH and LH) and steroid hormones, estrogen and progestagen, in the peripheral plasma of the mare during the estrous cycle. The horizontal bar represents estrus and OV indicates the day on which ovulation occurs.

smaller, peak associated with ovulation.[5] LH levels, on the other hand, remain low throughout diestrus but rise steadily during estrus to reach a peak at the time of, or up to, 1 to 3 days after ovulation (Figure 1). This protracted release of LH during estrus in the mare contrasts sharply with the rapid, more pulsatile release of LH which occurs some 24 to 30 hr prior to ovulation in other species,[6] and it is thought likely to be responsible for at least some of the twin and early diestrous ovulations which occur in some mares.[7]

The marked variability which exists both in the duration of estrus and in the occurrence of ovulation in relation to estrous signs precludes any reference to estrus as a pivotal point upon which to calculate the age of embryos for recovery purposes. The time of ovulation is the only meaningful parameter in this regard, and day 0 is usually taken as the day on which ovulation is first detected as having occurred. However, since donor mares are generally only examined once daily, either by rectal palpation of the ovaries or daily measurements of peripheral plasma progesterone levels, or a combination of both methods, any estimate of the time of ovulation is subject to an inherent error which can be as great as 24 hr. This is reflected in the considerable variation which is found in the stage of development of horse blastocysts recovered on days 7 or 8 for embryo transfer.[8]

## III. SYNCHRONIZATION OF OVULATION

L.E.A. Rowson and colleagues demonstrated that the success rate from surgical embryo transfer in sheep and cattle is highest when ovulation is exactly synchronized in donor and recipient animals, but begins to drop off sharply when the degree of synchrony exceeds 24 hr either way in the cow or 48 hr either way in the ewe.[9] Only one published study, that of Oguri and Tsutsumi,[10] has seriously examined the question of synchrony in relation to equine embryo transfer. In transfers of 79 horse blastocysts using two different nonsurgical techniques, initial conception rates of 63%, 53%, 41%, 38%, and 0%, and live foal rates of 50%, 35%, 30%, 19%, and 0%, respectively, were obtained when the degree of synchrony between recipient and donor mares was, respectively, −2, −1, 0, +1, and +2 days. Although the situation was slightly confounded by the use of two different transfer methods, and the presence of half the number of animals in the −2- and +2-day groups compared with the other three groups, the figures nonetheless gave a clear indication that a degree of asynchrony as great as 48 hr or more is not disadvantageous in terms of pregnancy rate, provided the recipient has ovulated *after* the donor.

The critical period for maternal recognition of pregnancy in the mare occurs around day 14 after ovulation.[11] This important message from the embryo to prevent luteolysis presumably involves suppression of the normal, cyclical release of prostaglandin F by the endometrium at the end of diestrus,[12] and it may well be related to the large quantities of estrogens secreted by the blastocyst from as early as day 12.[13,15] It is conceivable that handling, changes in temperature, and any other adverse conditions which the blastocyst experiences during the nonsurgical recovery and transfer process may temporarily halt its development. A delayed blastocyst placed in the ovulation-synchronized uterus could become asynchronous in terms of the time-related interchange of signals between itself and the endometrium which are necessary to maintain pregnancy. Thus, when the critical period of day 14 of diestrus is reached in the recipient mare, the blastocyst will be insufficiently advanced in its development to be able to secrete enough estrogen, or whatever other factors are needed to suppress luteolysis, to prevent the mare returning to estrus.

The development of efficient methods to control and synchronize estrus and ovulation in the large domestic species has provided a keystone for the successful commercial exploitation of embryo transfer in cattle.[9] The horse is no exception and the need to be able to control ovulation in groups of donor and recipient mares is perhaps even greater than in the cow, due to the much longer period of estrus and the marked individual variation between mares in the time from the onset of estrous behavior to the occurrence of ovulation. The ability of a single injection of a relatively low dose of prostaglandin F or various prostaglandin analogues to induce rapid luteolysis in the diestrous mare[16,17] has provided the basis of the treatment regimes currently being employed to control ovulation in mares. Two injections, each of 250 g, of the prostaglandin analogue fluprostenol (Equimate, ICI Pharmaceuticals Division, Cheshire, U.K.), given 14 or 15 days apart to a group of randomly cycling mares, results in over 90% of the mares displaying estrus by 6 days after the second injection (day 20) and approximately 75% of the mares ovulating during a 4-day period between days 20 and

24.[18] The success of this double prostaglandin synchronization schedule, with or without injection of human Chorionic Gonadotrophin (hCG) given on days 7 and 20 after the first prostaglandin treatment to hasten ovulation in each prostaglandin-induced estrus, has been confirmed.[19,20] More recently, Palmer[21] showed that daily administration of the orally-active synthetic progestagen, allyl trenbolone (Regumate, Roussel Uclaf, Paris), to cycling mares for 9 to 10 days, accompanied by treatment with fluprostenol on the last day of progestagen treatment and hCG a further 10 days later, likewise results in synchronized ovulation in about 75% of the treated animals.

As previously mentioned, the difficulty of inducing superovulation in mares by the administration of exogenous gonadotrophins seriously limits the yield of embryos from any particularly desirable donor animal. The best solution to the problem at present is to increase the frequency of ovulation in selected donors by repeatedly shortening their estrous cycle with prostaglandins. Such a method is employed routinely in the author's laboratory with good results and without any apparent side effects. Donor mares are injected with 250 $\mu$g of the prostaglandin analogue, chloprostenol (Estrumate, ICI Pharmaceuticals Division) immediately after undergoing nonsurgical embryo recovery on day 7 or 8 after ovulation. This induces a return to estrus in 2 to 3 days when each mare can again be mated with the required stallion and should ovulate within 4 to 5 days. After a further 7 days the mares are ready to be flushed again, and in this way the interval between successive embryo-recovery attempts is shortened from the expected 21 to 22 days of the normal estrous cycle to about 13 to 15 days. Repeated shortening of the cycle in this manner does not appear to lower fertility, and from one Pony mare flushed a total of six times during a period of 83 days in 1979, five normal blastocysts were obtained.

## IV. SUPEROVULATION IN THE MARE

Pregnant Mare Serum Gonadotrophin (PMSG), the high molecular weight glycoprotein hormone which uniquely expresses both FSH-like and LH-like biological activities,[22,23] remains the most widely used and effective gonadotrophin for stimulating superovulation in cattle, sheep, goats, and pigs. It is, therefore, somewhat ironic that this particular hormone, which is present in high concentration in the blood of mares between 40 and 120 days of pregnancy,[24,25] is completely ineffective with regard to stimulating follicular development when given to cycling and pregnant mares, even in repeated doses as high as 150,000 i.u.[4,26] The studies of Stewart and Allen[27,28] demonstrated that the gonadal gonadotrophin receptors in equids, as distinct from those in other species, have somehow evolved the ability to differentiate between the binding properties of their own chorionic gonadotrophin (PMSG) and those of their pituitary gonadotrophins, FSH and LH. That such an unusual mechanism allowing differential recognition of, and response to, different types of gonadotrophin molecules should have evolved in the mare is perhaps not surprising when one considers that at the time of peak PMSG production around day 60 of gestation, a total quantity in excess of 1 million i.u. PMSG is likely to be circulating in her blood. Were such a protective system not to exist, her ovaries would be hyperstimulated each time she became pregnant.

The one advance with regard to superovulation in the mare has come from a number of experiments undertaken in Wisconsin in recent years. Douglas et al.[29] induced estrus and ovulation in 25 of 29 seasonally anestrous Pony mares by giving daily injections for 14 consecutive days of a crude extract of equine pituitary glands. Based on rectal palpation of the ovaries, they noted that 14 of the 25 treated mares which ovulated had two or more ovulations. Subsequently, the same group[30] obtained a similar proportion of double and triple ovulations when giving the same type of course of daily injections of equine pituitary extract, with or without a subsequent injection of hCG, for 14 days to seasonally anestrous mares and for 6 consecutive days, between either days 11 to 16 of diestrus or days 1 to 6 of estrus, to cycling mares. Once again, however, the ovulation figures were based on rectal palpation only and were not supported by any attempts at embryo recovery. Furthermore, the repeated injections of the crude pituitary extract caused the death of one mare and gave rise to injection site abscesses in many of the others.

More recently, Douglas[31] used a commercially available equine pituitary extract (Pitropin, Biological Specialties, Wisconsin) to similarly treat cycling mares daily for 6 days during mid-diestrus or late diestrus/early estrus. As in the previous experiments, ovulation rates in the treated and control mares were assessed by rectal palpation only and on this basis, six of the eight treated mares were considered to have had between two and four ovulations each during the estrus following treatment. However, when they were subjected to nonsurgical flushing of the uterus between days 8 and 16 after the induced ovulation, the embryo recovery rate was only 35% of that expected (based on ovulation rate) in the eight treated animals compared with 67% of that expected in 20 untreated control mares. Similarly, Woods and Ginther[32] found that the conception rate was markedly reduced compared with that in untreated controls, in mares treated daily for 14 days with equine pituitary extract to induce single and/or multiple ovulations early in the breeding season (31% vs. 60%), and that only one conceptus was present on day 50 after ovulation in all 8 of the 27 treated mares which conceived during the estrus in which two or more ovulations had been diagnosed per rectum.

The incidence of spontaneous twin ovulations in the mare is strongly breed-dependent, ranging from as high as 20% in the larger, more modern breeds such as the Thoroughbred to zero in smaller, more primitive breeds such as the Welsh mountain pony. Although the Wisconsin studies have demonstrated that administration of equine pituitary gonadotrophins in sufficient quantity may induce some degree of multiple ovulation in a proportion of mares, it seems unlikely that this type of therapy would be advantageous for routine use in a commercial equine embryo transfer program. The cost and impracticality of the treatment regime, the variability in response rate, and the significant reduction in pregnancy rate in treated mares would combine to outweigh any possible advantages. Rather, it seems that some completely novel form of treatment is required that will somehow by-pass the fundamental evolutionary resistance of the mare's ovaries to gonadotrophic stimulation.

## V. EMBRYO RECOVERY

### A. Nonsurgical Embryo Recovery

Certain features of the anatomy of the mare's reproductive tract make the whole process of nonsurgical uterine flushing much simpler than its equivalent in the cow. The mare's cervix is relatively short and is more easily distended in diestrus than that in the cow, so allowing the easy insertion of a much larger diameter catheter. Furthermore, the horns of the mare's uterus are not coiled like those of the cow and they diverge from each other at a much greater angle, thereby more easily permitting the passage of a flushing catheter from the body of the uterus into either horn. Two main methods of nonsurgical flushing have been employed to date in the mare.

#### *1. Single-Horn Flushing*

For this technique, a rigid instrument[33,35] or a flexible rubber or polythene two-way catheter[8,36,37] is passed through the cervix into the body of the uterus. With the operator's arm in the rectum the catheter is then manipulated into the ipsilateral horn of the uterus so that the inflatable rubber cuff is situated at the base of the horn. The

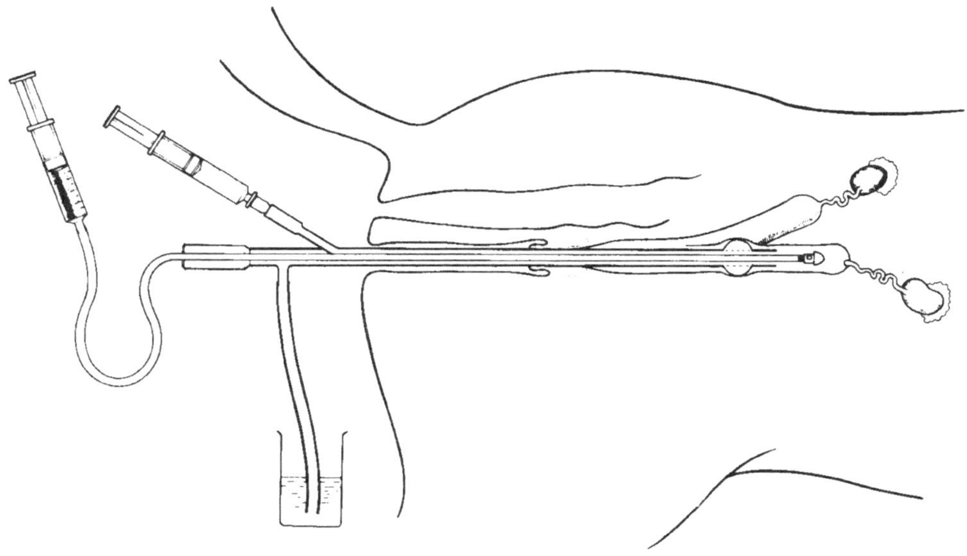

FIGURE 2. The older method of single uterine horn, nonsurgical embryo recovery using a rigid, metal, three-way flushing apparatus manipulated into the base of the ipsilateral uterine horn with the operator's arm in the mare's rectum.

FIGURE 3. Two-way 18-French gauge human degustation tube catheter with inflatable cuff used for both single horn and whole uterus, nonsurgical embryo flushing in mares and donkeys.

cuff is inflated with 30 to 40 mℓ of air or water to isolate the lumen of the horn from the remainder of the uterus, and 50 to 60 mℓ of medium is flushed rapidly into the horn using a catheter-tipped 60-mℓ syringe attached directly to the free end of the catheter (Figure 2). While massaging the fluid-filled uterine horn per rectum the medium is then either positively withdrawn into the same syringe or is allowed to flow back out through the catheter by gravity into a suitable collecting vessel (Figure 3). The whole process is then repeated twice.

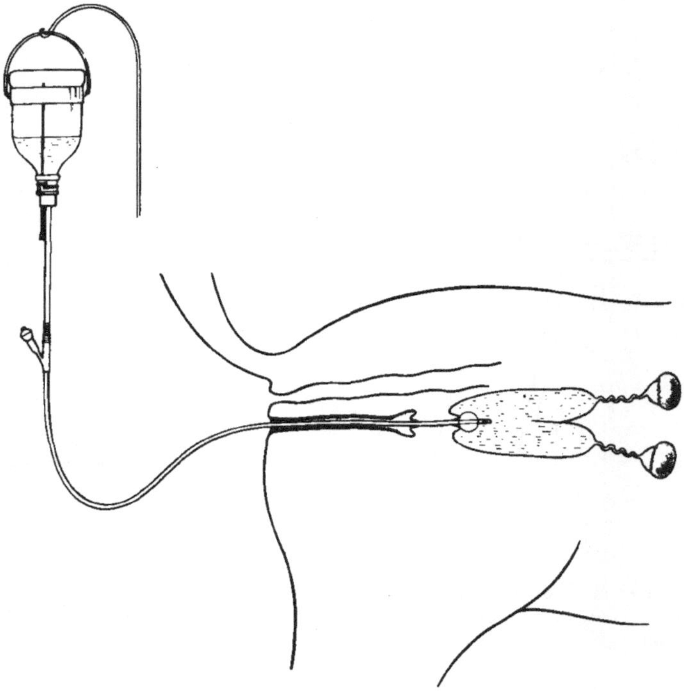

FIGURE 4. The whole uterus, nonsurgical embryo flushing technique using the flexible two-way 18-FG catheter shown in Figure 3 inserted into the body of the uterus and withdrawn so that the inflated cuff occludes the internal os of the cervix. Some 300 to 500 m*ℓ* of flushing medium is run into the uterus under gravity.

*2. Whole-Uterus Flushing*

This method, which has been used more widely in recent years,[10,31,38,39] simply involves passing the flushing instrument, either the rigid structure[38] or the flexible two-way catheter with inflatable cuff, through the cervix into the body of the uterus. The cuff is inflated and a slight backward tension maintained on the catheter so that the cuff is kept adjacent to, and hence occludes, the internal os of the cervix. The flushing medium (300 to 800 m*ℓ* depending on the size of the mare) is run into the uterus under gravity until the lumen is distended, and is then allowed to run back out into suitable collecting vessels (Figure 4). The whole process is repeated twice so that eventually a total of 900 to 2000 m*ℓ* of medium has been infused into the uterus in three stages.

The technique of flushing the whole uterus is much easier to carry out than the single-horn method since it requires only the simple insertion of the catheter through the cervix without any need for manipulation *per rectum*. Furthermore, it seems to give a much higher recovery rate of embryos than that obtainable by flushing only the ipsilateral uterine horn. In earlier studies[34,35] in which a rigid instrument (Figure 2) was used to flush the ipsilateral uterine horn in fertile Pony mares, recovery rates of only 49% and 40% were achieved. Likewise, Tischner[8] recovered only 11 embryos from 21 mares (52%) by the ipsilateral horn method, and in the author's laboratory during the 4-year period from 1976 to 1980, a total of 54 single-horn flushes of Pony mares and jenny donkeys which had been covered by either a horse stallion or a Jack donkey yielded 26 intraspecies and hybrid embryos, to give an overall recovery rate of only 48%.

In their later papers the Japanese workers[10,38] reported greatly improved embryo recovery rates when using a similar, rigid, three-way flushing instrument to that employed in their original study, but this time passing the instrument only through the cervix and infusing the whole uterus with a total of 1500 mℓ of medium in three "lots" each of 500 mℓ. In the first experiment they recovered 18 blastocysts from 20 attempted flushings of Pony and half-breed mares between days 6 and 7 after ovulation. In the subsequent and much larger study[38] involving the recovery of a total of 148 embryos from 200 attempted flushes between days 6 and 9 after ovulation, they made a direct comparison between the single-horn and whole-uterus flushing methods. Whereas the recovery rate from single ipsilateral horn flushing was only 44%, it rose to 87% (121 embryos) in the case of the whole uterus. The whole uterus flushing method was also adopted in the author's laboratory during the 1981 breeding season when attempting to recover donkey blastocysts for surgical transfer to mares. The same flexible plastic, 18 French gauge human degustation tube catheter used for the single horn flushes (see Figure 3) was employed and three "lots", each of 300 to 500 mℓ, of medium were infused into the uterus under gravity and allowed to flow out again as and when the uterus contracted to expel it (Figures 4 and 5). In this manner, a total of 27 donkey and 8 horse blastocysts was recovered from the attempted flushing of 37 jenny donkeys and 11 Pony mares between days 6 and 8 after ovulation, giving an overall recovery rate of 72%; this compares favorably with an average recovery rate of only 40 to 50% obtained during the previous 4 years when using the ipsilateral horn flushing method. A similar recovery rate was reported by Douglas,[31] who obtained 14 embryos from the attempted whole-uterus flushing of 20 Pony mares between days 8 and 16 after ovulation.

From the available figures, and especially those provided by the Japanese workers, it is apparent that repeated infusions of the whole uterus gives a marked improvement in embryo recovery compared with what can be achieved by flushing only the ipsilateral uterine horn. In examining the possible reasons for this superiority, it may be noted that a high rate of transuterine migration of the conceptus (±50%) occurs spontaneously in all breeds of horses[4,40,41] with a strong probability in lactating mares for the conceptus to implant in the uterine horn contralateral to that in which the previous conceptus implanted.[42] If the blastocyst is destined to migrate, for whatever reason, it appears to do so very soon after entering the uterus on day 5 to 6, so that it has already vacated the ipsilateral horn by day 7 or 8 when that horn is flushed by the single-horn method.

**B. Surgical Embryo Recovery**

Since the fertilized horse egg does not enter the uterus until days 5 to 6 after ovulation,[10,43] embryo recovery at any stage prior to day 6 (late morula) necessitates entry into the abdomen via a ventral mid-line laparotomy under general anesthesia followed by flushing of the ipsilateral oviduct with 30 to 40 mℓ of medium. The flushing technique is similar to that described for the cow by Rowson et al.,[44] although two factors can complicate the process in the mare. Firstly, the ovarian ligament on the right side of the animal is considerably shorter than that on the left so that, in nulliparous Pony mares especially excessive traction is necessary to exteriorize the right ovary and oviduct sufficiently through a mid-ventral incision to enable the oviduct to be flushed. This interferes with the flushing technique and carries the risk of rupture of the ovarian ligament with resulting hemorrhage. Secondly, the uterotubal junction in the mare is in the form of a firmly occluded papilla which protrudes into the uterine lumen.[45] While this will permit fluids to flow from the oviduct into the uterus, a very effective valve-like action completely prevents the reverse passage of fluid from the uterus into the oviduct. Thus, it is not possible to flush mares by injecting medium into the uterine lumen and then forcing it by manipulation into the oviduct as it is in the cow.

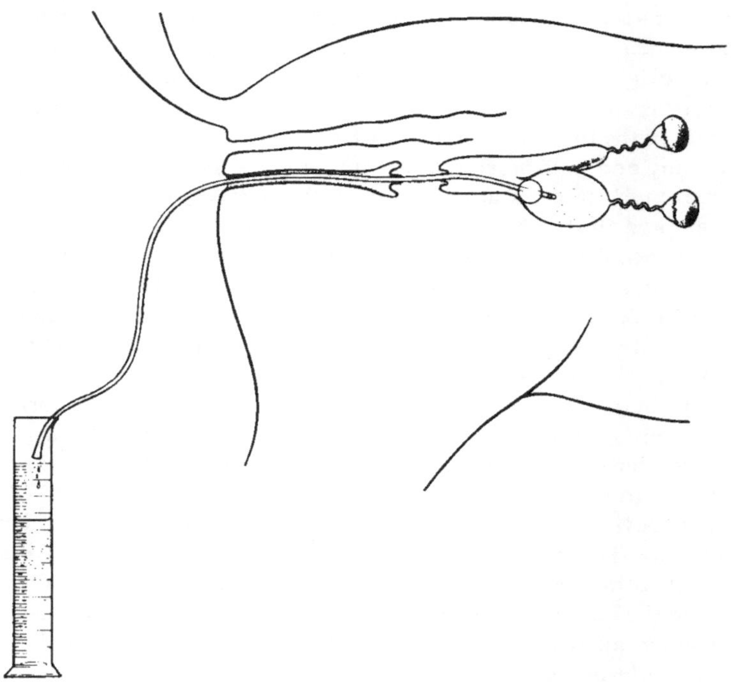

FIGURE 5. The recovery of flushing medium under gravity and uterine contraction following the infusion of 60 to 100 mℓ flushing medium into the ipsilateral uterine horn via a catheter-tipped syringe connected to the flexible 18-FG catheter.

The method used successfully for many years in the author's laboratory[33,34,46] involves the insertion of a 5- to 7-cm length of firm, polystyrene cannula (5 mm O.D.) into the dilated fimbria of the oviduct. This is connected by a short length of flexible tubing to a 20-mℓ syringe containing the flushing medium. The cannula is held in place in the oviduct, and back-flow of fluid prevented, by firm manual pressure. A 1-cm longitudinal incision is made in the outer serosa and myometrium of the ipsilateral uterine horn about 3 cm posterior to the uterotubal junction. The endometrium is pierced with the blades of a pair of rounded scissors and the fluted end of a short length of glass Pasteur pipette, previously shaped in a bunsen burner flame, is inserted into the incision and held firmly against the utero-tubal junction by manual pressure. The other end of the glass tube is tapered and connected to a length of soft rubber tubing which passes to the egg collection dish (Figures 6 and 7).

Some 30 to 40 mℓ of an enriched phosphate buffered saline medium (PBS)[47] is forced through the oviduct and collected into two or three dishes. Considerable pressure on the syringe plunger is often required in the initial stages of the flush to force the fluid through, and hence dilate, the highly tortuous oviduct and the sphincter of the utero-tubal papilla. Pressure is also required to force through the oviduct the spiral-shaped or globular gelatinous structures which frequently occur in, and partly occlude the lumina of equine oviducts[35,48] and which van Niekerk and Gerneke[49] described as originating within the lumen of the maturing follicles and being shed into the oviduct at ovulation. The need for pressure to start the medium flowing through the oviduct carries the slight risk of rupture of the oviduct, and the chance of doing so is greatly increased when trying to flush the right oviduct in maiden mares due to the tension needed to lift the ovary up to the abdominal incision in order to perform the flush.

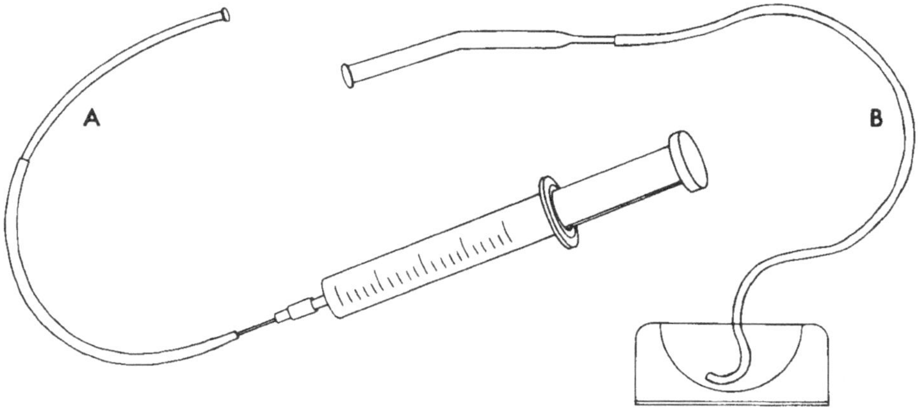


FIGURE 6. Apparatus used for surgical embryo recovery in equids. For flushing, the rigid plastic cannula (A) is inserted into the ovarian end of the oviduct and held in position as shown in Figure 7. It connects via a flexible polythene catheter to the syringe containing the flushing medium. For collection the fluted end of a Pasteur pipette (B) is inserted into the lumen of the uterus and held against the papilla of the utero-tubal junction as shown in Figure 7. The other end of the pipette is connected to flexible polythene tubing which then passes to the embryo collection dish.

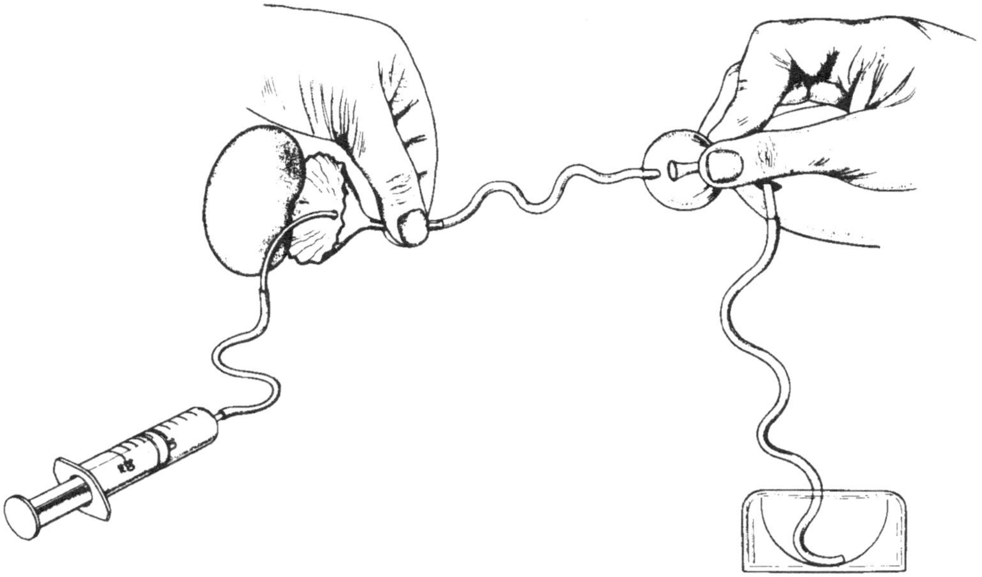

FIGURE 7. Illustration of the technique of surgical embryo recovery in equids. The rigid plastic cannula is held between thumb and forefinger in the ampulla of the oviduct and the fluted end of the glass pipette is held also by digital pressure in the tip of the ipsilateral uterine horn adjacent to the utero-tubal papilla.

A high recovery rate of embryos and/or unfertilized eggs can be achieved by use of the surgical method to flush young, fertile mares. For example, Allen and Rowson[34] obtained 23 embryos from 30 Welsh Pony mares (77% recovery rate) flushed between days 1 and 6 after ovulation. In five mares a recently ovulated but unfertilized egg was found in the flushings, and in the remaining two animals neither an embryo nor an unfertilized egg was recovered, thus giving an overall total recovery rate of 93%. In a more recent experiment[50] surgical flushing of the left oviduct of 39 mares between 1 and 3 days after ovulation yielded 28 embryos between the one- and eight-cell stages and four freshly ovulated but unfertilized eggs. Many old, unfertilized eggs from previous cycles were also recovered in most flushes and on only seven occasions, including one in which the flushing could not be carried out due to rupture of the oviduct, were no eggs recovered.

Adhesions are an occasional complication of surgical embryo recovery in the mare, and these occur most commonly between the uterine incision used for insertion of the collecting tube and the adjacent ovary. They are normally quite easy to break down and only very rarely are they sufficiently extensive to occlude the oviduct or otherwise impair ovarian/tubal function on that side. Careful suturing of the uterine incision and removal of blood clots from the surface of the uterus, ovaries, and other adjacent organs at the conclusion of the flush minimize the occurrence of adhesions.

## VI. EMBRYO TRANSFER

### A. Nonsurgical Embryo Transfer

This is a very straightforward technique in the mare whereby the embryo, in 0.2 to 0.5 mℓ of medium, is drawn into the lumen of a rigid plastic cattle insemination pipette or similar device using a syringe attached to the other end of the pipette. A 1- to 2-mℓ cushion of air is drawn into the pipette before the medium so as to ensure complete expulsion of the medium and embryo when the plunger of the syringe is depressed. With a gloved hand protecting the end of the pipette as it is passed into the vagina, one finger is used to open the external os of the cervix and direct the catheter into the uterus. The embryo may then be deposited either in the body of the uterus or, by manipulation with the operator's arm in the rectum, in the ipsilateral horn.

Just as nonsurgical recovery of equine embryos has given rise to marked differences in recovery rates between research groups, so the conception rates resulting from nonsurgical transfers have varied considerably. In their original study, Oguri and Tsutsumi[35] failed to obtain any pregnancies when transferring ten blastocysts and one morula via the cervix to the ipsilateral uterine horn of precisely synchronized recipient mares. In a later study,[10] however, they achieved a 40% pregnancy rate following the nonsurgical transfer of 15 6-day old blastocysts to synchronized recipient mares. For this purpose they used an apparatus which allowed the embryo to be placed in the uterine lumen via a long needle passing through the anterior fornix of the vagina and thence through the wall of the uterus, thereby avoiding passage through the cervix. In a small study involving Pony mares, Allen and Rowson[34] obtained five pregnancies from the nonsurgical transfer of seven blastocysts via the cervix. Transfer was carried out with a disposable plastic cattle inseminating pipette as described above, and the embryo was deposited in the body of the uterus without any attempt to direct the pipette into either of the uterine horns.

In the recent Japanese study[38] 82 blastocysts were transferred nonsurgically to unmated recipient mares which ranged from +2 to −2 days as expressed in terms of synchrony of ovulation. Of these, 45 were transferred transcervically as described above and the remaining 37 were transferred via the cervix and deposited in the middle of the ipsilateral uterine horn, presumably after positioning the pipette by manipulation *per rectum*. No significant difference in conception rate resulted from the two different transfer methods, the figures being 40% by transcervical transfer and 44% by transfer via the cervix. However, some pregnancy losses occurred in both groups of recipient mares such that the initial figure of 42% for overall conception rate was reduced to 33% in terms of the birth of live foals.

Other workers attempting nonsurgical embryo transfer in mares in recent years have obtained disappointing conception rates of <30%, even when transferring apparently normal blastocysts to healthy, fertile, and closely synchronized recipients.[51] However, all report having attempted to deposit the embryo in the ipsilateral horn, rather than just the body, of the uterus by manipulating the uterus with a hand placed in the rectum. On the other hand, a conception rate of ±60% from nonsurgical transfer of equine blastocysts has been maintained consistently by simply passing the catheter through the cervix and depositing the embryo in the body of the uterus with a minimum of interference and manipulation.[52]

There is obviously a need for more detailed investigation into the best technique for nonsurgical transfer of equine embryos and the reasons for the disappointing conception rates achieved by the methods used to date. The small amount of evidence available seems to suggest that correct positioning of the transferred blastocyst within the uterus is far less important than the need to carry out the transfer as rapidly as possible with the minimum handling and manipulation of the recipient's reproductive tract. Hurtgen and Ganjam[53] demonstrated that manipulation of the reproductive tract of normal, diestrous mares, either by taking an endometrial biopsy or simply dilating the cervix on days 6 or 7 after ovulation, results in luteolysis and a premature return to estrus in the majority of animals. Furthermore, measurement of myometrial electrical activity in mares throughout the estrous cycle by means of chronically implanted myometrial electrodes showed that the equine myometrium is very much more active and responsive to external stimuli during diestrus than that of the cow, pig, or sheep.[54] Thus, it seems reasonable to suggest that the increased manipulation required to place a blastocyst in the horn, rather than just the body, of the uterus of the recipient mare may be sufficient to stimulate release of prostaglandins or other hormones which disrupt her cycle, and hence her intrauterine environment, to such an extent that embryonic development is also disturbed and the normal mechanisms involved in the maternal recognition of pregnancy are unable to occur.

### B. Surgical Embryo Transfer

The method used for surgical transfer of embryos in the mare is very similar to that described in cattle.[44] The ipsilateral uterine horn is gently exteriorized through a midventral laparotomy incision, and a small hole is made in the wall of the uterus and endometrium about one third the way down the horn by forcing through a blunted 18-gauge needle, the tip of which is filled with solder and rounded off on a grindstone. Such forced entry into the uterus greatly reduces the occurrence of endometrial hemorrhage into the lumen compared with passing a normal sharpened needle. Furthermore, application of solder to the shaft of the needle to give it a slight cone shape allows the construction of a smaller or larger hole in the wall of the uterus depending upon the stage of development, and, hence, diameter, of the blastocyst being transferred, which in turn dictates the bore size and overall diameter of the pipette used to make the transfer. The embryo is taken up in a minimal volume of medium (0.05 to 0.1 m$\ell$) in the tip of a Pasteur pipette, attached by a short length of rubber tubing to a 1-m$\ell$ tuberculin syringe. The tip of the pipette is eased through the hole in the uterine wall and passed posteriorly in the uterine lumen for 2 to 3 cm before gently expelling the embryo.

Surgical transfer of equine embryos gives a high pregnancy rate provided that transfers are carried out >3 days after ovulation. In earlier experiments, Allen and Rowson[34] obtained pregnancies in only 7 of 16 mares (44%) to which a single embryo was transferred surgically between days 1 and 6 after ovulation; embryos recovered on days 1 to 3 were transferred to the ipsilateral oviduct, whereas those recovered on days 4 to 6 were transferred to the ipsilateral uterine horn. However, analysis of the results showed that whereas only one of eight embryos transferred on days 1 to 3 continued to develop, six of eight (75%) transferred to the uterus on days 4 to 6 gave rise to pregnancies.

Since these original experiments, surgical transfer of equine embryos has been performed routinely in the author's laboratory on days 6 to 8 after ovulation. Most of the embryos have been recovered nonsurgically from the donor mare 1 to 3 hr previously and maintained at 30°C in a small volume (2 to 5 mℓ) of fresh PBS while awaiting transfer. Very few intraspecific transfers between horses have been carried out, the great majority having involved the transfer of interspecific hybrid mule (♀ horse × ♂ donkey) or hinny (♀ donkey × ♂ horse) embryos to a synchronized recipient of the opposite species (donkey and horse, respectively). More recently this has been extended to include reciprocal interspecies transfer of horse (♀ horse × ♂ horse) and donkey (♀ donkey × ♂ donkey) embryos to unmated synchronized recipients of the other species. Yet despite the xenogeneic nature of these transfers, the conception rate has remained relatively high. For example, during 1979 to 1981 inclusive, surgical transfer of eight horse embryos to unmated jenny donkeys resulted in five pregnancies (62%), and transfer of a total 34 donkey embryos to unmated Pony mares during the 3-year period resulted in 24 pregnancies (71%).

These figures seem to provide sufficient evidence to permit the conclusion that provided (1) the transfer is carried out between days 6 and 8 after ovulation, (2) the embryo is transferred within 3 hr of recovery and is maintained pending transfer at 30 to 33°C in a small volume of fresh, serum-enriched medium, and (3) transfer is via a small hole in the wall of the anterior portion of the uterine horn in the absence of endometrial bleeding, a pregnancy rate >70% can be achieved.

## VII. STORAGE, TRANSPORT, AND MANIPULATION OF EQUINE EMBRYOS

Very little information has been published concerning the storage and transport of equine embryos. In 1975 the author and colleagues[55] surgically recovered six morulae from Welsh Pony mares in Cambridge, England and transported them by car to Krakow, Poland in the ligated oviducts of two rabbits, the journey lasting 33 hr. Five of the embryos had continued to develop normally in the rabbits and four of these were transferred (three surgically and one nonsurgically) to Polish recipient mares in which ovulation had been synchronized to within ±48 hr by the prior use of prostaglandin and hCG; the fifth embryo was accidentally lost. Three pregnancies developed from which one foal aborted at day 292 of gestation and two colt foals were born live at term.

Isolated attempts in Cambridge during the past 5 years to freeze surplus equine morulae and early blastocysts in liquid nitrogen have so far proved unsuccessful.[56] A freezing method employed routinely for sheep embryos[57] has been used, and although the embryos have, for the most part, appeared viable and relatively undamaged when thawed, transfer of nine of them (six surgical, three nonsurgical) to suitable recipients has failed to result in any live foals; three mares were suspected of having become pregnant but, if so, the conceptus did not develop beyond day 40.

The resistance of the mare to superovulation and the resulting difficulty in obtaining sufficient numbers of embryos for investigation severely limits the amount of useful experimental study which can be applied to determining the optimum cooling and freezing requirements of equine embryos. This problem, combined with the refusal of most horse breed registration authorities to countenance the use of straight-forward embryo transfer, let alone the long-term storage of embryos, seems destined to hamper significant advances in this area in the future.

Mention is made elsewhere (see Chapter 11) to the production of two sets of monozygotic horse twins by use of the technique of micromanipulation. During the breeding seasons of 1979 and 1980, a total of eight two- to eight-cell horse embryos, recovered surgically from the oviducts of Welsh Pony mares between 1 and 3 days after ovulation, were separated into individual blastomeres, reconstituted in varying proportions of the original cell number in empty porcine zonae pellucidae (4 × 2/4; 1 × ⅓; 2 × ¼; 1 × 4/8), embedded in a small chip of agar gel and the whole placed in the ligated oviducts of anestrous ewes for 3 to 5 days as described by Willadsen.[58] Of the total of 20 divided embryos made in this manner and placed in sheep, 18 (90%) had continued to develop normally to the morula or blastocyst stage when subsequently recovered from the ewe; these were transferred surgically to the uteri of synchronized recipient mares, some of which had been the donor of the original embryo 4 to 5 days previously. A total of ten pregnancies resulted, and in five instances only one half of the original embryo continued to develop, the other half failing to progress in its recipient mare; in all three cases where the donor also served as the recipient for one half of its manipulated embryo, pregnancy did not result. Two sets of monozygotic twins were born alive, one pair of colt foals in 1980 and a pair of filly foals in 1981. Both these sets of identical twins originated from separation of the two four-cell embryos into four × one-cell embryos. With the first of these, undertaken in 1979, only three of the reconstituted one-quarter embryos continued to develop in the sheep and were, therefore, retransferred to mares. With the second manipulation, undertaken in 1980, all four reconstituted embryos progressed to the morula/blastocyst stage in the ewe and were retransferred to recipient mares. Although three pregnancies became established, one of these subsequently failed and was resorbed between days 43 and 51 after ovulation.

These preliminary experiments, although few in number, nevertheless demonstrated quite clearly that the equine embryo, like that of the ewe[58] and cow,[59] remains pluripotent until at least the eight-cell stage and is not destroyed by manipulation and a reduction in overall cell number. Since only a small proportion (<3%) of mares conceive twins naturally and since those that do almost always abort between 5 and 9 months of gestation due to competition between the twin placentae for the available surface area of endometrium,[60] the chances of the live birth of spontaneously occurring identical twin horses is extremely remote and has not yet been reported in the literature. The technique of embryo micromanipulation, therefore, provides a useful tool for the artificial production of multiple monozygotic equine offspring that are genetically identical and, hence, immunologically compatible. These would have obvious potential value as experimental animals.

## VIII. INTERSPECIES TRANSFER OF EQUINE EMBRYOS

Equidae is one of the few genera of mammals in which mating between separate species which have widely differing chromosome complements results in the birth of live hybrid offspring.[67] This is perhaps surprising in view of the pronounced maternal sensitization against the paternally-derived histocompatibility antigens of her fetus which occurs as the result of the invasion of specialized trophoblast cells into the maternal endometrium to form the PMSG-secreting endometrial cups.[62]

Allen and Rowson[33] reported the successful transfer of an interspecific mule embryo (♀ horse × ♂ donkey) to a jenny donkey and vice versa, a hinny embryo (♀ donkey × ♂ horse) to a horse mare. More recently, attempts to transfer intraspecific embryos between horses and donkeys were surprisingly successful.[26] From a total of eight horse embryos (♀ horse × ♂ horse) recovered nonsurgically at the blastocyst stage between days 6 and 8 after ovulation and transferred surgically to the uteri of recipient jenny donkeys, five established pregnancies resulted. Two of these pregnancies were subsequently terminated surgically on days 63 and 71 of gestation for histological examination of the endometrial cups and other fetal tissues,[26] one conceptus was aborted spon-

FIGURE 8. Three-day-old horse (*E. caballus*) foal with its recipient donkey (*E. assinus*) mother. The foal was born on May 8th, 1980 at day 346 of gestation after surgical transfer on day 7 after ovulation.

taneously on day 82, and the other two were carried normally to term and born live at 346 and 364 days of gestation, respectively (Figure 8). With the reciprocal interspecies transfer (♀ donkey × ♂ donkey → ♀ horse) an even higher success rate was achieved; the attempted transfer of a total of 34 donkey embryos into synchronized recipient horse mares resulted in 24 established xenogeneic pregnancies, giving a transfer success rate of 71%.

The horse conceptuses in jenny donkeys gave rise to large and active endometrial cups with resulting high concentrations of PMSG in the maternal circulation, and they seemed readily able to implant and continue normally to term. This was not the case with the transferred donkey conceptuses in horse mares, however, and of the 24 such pregnancies established only one resulted in the birth of a live foal on day 346 (Figure 9). In 12 other cases abortion or resorption of the already dead conceptus occurred between 47 and 105 days of gestation and in the remaining 11 recipient mares, surgical removal of the conceptus between days 60 and 87 of gestation showed definite evidence of placental and fetal damage resulting from a generalized maternal cell-mediated immune attack against the xenogeneic conceptus. In marked contrast to the "horse-in-donkey" pregnancies, none of the "donkey-in-horse" pregnancies showed any endometrial cup development due to the complete failure of the donkey chorionic girdle to invade the horse endometrium between days 35 and 40 of gestation.[26]

FIGURE 9. Two-day-old donkey (*E. assinus*) foal with its recipient horse (*E. caballus*) mother. The foal was born on May 20th, 1980 at day 346 of gestation. Surgical transfer had been carried out on day 8 after ovulation.

As mentioned above, it is perhaps surprising that the pregnancy rate resulting from xenogeneic transfer between two different equine species is so high — at least equal to that obtainable from normal within-species transfer. This is particularly so in view of the very high subsequent failure rate of the one type of transfer (donkey conceptus in horse mare) as a result of what appears to be intense maternal immune rejection of the conceptus. The equine placenta is diffuse, noninvasive, and epitheliochorial in type,[63] and intimate and stable contact between trophoblast and maternal endometrium does not occur until as late as day 45 after ovulation.[64] Prior to this time, the spherical conceptus lies unattached in the uterine lumen and is held stationary in one or other of the uterine horns by the marked contraction of the myometrium and resulting increase in uterine tone which occurs from about day 18 onwards.[65,66] As discussed previously, a high rate of transuterine migration of the blastocyst occurs in the mare,[41] and the conceptus may move from one uterine horn to the other as late as day 35 after ovulation.[40] This apparent lack of any need for close physical contact with maternal tissues, coupled with the high early survival of the xenogeneic embryo in the uterus of another species, seems to indicate that the young equine conceptus is relatively nonspecific in its requirements. It is only after about day 40 of gestation, when the immunological die of the endometrial cup reaction has been cast, that the need for an intimate and stable relationship with its maternal host begins to seriously concern the embryo. Whether the initial survival is possible as a result of either or both the large quantities of estrogens which the young conceptus manufactures,[15] or the amorphous, shell-like layer of material which surrounds it after hatching from the zona pellucida,[67] remains an interesting question for future investigation.

## IX. CONCLUSIONS

Many features of the reproductive anatomy and physiology of the mare make the

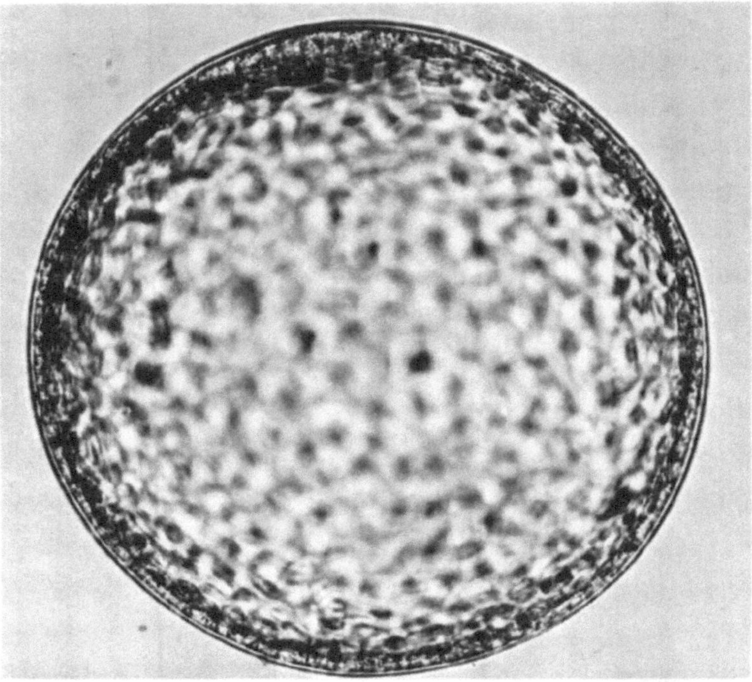

FIGURE 10. Expanded horse blastocyst on day 8 after ovulation. This embryo is just visible to the naked eye and is at an ideal stage for surgical transfer.

technique of embryo transfer somewhat simpler in equids than it is in other large domestic species. The equine blastocyst is robust, readily visible to the naked eye by day 8 after ovulation (Figure 10), and rather catholic in its environmental requirements. It is easily recovered by nonsurgical means and it survives well, if transferred correctly, even in an imperfectly synchronized recipient. These features, coupled with our desire to reproduce a whole variety of characters of excellence in the various breeds of horses, would seem to favor the rapid development and exploitation of all facets of the recovery, storage, transport, and transfer of equine embryos.

But the two major stumbling blocks, namely the resistance of the mare's ovaries to exogenous gonadotrophins and the resistance of breed registration authorities to allow use of the technique, must first be overcome if any real progress is to be made in the future. While it seems reasonable to suppose that determined research effort could eventually overcome the first of these hurdles, it is more difficult to foresee any real change in the politics which govern the second. In the meantime, horse breeding as a whole remains largely deprived of the use of a potentially valuable management tool.

## REFERENCES


1. **Andrews, F. N. and McKenzie, F. F.**, Estrus, ovulation and related phenomena in the mare, *Res. Bull. Univ. Mo. Agric. Stn.*, 329, 1941.
2. **Osborne, V. E.**, An analysis of the pattern of ovulation as it occurs in the annual reproductive cycle of the mare in Australia, *Aust. Vet. J.*, 42, 149, 1966.
3. **Heinze, H.**, Pelviscopy in the mare, *J. Reprod. Fert.*, Suppl. 23, 319, 1975.

4. **Day, F. T.**, Clinical and experimental observations on reproduction in the mare, *J. Agric. Sci. (Camb.)*, 30, 244, 1940.
5. **Evans, M. J. and Irvine, C. H. G.**, Serum concentrations of FSH, LH and progesterone during the oestrous cycle and early pregnancy in the mare, *J. Reprod. Fert.*, Suppl. 23, 193, 1975.
6. **Geschwind, I. I.**, Dynamics of pituitary gonadotrophin secretion, *J. Anim. Sci.*, 34 (Suppl. 1), 19, 1972.
7. **Stabenfeldt, G. H., Hughes, J. P., Evans, J. W., and Geschwind, I. I.**, Unique aspects of the reproductive cycle of the mare, *J. Reprod. Fert.*, Suppl. 23, 155, 1975.
8. **Tischner, M.**, Evaluation of deep frozen semen in stallions, *J. Reprod. Fert.*, Suppl. 27, 53, 1979.
9. **Rowson, L. E. A.**, The role of reproductive research in animal reproduction, *J. Reprod. Fert.*, 26, 11, 1971.
10. **Oguri, N. and Tsutsumi, Y.**, Non-surgical egg transfer in mares, *J. Reprod. Fert.*, 41, 313, 1974.
11. **Hershman, L. and Douglas, R. H.**, The critical period for maternal recognition of pregnancy in Pony mares, *J. Reprod. Fert.*, Suppl. 27, 395, 1979.
12. **Neely, D. P., Kindahl, H., Stabenfeldt, G. H., Edqvist, I.-E., and Hughes, J. P.**, Prostaglandin release patterns in the mare: physiological, pathophysiological, and therapeutic responses, *J. Reprod. Fert.*, Suppl. 27, 181, 1979.
13. **Flood, P. F., Betteridge, K. J., and Irvine, D. S.**, Oestrogens and androgens in blastocoelic fluid and cultures of cells from equine conceptuses of 10 to 22 days gestation, *J. Reprod. Fert.*, Suppl. 27, 413, 1979.
14. **Zavy, M. T., Mayer, R., Vernon, M. W., Bazer, F. W., and Sharp, D. C.**, An investigation of the uterine luminal environment of non-pregnant and pregnant Pony mares, *J. Reprod. Fert.*, Suppl. 27, 403, 1979.
15. **Heap, R. B., Hamon, M., and Allen, W. R.**, Studies on oestrogen synthesis in the equine conceptus, *J. Reprod. Fert.*, Suppl. 32, in press, 1982.
16. **Douglas, R. H. and Ginther, O. J.**, Effect of prostaglandin F2$\alpha$ on length of diestrus in mares, *Prostaglandins*, 2, 265, 1972.
17. **Allen, W. R. and Rowson, L. E. A.**, Control of the mare's oestrous cycle by prostaglandins, *J. Reprod. Fert.*, 33, 539, 1973.
18. **Palmer, E. and Jousset, B.**, Synchronization of oestrus in mares with a prostaglandin analogue and hCG, *J. Reprod. Fert.*, Suppl. 23, 269, 1975.
19. **Hyland, J. H. and Bristol, F.**, Synchronization of oestrus and timed insemination in mares, *J. Reprod. Fert.*, Suppl. 27, 251, 1979.
20. **Voss, J. L., Wallace, R. A., Squires, E. L., Pickett, B. W., and Shideler, B. K.**, Effects of synchronization and frequency of insemination on fertility, *J. Reprod. Fert.*, Suppl. 27, 257, 1979.
21. **Palmer, E.**, Reproductive management of mares without detection of oestrus, *J. Reprod. Fert.*, Suppl. 27, 263, 1979.
22. **Stewart, F., Allen, W. R., and Moor, R. M.**, Pregnant mare serum gonadotrophin: ratio of follicle stimulating hormone and luteinizing hormone activities measured by radioreceptor assay, *J. Endocrinol.*, 71, 371, 1977.
23. **Papkoff, H.**, Chemical and biological properties of the subunits of pregnant mare serum gonadotrophin, *Biochem. Biophys. Res. Comm.*, 58, 397, 1974.
24. **Cole, H. H. and Hart, G. H.**, The potency of blood serum of mares in progressive stages of pregnancy in effecting the sexual maturity of the immature rat, *Am. J. Physiol.*, 93, 57, 1930.
25. **Allen, W. R.**, The immunological measurement of pregnant mare serum gonadotrophin, *J. Endocrinol.*, 43, 593, 1969.
26. **Allen, W. R.**, Immunological and endocrinological studies of endometrial cups during equine pregnancy, symposium in honour of E. C. Amoroso, *J. Reprod. Fert.*, (Suppl. in press), 1982.
27. **Stewart, F. and Allen, W. R.**, The binding of FSH, LH and PMSG to equine gonadal tissues, *J. Reprod. Fert.*, Suppl. 27, 431, 1979.
28. **Stewart, F. and Allen, W. R.**, Biological functions and receptor binding activities of equine chorionic gonadotrophins, *J. Reprod. Fert.*, 62, 527, 1981.
29. **Douglas, R. H., Nuti, L., and Ginther, O. J.**, Induction of ovulation and multiple ovulation in seasonally-anovulatory mares with equine pituitary fractions, *Theriogenology*, 2, 133, 1974.
30. **Lapin, D. R. and Ginther, O. J.**, Induction of ovulation and multiple ovulations in seasonally anovulatory and ovulatory mares with an equine pituitary extract, *J. Anim. Sci.*, 44, 834, 1977.
31. **Douglas, R. H.**, Review of induction of superovulation and embryo transfer in the equine, *Theriogenology*, 11, 33, 1979.
32. **Woods, G. L. and Ginther, O. J.**, Pregnancy rate and incidence of multiple foetuses in mares with induced multiple ovulations, *J. Reprod. Fert.*, Suppl. 32, in press, 1982.
33. **Allen, W. R. and Rowson, L. E. A.**, Transfer of ova between horses and donkeys, Proc. 7th Int. Congr. Anim. Reprod. A. I., Munich, 1972, 484.

34. **Allen, W. R. and Rowson, L. E. A.**, Surgical and non-surgical egg transfer in horses, *J. Reprod. Fert.*, Suppl. 23, 525, 1975.
35. **Oguri, N. and Tsutsumi, Y.**, Non-surgical recovery of equine eggs, and an attempt at non-surgical egg transfer in horses, *J. Reprod. Fert.*, 31, 187, 1972.
36. **Allen, W. R.**, unpublished data, 1978.
37. **Vogelsang, S. G., Sorensen, J. D., Potter, G. D., Burns, S. J., and Kraemer, D. C.**, Fertility of donor mares following non-surgical collection of embryos, *J. Reprod. Fert.*, Suppl. 27, 383, 1979.
38. **Oguri, N. and Tsutsumi, Y.**, Non-surgical transfer of equine embryos, *Arch. Androl.*, 5 (Abstr. 102), 108, 1980.
39. **Squires, E. L.**, Equine embryo transfer, *Equine Vet. Data (Wildomer, Calif.)*, 1, 201, 1980.
40. **Bain, A. M. and Howey, W. P.**, Ovulation and transuterine migration of the conceptus in Thoroughbred mares, *J. Reprod. Fert.*, Suppl. 23, 541, 1975.
41. **Butterfield, R. M. and Mathews, R. G.**, Ovulation and the movement of the conceptus in the first 35 days of pregnancy in Thoroughbred mares, *J. Reprod. Fert.*, Suppl. 27, 447, 1979.
42. **Allen, W. E. and Newcombe, J. R.**, Relationship between early pregnancy site in consecutive gestations in mares, *Equine Vet. J.*, 13, 51, 1981.
43. **Hamilton, W. J. and Day, F. T.**, Cleavage stages of the ova of the horse with notes on ovulation, *J. Anat.*, 79, 127, 1945.
44. **Rowson, L. E. A., Moor, R. M., and Lawson, R. A. S.**, Fertility following egg transfer in the cow: effect of method, medium and synchronization of oestrus, *J. Reprod. Fert.*, 18, 511, 1969.
45. **Sisson, S. and Grossman, J. D.**, *The Anatomy of the Domestic Animals*, W. B. Saunders, Philadelphia, 1953.
46. **Allen, W. R., Pashen, R. L., and Willadsen, S. M.**, Successful interspecies embryo transfer between the horse (Equus caballus) and the donkey (Equus assinus), unpublished data, 1982.
47. **Whittingham, D. G.**, Fertilization of mouse eggs in vitro, *Nature (London)*, 220, 592, 1968.
48. **Onuma, H. and Ohnami, Y.**, Retention of tubal eggs in mares, *J. Reprod. Fert.*, Suppl. 23, 507, 1975.
49. **van Niekerk, C. H. and Gerneke, W. H.**, Persistence and parthenogenetic cleavage of tubal ova in the mare, *Onderstepoort J. Vet. Res.*, 31, 195, 1966.
50. **Willadsen, S. M., Pashen, R. L., and Allen, W. R.**, unpublished data, 1982.
51. **Squires, E. L., Sharp, D. C., and Tischner, M.**, personal communications, 1981.
52. **Douglas, R. H.**, personal communication, 1981.
53. **Hurtgen, J. P. and Ganjam, V. K.**, The effect of intrauterine and cervical manipulation on the equine oestrous cycle and hormone profiles, *J. Reprod. Fert.*, Suppl. 27, 191, 1979.
54. **Taverne, M. A. M., van der Weyden, G. C., Fontijne, P., Dieleman, S. J., Pashen, R. L., and Allen, W. R.**, In vivo myometrial electrical activity in the cyclic mare, *J. Reprod. Fert.*, 56, 521, 1979.
55. **Allen, W. R., Stewart, F., Trounson, A. O., Tischner, M., and Bielanski, W.**, Viability of horse embryos after storage and long-distance transport in the rabbit, *J. Reprod. Fert.*, 47, 387, 1976.
56. **Willadsen, S. M., and Allen, W. R.**, unpublished data, 1980.
57. **Willadsen, S. M., Polge, E. J. C., Rowson, L. E. A., and Moor, R. M.**, Deep freezing of sheep embryos, *J. Reprod. Fert.*, 46, 151, 1976.
58. **Willadsen, S. M.**, A method for culture of micromanipulated sheep embryos and its use to produce monozygotic twins, *Nature (London)*, 277, 298, 1979.
59. **Willadsen, S. M. and Polge, E. J. C.**, Attempts to produce monozygotic quadruplets in cattle by blastomers separation, *Vet. Rec.*, 108, 211, 1981.
60. **Jeffcott, L. B. and Whitwell, K. E.**, Twinning as a cause of foetal and neonatal loss in the Thoroughbred mare, *J. Comp. Pathol.*, 83, 91, 1973.
61. **Gray, A. P.**, *Mammalian Hybrids*, Commonwealth Agricultural Bureaux: Farnham Royal, Bucks, England, 1954, 45.
62. **Allen, W. R.**, Immunological aspects of the equine endometrial cup reaction, in *Immunobiology of Trophoblast*, Edwards, R. G., Howe, C. E. S., and Johnson, M. H., Eds., Cambridge University Press, London, 1975, 217.
63. **Amoroso, E. C., Placentation, in** *Marshall's Physiology of Reproduction*, Vol. 2, 3rd ed., Parkes, A. S., Ed., Longmans Green, London, 1952, 127.
64. **Samuel, C. A., Allen, W. R., and Steven, D.**, Ultrastructural development of the equine placenta, *J. Reprod. Fert.*, Suppl. 23, 575, 1975.
65. **van Niekerk, C. H.**, The early diagnosis of pregnancy, the development of the foetal membranes and nidation in the mare, *J. S. Afr. Vet. Med. Assoc.*, 36, 483, 1965.
66. **van Niekerk, C. H. and Allen, W. R.**, Early embryonic development in the horse, *J. Reprod. Fert.*, Suppl. 23, 495, 1975.
67. **Flood, P. F. and Betteridge, K. J.**, Horse embryos from 3 to 17 days; transmission electron microscopy, *J. Reprod. Fert.*, Suppl. 32, in press, 1982.

Chapter 9

# OVUM RECOVERY AND EMBRYO TRANSFER IN PRIMATES*

**W. Richard Dukelow**

## TABLE OF CONTENTS



* This chapter is dedicated to the late Dr. Pierre Soupart of Vanderbilt University in recognition of his advice, friendship, and inspiration to this author.

# I. INTRODUCTION

In July of 1978 the world received news of the birth of a child following laparoscopic ovum recovery, in vitro fertilization, and transfer of the embryo. This seemingly simple event, repeating what had been achieved in several species of laboratory animals, represents both the culmination of nearly 10 years of active research and the start of an era in primate reproductive biology. Indicative of the activity in this field is the fact that as of the present time (May 1981) human pregnancies have been reported from at least four laboratories, and six children have been born by the procedure with several other pregnancies in progress.

With nonhuman primates progress has been less rapid. In vitro fertilization has been achieved but no pregnancies have yet resulted from transfer of these embryos. Transfer of in vivo fertilized embryos, however, has resulted in pregnancy.

In the present review emphasis has been placed on the embryo handling and transfer procedure with some discussion of the techniques of embryo procurement including in vitro fertilization. Less emphasis has been placed on techniques of ovum maturation prior to recovery since this subject has been reviewed elsewhere for both human[1-7] and nonhuman[1,8-10] primates.

# II. NONHUMAN PRIMATES

## A. Induction of Follicular Growth and Ovulation

Morphological development of the ovary has been described extensively in the rhesus monkey[11] and reviewed for other primate species.[12] Nonhuman primate development correlates well with man except for some of the prosimians. Follicular development on the ovarian surface has been laparoscopically documented during the final 24 to 36 hr before ovulation in the cynomolgus macaque.[13-15] Examination of this development allows prediction of the estimated time of ovulation and is useful for timed-mating procedures.

The actual ovulation process has been described in cynomolgus macaques,[16] the Japanese macaque,[17] the bonnet macaque,[18] the squirrel monkey,[19] and the baboon.[20] With the exception of the bonnet macaque, ovulation is normally preceded by the release of clear follicular fluid and then expulsion of the cumulus cell mass containing the ovum. The follicle collapses and some bleeding is observed. The ovulation process requires about 30 sec and is not the explosive phenomenon observed in rodents. The time of ovulation varies greatly in macaque monkeys as it does in humans. In the cynomolgus macaque the day of ovulation can be described, relative to cycle length, by the regression equation of $Y = 0.31\ X + 5.12$ where $X$ = cycle length. Despite such calculations, wide variations can occur and ovulation has been observed as early as day 12 of 37-day cycle and as late as day 17 of a 24-day cycle.[21]

Ovulation induction in nonhuman primates has been reviewed[22,23] and will not be dealt with extensively here. Basically the standard human ovulation induction protocols are effective in nonhuman primates.[22,24] Some seasonal effects are noted, as in the squirrel monkey.[25] The seasonal effect seems to relate to the minimal effective dose of follicle stimulating hormone (FSH), as ovulation can be induced by extending the time of FSH injection or increasing the dose.[26] The minimal effective dose of human chorionic gonadotrophin has been determined for the squirrel monkey to be between 100 and 250 i.u.[27] Ovulations resulting from these procedures produce viable ova capable of being fertilized either in vivo or in vitro.

Follicular puncture and aspiration, either by laparotomy or laparoscopy,[31] is the commonly accepted procedure to obtain ova for study or in vitro fertilization procedures. To accomplish this a 25-gauge needle (½ to 1 in. long) is passed through the skin (in the case of laparoscopy) and abdominal wall and pushed gently (bevel side down) into the follicle where slight suction is applied with the aspirating syringe. One of the major problems with in vitro fertilization is obtaining oocytes at the metaphase II stage of meiosis, characterized by the presence of a polar body. Such oocytes are mature and can be penetrated by spermatozoa. At least 30 hr of in vitro culture are required to mature rhesus monkey oocytes, and such maturation has been documented in the absence of gonadotrophins. The fertility of oocytes matured in vitro was tested by Suzuki and Mastroianni,[9] who obtained a total of six fertilized rhesus oocytes out of 31 (19.4%) from rhesus oviducts after allowing 48 hr for in vitro maturation and followed by oocyte transfer to mated females. Their in vitro maturation procedure yielded 55.3% matured oocytes. Shorter times (22 to 36 hr) for in vitro maturation have been reported for cynomolgus macaques.[32] There also may be differences in the maturation rate due to seasonal differences in the animals' reproductive cycle.[10] The follicular environment, particularly the estrogen level, may facilitate oocyte maturation in vitro during the breeding season.

### B. In Vitro and Xenogenous Fertilization

Among all primate species, in vitro fertilization has been reported in only baboons, squirrel monkeys, and humans. Among the nonhuman primates this work has been confined to only three laboratories. In 1972 D. C. Kraemer (personal communication) reported the development of five baboon oocytes recovered by laparotomy and fertilized in vitro with sperm incubated with cervical mucus, to at least the two-pronuclear stage. One of the embryos reached the six-cell stage and, upon staining, all six cells were found to contain nuclei.

In that same year two reports appeared simultaneously which described successful in vitro fertilization of squirrel monkey oocytes.[33,34] In an expanded report of one of these studies, 22 mature oocytes were recovered and 11 of these had spermatozoa in the perivitelline space, extrusion of the second polar body or pronuclear formation. Six ova cleaved to the two-cell stage.[35] From the other original report, expanded reports appeared in 1975 and 1979.[29,30] To 1979 a level of 35% in vitro fertilization had been achieved and embryos developed to the eight-cell stage. The optimum medium for fertilization was found to be TC-199 with 20% agamma newborn calf serum and 72 $\mu$g pyruvate per m$\ell$. The technique has been used to determine the fertility of captive males and it has been determined that the fatted male condition is not necessary for the production of fertile spermatozoa. Although the spermatozoan concentration and motility do fluctuate with the season, spermatozoa can be obtained throughout the year that are capable of fertilizing squirrel monkey ova in vitro.

More recent refinements in techniques have raised the in vitro fertilization rate to approximately 60%, and with the addition of 1 to 10 $\mu M$ dbcAMP to the fertilization medium, fertilization rates of 90% have been achieved.[36]

As an alternative method for the production of nonhuman primate embryos, xenogenous fertilization has been proposed. This procedure involves the introduction of laparoscopically recovered follicular oocytes and spermatozoa of the same species, into the oviductal lumen of the pseudopregnant rabbit. Fertilization does occur and the embryos can be developed to at least the eight-cell stage. Recovery and fertilization rates of 55 and 36%, respectively, have been reported for the squirrel monkey.[37]

### C. Embryo Recovery and Transfer

Nonhuman primate embryo recovery has primarily been studied in the baboon, rhesus, and squirrel monkey.

In their classic studies of the baboon embryo, Hendrickx and Kraemer[38-40] used an *in situ* flushing procedure (Figure 1). Using this procedure the vagina was first cleaned

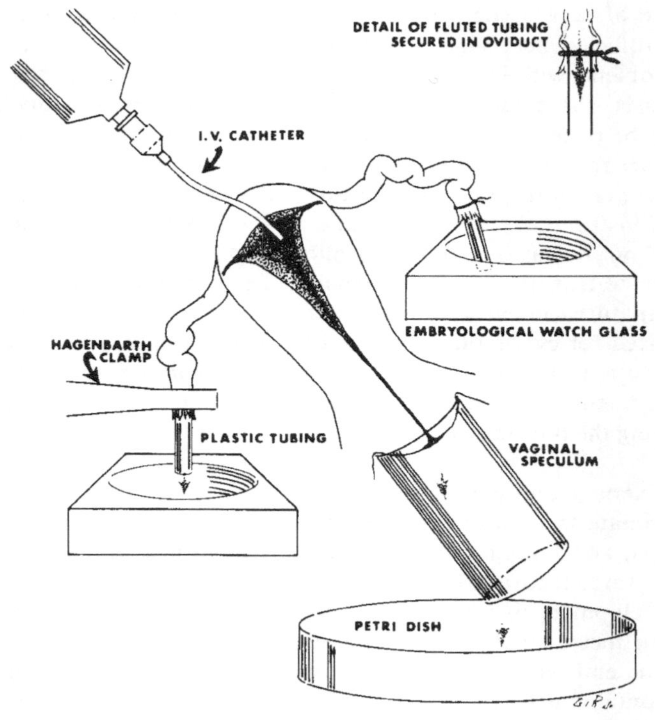


FIGURE 1. Technique used to collect preimplantation baboon embryos. (Reprinted from Hendrickx, A. G., Embryology of the Baboon, *University of Chicago Press,* Chicago, 1971. With permission.)

with saline and a glass speculum inserted. The reproductive tract was then exposed by a midventral incision and fluted plastic catheters were placed about 3 mm into the infundibular end of the oviduct and held in place with a silk suture. An 18-gauge plastic catheter was inserted through the uterine fundus into the uterine lumen. The cervix was then clamped shut and the medium (Hams F10 and 20% fetal calf serum) was forced through the uterus and out the oviduct into watch glasses. After this procedure the cervical clamp was removed and the fluid was forced through the uterus and cervix through the vaginal speculum and to collecting dishes.

To collect postimplantation embryos the procedure illustrated in Figure 2 was used. An 8-cm midline incision was made and a Balfour retractor used to expose the uterus. The implantation site was identified by its increased vascularity and it could be easily recognized as early as the 20th day of pregnancy. A sponge forceps was placed at the cervix to restrict uterine blood supply. A small uterine incision was made opposite the implantation site and this was slowly enlarged. The connective tissue septae were cut with scissors and the friable blood vessels detached by blunt dissection with a spatula. Eventually the endometrial sac containing the chorionic vesicle was excised anteriorly to the cervix uteri and the incision closed.

Transcervical approaches have also been used in attempts to collect baboon embryos. Kraemer et al.[41] used such a procedure modeled after similar procedures in humans.[42] They collected embryos on 5 of 26 attempts but found none of the approaches sufficiently efficient for practical application. More recent work[43,44] has involved the use of a continuous flow collection device (1 to 2 m$\ell$/min) using an Issacs endometrial cell sampler modified for continuous flow. Using this procedure recovery

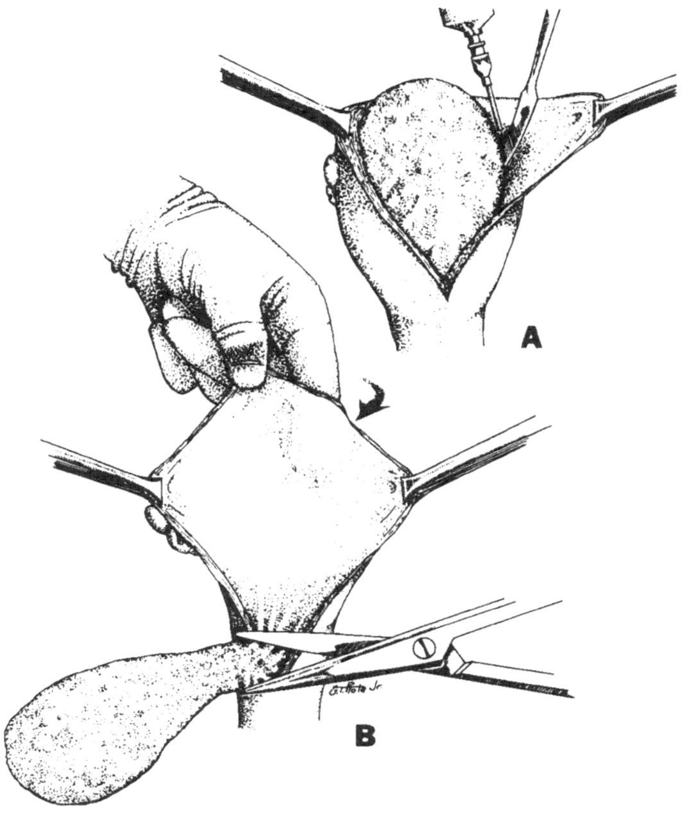


FIGURE 2. Illustration of technique used to collect baboon embryos after implantation. (A) A spatula is used to separate the endometrium from the myometrium; (B) the uterus is inverted (arrow) and the endometrial sac, containing the embryo, is cut at the cervix uteri. (Reprinted from Hendrickx, A. G., *Embryology of the Baboon*, University of Chicago Press, Chicago, 1971. With permission.)

rates ranging from 35 to 58% have been obtained. This collection device is illustrated in Figure 3.

There has been only a single successful transfer of a nonhuman primate between two animals and that is the report of Kraemer et al.[45] An embryo was collected from a naturally ovulating, naturally mated female baboon on day 5 of gestation and transferred to a synchronized nonmated female by surgical means. The embryo was recovered by the procedure shown in Figure 1. It was in the morula stage of development at transfer to the uterine lumen in a volume of 1 $\mu\ell$. Pregnancy was diagnosed on day 20, and a male offspring delivered on day 174 (Figure 4).

In rhesus monkeys a retrograde flushing procedure has been developed to obtain ova from the oviducts.[46-48] This is patterned after an earlier procedure[49] and is illustrated in Figure 5. A cellulose nitrate tube is cut at an angle and inserted into the abdominal cavity. The fimbriated end of the oviduct is held near the tube. Near the tubouterine junction a ½-in., 25-gauge needle is inserted into the oviduct and 2 to 3 $m\ell$ of medium is flushed towards the infundibulum.

To recover uterine embryos both oviducts are clamped near the tubouterine junction to prevent retrograde flushing (Figure 6). The cervix is clamped with a right-angle gall bladder clamp and a 19-gauge needle attached to the 10-$m\ell$ syringe barrel is inserted

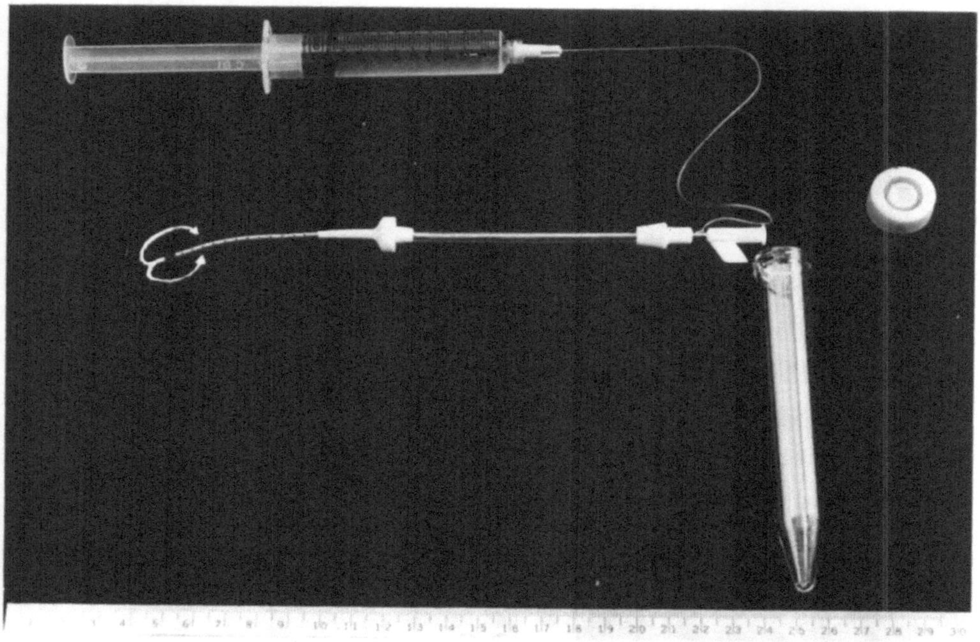

FIGURE 3. Continuous flow embryo collection device using an Isaacs endometrial cell sampler modified for continuous flow (arrows). (Centimeter scale at bottom of figure.) (Reprinted from Pope, C. E., Pope, V. Z., and Beck, L. R. *Biol. Reprod.*, 23, 657, 1980. With permission.)

FIGURE 4. Baboon infant produced by embryo transfer. At 63 days of age with the donor mother on the right and recipient mother on the left. (Reprinted from Kraemer, D. C., Moore, G. T., and Kramen, M. A., *Science*, 192, 1246, 1976. Copyright 1976 by the American Association for the Advancement of Science. With permission.)

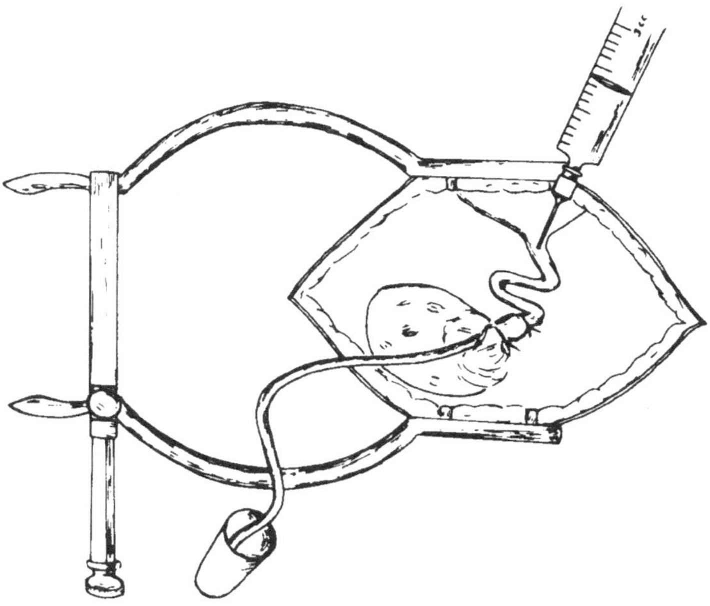


FIGURE 5. Diagrammatic representation of procedure for retrograde flushing of the rhesus monkey oviduct for recovery of tubal ova. (Reprinted from Brackett, B. G., *Methods in Mammalian Reproduction,* Daniel, J. C., Jr., Ed., Academic Press, New York, 1973, 333. With permission.)

into the uterine lumen from the ventral surface. A second 19-gauge needle on a 10-m$\ell$ syringe filled with medium is inserted into the fundus of the uterus. The uterus is then flushed by forcing the medium into the lumen and recovering the flushing in the second syringe barrel. Using these procedures ovum recovery rates ranging from 33 to 75% have been reported.

Another surgical flushing procedure for rhesus monkeys has been reported which involves flushing of the uterus with two bent 17-gauge fenestrated needles attached by polyethylene tubing to 5-m$\ell$ syringes.[50-52] One needle is placed into the uterine lumen from the ventral surface just anterior to the internal os and the second is placed at the fundus. About 4 m$\ell$ of medium is then introduced through the supracervical needle while simultaneously aspirating the fluid into the receiving syringe at the fundus. In about 22 flushes with this technique 93% of the fluid was collected and nine embryos and two unfertilized ova were recovered.

Although successful transfer of rhesus embryos between animals has not been accomplished, pregnancies have resulted by transfer of embryos back to the donor animal.[53] Thirteen naturally cycling animals were subjected to laparotomy and oviductal flushing between days 12 and 18 of their cycle. Dulbecco's phosphate-buffered saline was used as the flushing medium. The embryos were examined and then taken up in a fine nylon cannula for transfer (through a 19-gauge needle) into the fundus of the uterus. The embryos, which were expelled in 10 $\mu\ell$ of fluid,[54] were transferred within 30 min of recovery. Some embryos were transferred to the oviduct and this was accomplished by introducing the tip of the nylon cannula through the tubal ostium to the point of the first marked flexure of the ampulla. Each embryo was expelled in about 0.5 $\mu\ell$ of fluid and a similar volume of air. From ten embryos, ranging from 1 to 22 cells, transferred to oviducts, eight pregnancies resulted. Of three transfers to uteri (one five-cell embryo and two seven- to eight-cell embryos), none resulted in pregnancy.

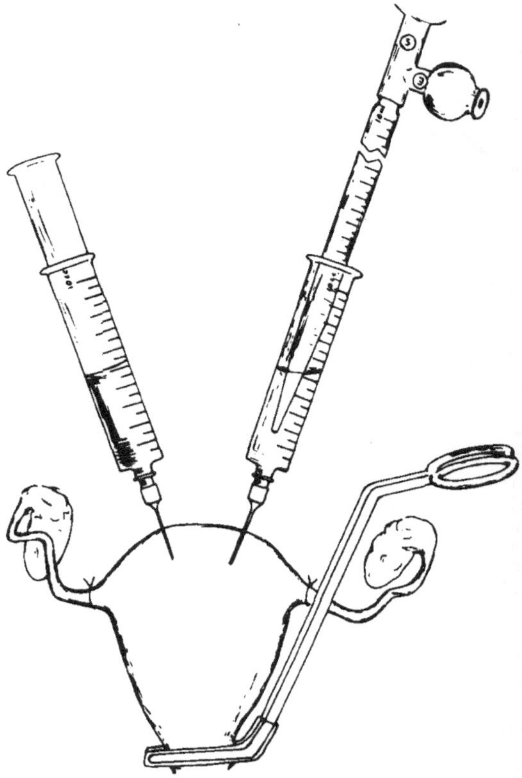

FIGURE 6. Diagrammatic representation of procedure for uterine lavage in the rhesus monkey. (Reprinted from Brackett, B. G., *Methods in Mammalian Reproduction*, Daniel, J. C., Jr., Ed., Academic Press, New York, 1973, 333. With permission.)

A recent study has demonstrated the possibility of recovery of rhesus follicular oocytes and subsequent transfer to the isthmal region of the oviduct (thus by-passing possible blocked oviducts, a common infertility problem in women).[55] The oocytes were recovered laparoscopically, examined and matured, and then transferred on the ipsilateral side through a 20-gauge needle about 1 to 2 cm from the tubal-uterine junction. The animal was mated the day before and the day after the procedure. Of 31 such attempts, 5 pregnancies resulted, none of which was ectopic.

In the squirrel monkey laparoscopic recovery of follicular oocytes has been described and used extensively.[30,31] These procedures have been adapted to many uses including intrafollicular injections and topical application of compounds to the ovarian surface.[31]

Laparoscopic techniques have also been used to recover uterine blastocysts from the squirrel monkey. The animal is first subjected to an ovulation induction regimen such as that shown in Figure 7.[24] The female is exposed to the male for a period of 12 to 24 hr and a preliminary laparoscopy is performed about 36 hr after the hCG to determine if ovulation has occurred. If positive, the uterine flushing procedure illustrated in Figure 8 is then carried out on day 4, 5, or 6, depending on the stage of embryo desired.[56] Using this laparoscopic technique the uterus is manipulated with the Verres cannula and grasped with the alligator forceps to steady the uterus. A 25-gauge, 5/8-in. needle

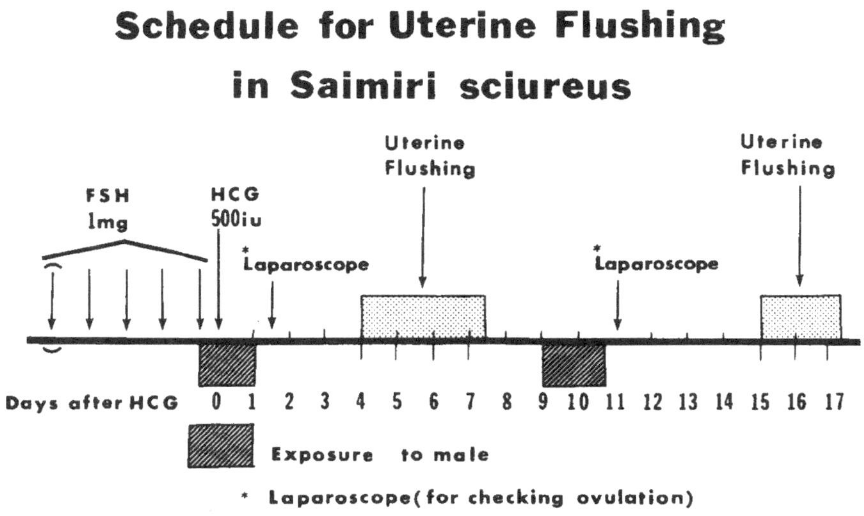


FIGURE 7. Follicular growth induction schedule for laparoscopic uterine flushing in the squirrel monkey. (Reprinted from Ariga, S. and Dukelow, W. R., Recovery of Preimplantation Blastocysts in the Squirrel Monkey by a Laparoscopic Technique. *Fert. Steril.*, 28, 577, 1977. With permission.)

is inserted through the lower abdominal wall and the uterine fundus. Prewarmed medium (2 to 3 mℓ) is then flushed through the uterus and recovered through a polyeth ylene catheter previously inserted into the external cervical os through a cleansed vagina. Using this procedure an average of 73% of the flushed fluid has been recovered and two preimplantation blastocysts and six ova have been recovered for study following 58 attempts.[57] One of the blastocysts is shown in Figure 9.

A similar laparoscopic transfer procedure has been described for placing in vitro fertilized embryos into the ovarian end of the oviduct of squirrel monkeys.[30,31] This procedure is illustrated in Figure 10. Using this technique the organs are first manipulated into position with the Verres cannula and the alligator forceps used to grasp the oviductal supporting mesentery. Then a micropipettor is inserted through a secondary point of entry through the skin and abdominal wall and the tip placed 1 to 2 mm into the infundibulum. Upon injection, 1 $\mu\ell$ of fluid is deposited with the embryos. Although no pregnancies have resulted from this procedure (with in vitro fertilized embryos),[30] dyes have been used to verify the site of deposition. Up to 3 $\mu\ell$ can be deposited without retrograde loss into the abdominal cavity.

## III. HUMANS

### A. Induction of Follicular Growth and Ovulation

Studies of ovum maturation have generally been of two types: (1) administration of pharmacologic agents to induce ovulation in women where anovulation represents the primary cause of infertility; and (2) recovery of preovulatory oocytes for subsequent maturation in vitro for use in subsequent in vivo or in vitro fertilization and transfer experiments. The latter area has received considerable attention in recent years because of the impracticality of direct study of oocyte maturation within the human ovary.

After puberty, levels of serum gonadotropins, prolactin, and the sex steroids influence follicular growth, ovulation, and corpus luteum formation. These also play a role in determining which follicle will mature and which will undergo atresia. Current evi-

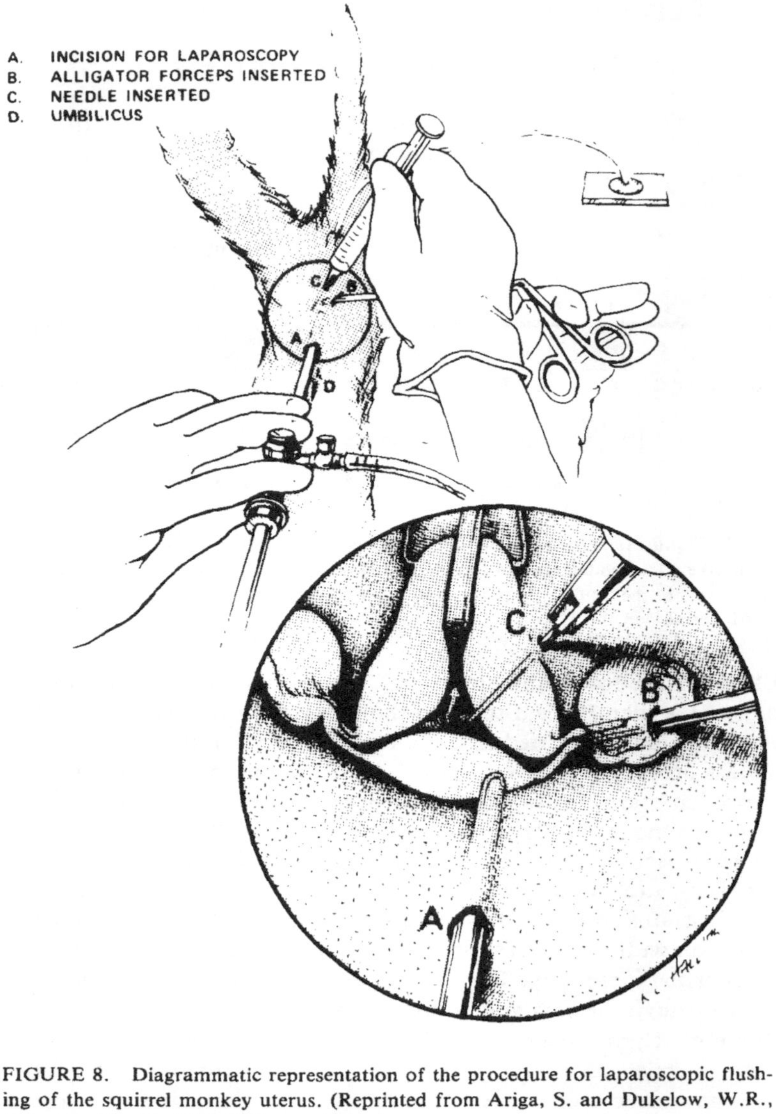


FIGURE 8. Diagrammatic representation of the procedure for laparoscopic flushing of the squirrel monkey uterus. (Reprinted from Ariga, S. and Dukelow, W.R., *Primates*, 18, 453, 1977. With permission.)

dence indicates strongly that the serum levels of the steroid hormones largely reflect the secretory activity of the preovulatory follicle. Within the follicle these steroids mediate the effects of gonadotropins and prolactin on follicles and ovum maturation.[58]

The changes that occur in the follicle just prior to ovulation are significant. If follicular fluid is aspirated from large preovulatory follicles the granulosa cells are removed in large sheets in contrast to small clumps or individual cells found in follicular fluid from nonovulatory follicles. The diameter of preovulatory follicles ranges from 2 to 4 cm with a volume of 3 to 14 m$\ell$. Aspiration is easiest 15 to 27 hr after the start of the LH surge. After 30 hr aspiration is more difficult due to follicle distention with hemorrhagic fluid. Levels of estradiol-17$\beta$ and progesterone are elevated in the serum of preovulatory women and the corresponding levels in follicular fluid are even higher.[59]

The preovulatory follicle is thin walled and vascularized with distinctive blood vessel patterns not unlike that observed in macaque monkeys.[14]

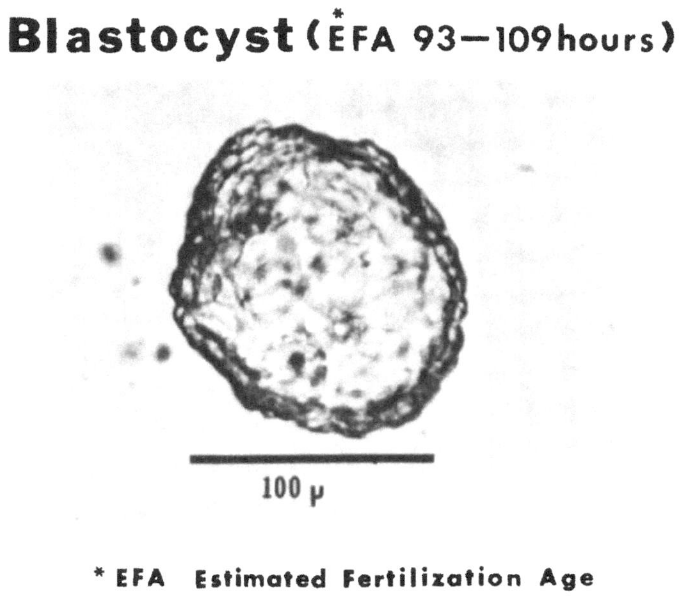


FIGURE 9. Squirrel monkey uterine preimplantation blastocyst recovered by laparoscopic flushing procedure. (Reprinted from Ariga, S. and Dukelow, W. R., Recovery of Preimplantation Blastocysts in the Squirrel Monkey by a Laparoscopic Technique. *Fert. Steril.*, 28, 577, 1977 With permission.)

Prediction of the time of ovulation has previously been based on hormonal changes, changes in the body temperature, or the physical properties of cervical mucus. These have been reviewed.[60] More recently, however, ultrasonic scanning of the ovaries has been used to detect preovulatory follicles in women.[61,62]

Occasional abnormalities of development are detected in maturing oocytes. Binucleate human oocytes have been recovered from large follicles[63] (2 ova out of 900 examined). While such oocytes can be assumed to be normally lost by atresia, they may be ovulated and a triploid zygote could result. Also, examination of such an oocyte after an in vitro fertilization attempt could lead to a false diagnosis of fertilization.

Although gonadotropins are commonly administered prior to ovum recovery, the simple release of the oocyte from the follicular environment often triggers oocyte maturation. In a study of 776 oocytes obtained from ovarian tissues after wedge resection, two matured in vivo and another had two vesicular nuclei. In a smaller sample incubated in culture medium for up to 42 hr, 42% showed germinal vesicle breakdown and 31% progressed to the metaphase II stage. Neither the stage of the cycle at which the ovaries were removed nor the addition of 10 µg progesterone had a significant effect on oocyte maturation.[6] First polar body extrusion in humans occurs 40 to 48 hr after start of culture.[1]

Early efforts with the use of human menopausal gonadotropin (hMG) and human chorionic gonadotropin (hCG) to induce follicular growth and ovum maturation for in vitro fertilization experiments did not result in successful pregnancies, and such attempts were abandoned in favor of timing oocyte recovery to the woman's natural cycle and monitoring the LH secretion by radioimmunoassay and immunological urinary tests.[65] It was found that a high rate of successful recoveries could be obtained when recovery was between 15 and 24 hr after the start of the natural LH surge. Four pregnancies did result (following in vitro fertilization) from this work and all four were

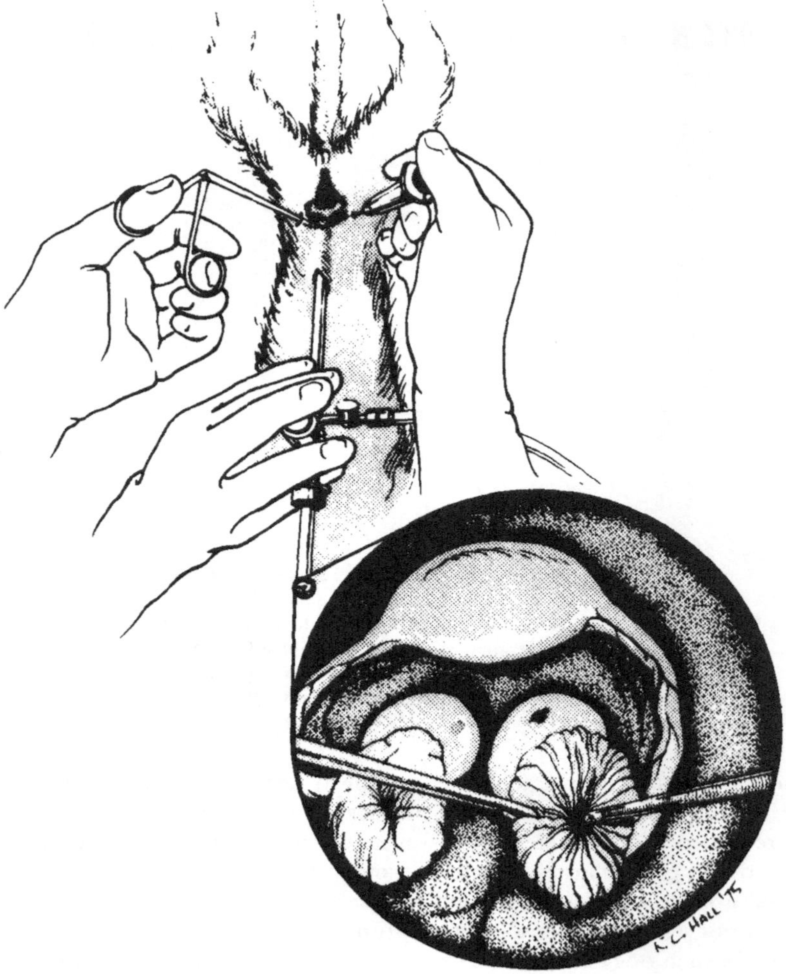


FIGURE 10. Laparoscopic embryo transfer procedure used with the squirrel monkey. (Reprinted from Dukelow, W. R., *Methods in Mammalian Reproduction*, Daniel, J. C., Jr., Ed., Academic Press, New York, 1978, 438. With permission.)

when the time interval between the start of the LH surge and the recovery was from 24 to 30 hr. These workers also reported a diurnal rhythm for the start of the LH surge. Most patients marked the beginning of the surge between 0600 and 1200 hr, especially during the winter months.[64] This rhythm remains to be confirmed.

Despite the subsequent failure to achieve a pregnancy after hMG-hCG induction and in vitro fertilization, much of our basic knowledge regarding follicular development immediately prior to ovulation comes from such treatment.[65-68] From these studies it was determined that the timing of ovulation could be controlled and that after recovery and in vitro culture human maturation occurred from 36 to 42 hr later. Such follicles also provided follicular material for many basic biochemical studies to ascertain the oocyte environment just prior to ovulation and maturation.

Using the natural cycle of the patient but administering 3000 i.u. hCG 36 to 38 hr before surgery, others have recovered nine ova of which five were subsequently fertilized.[69]

Recently clomiphene (50 to 100 mg daily) or human pituitary gonadotropin (hPG, 480 to 620 i.u. daily) have been given for 5 consecutive days starting on day 5 of the menstrual cycle. With the former treatment, 5000 i.u. hCG was given 19 to 36 hr before laparoscopy, whereas with the hPG treatment, 10,000 i.u. hCG was given immediately after the preovulatory estrogen peak with ovum collection 31 to 36 hr later. Numbers of ova recovered averaged one and four for the two groups, respectively. About 70% of all ova were in the process of maturation upon recovery.[70] An identical clomiphene-hCG treatment resulted in the development of 5.7 follicles per patient.[7] The mean duration of the luteal phase of these women was 13.4 days and increased significantly to 16.4 days when 1000 i.u. hCG was injected on the 2nd and 4th day after follicular aspiration. Of 33 oocytes inseminated 16 were fertilized and 5 cleaved in culture. Recent studies by this same research group have shown similar rates of in vitro fertilization in women receiving clomiphene alone, clomiphene with hCG, or by following the natural cycle. Recovery of mature oocytes from these three groups were 70, 100, and 65%, respectively, and some pregnancies occurred following embryo transfer in all but the natural cycling group.[72]

**B. In Vitro Fertilization**

The previous section of this chapter has dealt primarily with the maturation of human oocytes after in vivo treatment of the patient, ovum recovery, and in vitro incubation of the oocyte prior to fertilization.

The timing of the maturation in vitro is critical to success with in vitro fertilization, and many research workers have attempted to enhance maturation by altering the culture conditions. Edwards[1] observed germinal vesicles after 20 hr in culture and reported in vitro maturation after 40 to 48 hr. Germinal vesicle breakdown, metaphase I, and metaphase II occurred 25, 28 to 35, and 36 to 43 hr after oocyte recovery.[2] An in vitro maturation rate of 57.5% was reported with oocytes incubated in follicular fluid supplemented with either Hank's, Brinster's, or Bavister's media. When a chemically defined medium is used, Krebs'-Ringer bicarbonate with added glucose, lactate, and pyruvate, a 15% maturation was obtained.[73] These latter workers also reported that the maturation rate was higher with oocytes surrounded by cumulus cells than with cumulus-free oocytes.

Ultrastructural evaluation of matured oocytes indicated a 38.2% maturation rate after 37 hr or less of culture and 67.4% after 40 to 53 hr.[5] Comparison of the oocytes matured in vitro and in vivo showed no differences ultrastructurally and that in vitro maturation did not induce anomalies.

It has already been indicated that addition of progesterone (10 μg/mℓ) did not influence maturation. With addition of FSH, LH, and hCG to a basic 25% Bavister's medium, 75% follicular fluid resulted in a higher maturation rate (57.1%) than controls (28.0%).

Similarly, maturation of immature oocytes in Ham's F-10 medium was higher if 17β-estradiol was added for 4 hr and followed by 17α-hydroxyprogesterone.[75,76]

The most common modern procedure for recovering the oocytes from patients has been with laparoscopy, as first reported by Steptoe and Edwards;[65] various refinements and modifications have appeared during the past 10 years. An ovum recovery unit has been described consisting of a 25- or 27-gauge cartridge type double end dental needle and a 20-gauge hypodermic needle bent to a gentle curve not greater than 45° and attached to appropriate syringe and collecting vial.[77] This device has a reported recovery rate, *in situ,* of 25.9%. Other research workers used a 24 cm long, 20-gauge needle (with 45° bevel and a 15-gauge sheath) attached to a follicular fluid trap (5 mℓ capacity). Vacuum pressure was at 200 mm of mercury which was found to be optimal for oocyte collection. The recovery rate under these conditions was 45.6%.[78] Using a similar procedure but with vacuum pressure limited to 120 mm of mercury (alledgedly to prevent damage to the ovum) the recovery rate was 66.1%.[64] In a limited series of patients a 10-in., 20-gauge needle was used at a vacuum pressure of 90 to 100 cm of water, and this yielded 75% recovery (12 of 16).[79]

The most recent development for oocyte recovery consists of a Teflon®-protected needle 23 cm long with an internal diameter of 1.6 mm and an outside diameter of 2.1 mm. The needle is lined with a 19-gauge Teflon® tube giving an internal diameter of 1.0 mm. The Teflon® tubing is then passed to a collecting tube. In comparison with needles of course and fine bore, the Teflon®-lined apparatus yielded 100% recovery from 17 follicles in 13 patients.[80]

Since the discovery of spermatozoan capacitation in 1951 and the first successful in vitro fertilization, transfer of embryos, and parturition in the rabbit a few years later (1959), a large number of laboratory and domestic animals have been used in attempts to achieve this unique form of conception. These efforts have been documented in other chapters of this book. In primates this work has recently been reviewed by Blandau,[81] who correctly points out that most investigators in the field would agree that there is probably no area of reproductive biology in which it is more difficult to document the success of the procedure used. Initially success is measured in terms of ova penetrated and percentage of embryos that cleave to a certain stage. The presence of a spermatozoan midpiece and tail in the cytoplasm of the ovum is presumptive evidence of fertilization, but even this is criticized for possible errors in focusing the microscope. Transfer and implantation with delivery of living young was considered for many years the finite proof of in vitro fertilization, and, if coupled with chromosomal studies, was even more persuasive. It is rather ironic, after years of arguing the necessary criteria for proof of fertilization (which was not nearly so controversial when laboratory or domestic animals were involved!), that when the first human was born as a result of this procedure some knowledgeable scientists decried the fact that the child was a female and, thus, one could not rule out the possibility of parthenogenic cleavage! Fortunately the second such child born was male.

In 1944 a report appeared regarding isolation of 800 human follicular ova of which 138 were observed after exposure to human spermatozoa. Of these, two ova were described as being in the two-cell stage and two others at three cells.[82] These same embryos were described in a later publication.[83] A few years later, in Japan, 160 ova were collected from 74 patients and incubated with spermatozoa after culture in a basic salt solution with human serum added. Cleavage was reported in 20 ova with some reaching the morula stage.[84] Critical examination of some of these cleaving ova indicated that fragmentation may have occurred. Other reports over the years have appeared that suggest in vitro fertilization, but they are lacking either in scientific validation or in confirmation by acceptable scientific standards.

In 1966 a report appeared where attempts were made to fertilize human oocytes in vitro after maturation in culture. Spermatozoa were washed and incubated with pieces of human endosalpinx, cervical mucus, or transferred to the oviducts of rabbits or hormonally primed monkeys. The rate of fertilization was low, probably because of a lack of spermatozoan capacitation or knowledge of the exact requirements for ovum maturation.[85,86]

The first successful in vitro fertilization of human oocytes occurred in 1969 with seven ova fertilized of 34 mature oocytes.[4,87] Five of the fertilized oocytes, however, were polyspermic. In a subsequent study the spermatozoan midpiece and tail in four of six fertilized oocytes was observed 10.5 to 24 hr after insemination.[88] Shortly thereafter, human oocytes had been fertilized and developed to the 8- to 16-cell stage in vitro.[66]

The second group to achieve successful in vitro fertilization of human oocytes reported in 1971.[89] In this case oocytes were recovered from excised ovaries and the spermatozoa used had been preincubated, for capacitation purposes, in a monkey uterus. Of the 70 oocytes examined, 8 displayed cleavage to the 2- to 12-cell stage.

In 1973 a third group reported in vitro fertilization[75] with a single oocyte showing evidence of fertilization out of 11 mature oocytes exposed to spermatozoa. In subsequent reports by these same research workers,[76,77] 14 of 16 oocytes were fertilized, and, more importantly, electron microscopy was used to provide the first definitive proof of fertilization in vitro. These studies alo indicated a very rapid dismantling of the spermatozoan midpiece and tail once within the oocyte cytoplasm, indicating that the value of this criterion of fertilization existed for only the first 24 hr after spermatozoan penetration.

Preovulatory oocytes were aspirated from women 36 to 38 hr after hCG and exposed, in vitro, to washed ejaculated spermatozoa. Of 182 oocytes inseminated, 35 cleaved and 41 were apparently fertilized as judged by the appearance of two polar bodies and/or pronuclei.[70,71] Seventeen of the embryos ranged from the eight-cell to the morula stage of development. It was also reported that spermatozoan penetration, after mixture of the spermatozoa and oocytes, was much more rapid than previously reported, less than 2.5 hr.[90]

The same group also reported the secretion of hormones during culture by in vitro fertilized embryos and, specifically, the role of the cumulus cells in progesterone and prostaglandin secretion.[91]

### C. Embryo Recovery and Transfer

Procedures for the recovery of follicular oocytes for in vitro fertilization or for laboratory study have already been discussed. Other procedures have been used to obtain human uterine embryos for developmental studies. Such a technique to obtain unimplanted oocytes from the uterine cavity was described in 1972.[42] Two polyethylene tubes were tied together, the smaller 4 cm longer than the larger, and initially tried for oocyte recovery by a double flushing procedure. Unfortunately this device resulted in cervical irritation and subsequently a stainless steel cannula was designed. This device was 5 cm long and slightly curved, with a 3-mm bore. A Teflon® tube was inserted through a side arm which allowed introduction of the tube to the fundus. This procedure was used with 42 women and a total of eight oocytes was recovered. Only unfertilized oocytes were recovered on the 2nd and 3rd postovulatory days, whereas fertilized oocytes were recovered on the 4th and 5th days.[92]

A seven-cell human embryo was recovered from the proximal middle quarter of the oviduct by a procedure employing bilateral salpingectomy, flushing of the endometrial cavity, and flushing of four sections of the oviduct. The embryo was recovered 83 hr after intercourse and 77 hr after the first significant elevation of the LH level of the urine. Thus the embryo was estimated to be about 72 hr old. Based on this report, plus ultrastructural analysis of the embryo,[94] it was suggested that human embryos enter the uterine cavity when they contain 7 to 12 blastomeres.

Despite earlier unconfirmed reports of possible pregnancies resulting from in vitro fertilization and embryo transfer, it was not until 1976 that success with such a procedure was confirmed. This pregnancy followed follicular growth induction with hMG and hCG and deposition of the in vitro fertilized embryo into the uterine lumen in about 0.2 mℓ of medium. Despite blockage of the oviducts somewhat above the uterotubal junction, the embryo lodged in the oviduct and an ectopic pregnancy resulted which terminated at 10 weeks.[95] Subsequent reports from the same laboratory indicated that three pregnancies (only one ectopic) resulted from transfer of in vitro fertilized embryos following hMG-hCG treatment, none of which progressed beyond 10 weeks.[64] Upon altering the procedure to involve only the use of natural ovulatory cycles, monitored by radioimmunoassay and by immunological techniques to detect hCG, four pregnancies occurred of which two resulted in live births.[64,96] The first birth occurred on July 25, 1978 and represents the first successful in vitro fertilization and embryo transfer in humans.[97]

The method of in vitro fertilization[59,64] has been described elsewhere in the present paper. Transfer of the fertilized embryo was through the cervix without anesthesia. The embryo was drawn into a smooth polyethylene cannula (of 1.3-mm diameter) in culture medium using microscopic observation. The embryo was held about 2 cm from the end of the cannula so that release of 0.05 to 0.09 mℓ of the fluid would carry it into the uterine cavity. The cannula was passed through the washed cervix so that the point of expulsion would lie about 1 cm from the fundus. The contents were expelled from the cannula and it was then slowly withdrawn, verification of the closure of the cervix observed, and the patient was confined to bed for 10 hr. In four additional patients the embryo was transferred surgically with a Tuohy needle being passed through the wall of the fundus and into the uterine cavity. The embryo, within a polyethylene catheter, was passed through this cannula and expelled into the uterine lumen with 0.05 to 0.08 mℓ of culture fluid.[64]

A similar procedure[98] was employed to transfer five eight-cell embryos into the uterine cavity but without a successful pregnancy resulting.[70,71] Subsequently this same research group obtained a pregnancy following in vitro fertilization. An eight-cell embryo was transferred to the uterine lumen 74 hr after insemination. No premedication was given to the patient and a fine metal cannula, 1.7 mm in outer diameter, was inserted through the cervix. A nylon catheter filled with medium was loaded with the embryo 2 cm from the tip. When inserted through the cannula and located about 4 cm from the cervical os, the embryo was expelled. The child was born in June 1980.[99] A second pregnancy was also reported about this time.

By the end of 1979 six pregnancies had been reported resulting from in vitro fertilization and embryo transfer, and the mean efficiency of embryo transfer (based on the reports from England and Australia) was 10.7%.[100]

The most recent report on human in vitro fertilization and embryo transfer records the use of clomiphene citrate and hCG to induce ovulation with a subsequent in vitro fertilization rate of 87%. Of 16 patients receiving embryo transfers under this regimen, 6 showed elevated hCG levels (signs of early pregnancies), and 3 (10%) of these have confirmed pregnancies. Interestingly, this same group reports one pregnancy (out of seven attempts) using clomiphene without hCG. Two embryos were transferred to a single recipient and she was diagnosed pregnant with twins, which were eventually delivered by cesarian section on June 5, 1981.[72]

## ACKNOWLEDGMENT

The author acknowledges the help and kind assistance of Mrs. L. M. Cleeves in the preparation of this manuscript, and of Mr. P. J. Chan and F. J. DeMayo for their assistance in the review.

## ADDENDUM

Since this chapter was prepared, Trounson and Wood have published a review, entitled *Extracorporeal fertilization and embryo transfer,* which provides an up-to-date account of techniques employed with human material.[101]

# REFERENCES


1. **Edwards, R. G.,** Maturation in vitro of mouse, sheep, cow, pig, rhesus monkey and human ovarian oocytes, *Nature (London),* 208, 349, 1965.
2. **Edwards, R. G.,** Maturation in vitro of human ovarian oocytes, *Lancet,* 2, 926, 1965.
3. **Edwards, R. G., Donahue, R. P., Baramki, T. A., and Jones, H. W.,** Preliminary attempts to fertilize human oocytes matured in vitro, *Am. J. Obstet. Gynecol.,* 96, 192, 1966.
4. **Edwards, R. G., Bavister, B. D., and Steptoe, P. C.,** Early stages of fertilization in vitro of human oocytes matured in vitro, *Nature (London),* 221, 632, 1969.
5. **Zamboni, L., Thompson, R. S., and Moore-Smith, D.,** Fine morphology of human oocyte maturation in vitro, *Biol. Reprod.,* 7, 425, 1972.
6. **Shea, B. F., Baker, R. D., and LaTour, P. A.,** Human follicular oocytes and their maturation in vitro, *Fert. Steril.,* 26, 1975, 1975.
7. **Breitenecker, G., Friedrich, F., and Kameter, P.,** Further investigations on the maturation and degeneration of human ovarian follicles and their oocytes, *Fert. Steril.,* 29, 336, 1978.
8. **Suzuki, S. and Mastroianni, L.,** Maturation of monkey ovarian follicular oocytes in vitro, *Am. J. Obstet. Gynecol.,* 96, 723, 1966.
9. **Suzuki, S. and Mastroianni, L.,** The fertilizability of in vitro cultured monkey ovarian follicular oocytes, *Fert. Steril.,* 19, 500, 1968.
10. **Smith, D. M., Conaway, C. H., and Kerber, W. T.,** Influences of season and age on maturation in vitro of rhesus monkey oocytes, *J. Reprod. Fert.,* 54, 91, 1978.
11. **Koering, M. J.,** Cyclic changes in ovarian morphology during the menstrual cycle in *Macaca mulatta, Am. J. Anat.,* 126, 73, 1969.
12. **Koering, M. J.,** Comparative morphology of the primate ovary, *Contrib. Primat.,* 3, 38, 1974.
13. **Jewett, D. A. and Dukelow, W. R.,** Follicular morphology in *Macaca fascicularis, Folia Primat.,* 16, 216, 1971.
14. **Jewett, D. A. and Dukelow, W. R.,** Serial observations of follicular morphology near ovulation in *Macaca fascicularis, J. Reprod. Fert.,* 31, 287, 1972.
15. **Jewett, D. A. and Dukelow, W. R.,** Follicular observation and laparoscopic aspiration techniques in *Macaca fascicularis, J. Med. Prim.,* 2, 108, 1973.
16. **Rawson, J. M. R. and Dukelow, W. R.,** Observation of ovulation in *Macaca fascicularis, J. Reprod. Fert.,* 34, 187, 1973.
17. **Nigi, H.,** Laparoscopic observations of ovaries before and after ovulation in the Japanese monkey *(Macaca fuscata), Primates,* 18, 243, 1977.
18. **Barnes, R. D., Lasley, B. L., and Hendrickx, A. G.,** Midcycle ovarian histology of the bonnet monkey, *Macaca radiata, Biol. Reprod.,* 18, 537, 1978.
19. **Harrison, R. M. and Dukelow, W. R.,** Morphological changes in the *Saimiri sciureus* ovarian follicles detected by laparoscopy, *Primates,* 15, 305, 1974.
20. **Wildt, D. E., Doyle, L. L., Stone, S. C., and Harrison, R. M.,** Correlation of perineal swelling with serum ovarian hormone levels, vaginal cytology and ovarian follicular development during the baboon reproductive cycle, *Primates,* 18, 261, 1977.
21. **Dukelow, W. R.,** Ovulatory cycle characteristics in *Macaca fascicularis, J. Med. Primat.,* 6, 33, 1977.
22. **van Wagenen, G. W.,** Induction of ovulation in *Macaca mulatta, Fert. Steril.,* 19, 15, 1968.
23. **van Wagenen, G. W. and Simpson, M. E.,** *Postnatal development of the ovary in Homo sapiens and Macaca mulatta and induction of ovulation in the macaque,* Yale University Press, New Haven, Conn., 1973.
24. **Dukelow, W. R.,** Induction and timing of single and multiple ovulations in the squirrel monkey *(Saimiri sciureus), J. Reprod. Fert.,* 22, 303, 1970.
25. **Harrison, R. M. and Dukelow, W. R.,** Seasonal adaption of laboratory-maintained squirrel monkeys *(Saimiri sciureus), J. Med. Primat.,* 2, 277, 1973.
26. **Kuehl, T. J. and Dukelow, W. R.,** Ovulation induction during the anovulatory season in *Saimiri scireus, J. Med. Primat.,* 4, 23, 1975.
27. **Dukelow, W. R.,** Human chorionic gonadotropin: induction of ovulation in the squirrel monkey, *Science,* 206, 234, 1979.
28. **Jarosz, S. J., Kuehl, T. J., and Dukelow, W. R.,** Vaginal cytology, induced ovulation and gestation in the squirrel monkey *(Saimiri sciureus), Biol. Reprod.,* 16, 97, 1977.
29. **Kuehl, T. J. and Dukelow, W. R.,** Fertilization in vitro of *Saimiri sciureus* follicular oocytes, *J. Med. Primat.,* 4, 209, 1975.
30. **Kuehl, T. J. and Dukelow, W. R.,** Maturation and in vitro fertilization of follicular oocytes of the squirrel monkey *(Saimiri sciureus), Biol. Reprod.,* 21, 545, 1979.
31. **Dukelow, W. R.,** Laparoscopic research techniques in mammalian embryology, in *Methods in Mammalian Reproduction,* Daniel, J. C., Jr., Ed., Academic Press, New York, 1978, 438.

32. **Thibault, C., Gerard, M., and Menezo, Y.**, Nuclear and cytoplasmic aspects of mammalian oocyte maturation in vitro in relation to follicle size and fertilization, *Sperm Action. Prog. Reprod. Biol.*, 1, 233, 1976.
33. **Johnson, M. P., Harrison, R. M., and Dukelow, W. R.**, Studies on oviductal fluid and in vitro fertilization in rabbits and nonhuman primates, *Fed. Proc.*, 31, 369, 1972.
34. **Cline, E. M., Gould, K. G., and Foley, C. W.**, Regulation of ovulation, recovery of mature ova and fertilization in vitro of mature ova of the squirrel monkey *(Saimiri sciureus), Fed. Proc.*, 31, 360, 1972.
35. **Gould, K. G., Cline, E. M., and Williams, W. L.**, Observations on the induction of ovulation and fertilization in vitro in the squirrel monkey *(Saimiri sciureus), Fert. Steril.*, 24, 260, 1973.
36. **Chan, P. J., Hutz, R. J., and Dukelow, W. R.**, In vitro maturation and fertilization effects on primate embryonic mortality, *Fert. Steril.* (In press.)
37. **DeMayo, F. J., Mizoguchi, H., and Dukelow, W. R.**, Fertilization of squirrel monkey and hamster ova in the rabbit oviduct (xenogenous fertilization), *Science*, 208, 1468, 1980.
38. **Hendrickx, A. G.**, *Embryology of the Baboon*, University of Chicago Press, Chicago, 1971.
39. **Hendrickx, A. G. and Kraemer, D. C.**, Preimplantation stages of baboon embryos (*Papio* sp.), *Anat. Rec.*, 162, 111, 1968.
40. **Panigel, M., Kraemer, D. C., Kalter, S. S., Smith, G. C., and Heverling, R. L.**, Ultrastructure of cleavage stages and preimplantation embryos of the baboon, *Anat. Embryol.*, 147, 45, 1975.
41. **Kraemer, D. C., Flow, B. L., Schriver, G. M., Kinney, G. M., and Pennycock, J. W.**, Embryo transfer in the nonhuman primate, feline and canine, *Theriogenology*, 11, 51, 1979.
42. **Croxatto, H. B., Fuentealba, B., Diaz, S., Pastene, L., and Tatum, H. J.**, A single nonsurgical technique to obtain unimplanted eggs from human uteri, *Am. J. Obstet. Gynecol.*, 112, 662, 1972.
43. **Pope, C. E., Pope, V. Z., and Beck, L. R.**, Nonsurgical recovery of uterine embryos in the baboon, *Biol. Reprod.*, 23, 657, 1980.
44. **Pope, C. E., Pope, V. Z., and Beck, L. R.**, Nonsurgical recovery and in vitro culture of baboon embryos, *Arch. Androl.*, 5, 99, 1980.
45. **Kraemer, D. C., Moore, G. T., and Kramen, M. A.**, Baboon infant produced by embryo transfer, *Science*, 192, 1246, 1976.
46. **Brackett, B. G.**, Experimentation involving primate embryos, in *Methods in Mammalian Reproduction*, Daniel, J. C., Jr., Ed., Academic Press, New York, 1978, 333.
47. **Eddy, C. A., Garcia, R. G., Kraemer, D. C., and Pauerstein, C. J.**, Detailed time course of ovum transport in the rhesus monkey *(Macaca mulatta), Biol. Reprod.*, 13, 363, 1975.
48. **Batta, S. K., Stark, R. A., and Brackett, B. G.**, Ovulation induction by gonadotropin and prostaglandin treatments of rhesus monkeys and observations of the ova, *Biol. Reprod.*, 18, 264, 1978.
49. **Mastroianni, L. and Rosseau, C. H.**, Influence of the interuterine cell on ovum transport and sperm distribution in the monkey, *Am. J. Obstet. Gynecol.*, 93, 416, 1965.
50. **Hurst, P. R., Jefferies, K., Eckstein, P., and Wheeler, A. G.**, Recovery of uterine embryos in rhesus monkeys, *Biol. Reprod.*, 15, 429, 1976.
51. **Hurst, P. R., Jefferies, K., Eckstein, P., and Wheeler, A. G.**, An ultrastructural study of preimplantation uterine embryos of the rhesus monkey, *J. Anat.*, 126, 209, 1978.
52. **Hurst, P. R., Wheeler, A. G., and Eckstein, P.**, A study of uterine embryos recovered from rhesus monkeys fitted with intrauterine devices, *Fert. Steril.*, 33, 69, 1980.
53. **Marston, J. R., Penn, R., and Sivelle, P. C.**, Successful autotransfer of tubal eggs in the rhesus monkey *(Macaca mulatta), J. Reprod. Fert.*, 49, 175, 1977.
54. **Marston, J. H., Kelly, W. A., and Eckstein, P.**, Effect of an intrauterine device on uterine motility in rhesus monkey, *J. Reprod. Fert.*, 19, 321, 1969.
55. **Kreitmann, O. and Hodgen, C. D.**, Low tubal ovum transfer: an alternative to in vitro fertilization, *Fert. Steril.*, 34, 375, 1980.
56. **Ariga, S. and Dukelow, W. R.**, Non-surgical (laparoscopic) uterine flushing and egg recovery technique in the squirrel monkey *(Saimiri sciureus), Primates*, 18, 453, 1977.
57. **Ariga, S. and Dukelow, W. R.**, Recovery of preimplantation blastocysts in the squirrel monkey by a laparoscopic technique, *Fert. Steril.*, 28, 577, 1977.
58. **Ross, G. T. and Lipsett, M. B.**, Hormonal correlates of normal and abnormal follicle growth after puberty in humans and other primates, *Clin. Endocrinol. Metabol.*, 7, 561, 1978.
59. **Edwards, R. G., Steptoe, P. C., Fowler, R. E., and Baillie, J.**, Observations on preovulatory human ovarian follicles and their aspiration, *Br. J. Obstet. Gynaecol.*, 87, 769, 1980.
60. **Moghissi, K. S.**, Prediction and detection of ovulation, *Fert. Steril.*, 34, 89, 1980.
61. **Mastroianni, L.**, The role and value, present and future, of ultrasound in the detection of ovulation, *Fert. Steril.*, 34, 177, 1980.
62. **Queenan, J. T., O'Brien, G. D., Bains, L. M., Simpson, J., Collins, W. P., and Campbell, S.**, Ultrasound scanning of ovaries to detect ovulation in women, *Fert. Steril.*, 34, 99, 1980.

63. **Kennedy, J. F. and Donahue, R. P.**, Binucleate human oocytes from large follicles, *Lancet*, 1, 754, 1969.
64. **Edwards, R. G., Steptoe, P. C., and Purdy, J. A.**, Establishing full term human pregnancies using cleaving embryos grown in vitro, *Br. J. Obstet. Gynaecol.*, 87, 737, 1980.
65. **Steptoe, P. C. and Edwards, R. G.**, Laparoscopic recovery of preovulatory human oocytes after priming of ovaries with gonadotrophins, *Lancet*, 1, 683, 1970.
66. **Edwards, R. G., Steptoe, P. C., and Purdy, J. M.**, Fertilization and cleavage in vitro of preovulatory human oocytes, *Nature (London)*, 227, 1307, 1970.
67. **Edwards, R. G.**, Physiological aspects of human ovulation, fertilization and cleavage, *J. Reprod. Fert.*, Suppl. 18, 87, 1973.
68. **Edwards, R. G.**, Studies on human conception, *Am. J. Obstet. Gynecol.*, 117, 587, 1973.
69. **Lopata, A., McMaster, R., McBain, J. C., and Johnston, I. W. H.**, In vitro fertilization of preovulatory human eggs, *J. Reprod. Fert.*, 52, 339, 1978.
70. **Lopata, A., Brown, J. B., Johnston, I. W., Leeton, J., Talbot, J. M., and Wood, C.**, In vitro fertilization and embryo transfer in infertile patients treated with clomiphene and gonadotrophins, *Theriogenology*, 8, 155, 1977.
71. **Lopata, A., Brown, J. B., Leeton, J. F., Talbot, J. M., and Wood, C.**, In vitro fertilization of preovulatory oocytes and embryo transfer in infertile patients treated with clomiphene and human chorionic gonadotrophin, *Fert. Steril.*, 30, 27, 1978.
72. **Trounson, A. O., Leeton, J. F., Wood, C., Webb, J., and Wood, J.**, Pregnancies in human by fertilization in vitro and embryo transfer in the controlled ovulatory cycle, *Science*, 212, 681, 1981.
73. **Kennedy, J. F. and Donahue, R. P.**, Human oocytes: maturation in chemically defined media, *Science*, 164, 1291, 1969.
74. **Soupart, P. and Morgenstern, L. L.**, Human sperm capacitation and in vitro fertilization, *Fert. Steril.*, 24, 462, 1973.
75. **Soupart, P. and Strong, P. A.**, Ultrastructural observations on human oocytes fertilized in vitro, *Fert. Steril.*, 25, 11, 1974.
76. **Soupart, P. and Strong, P. A.**, Ultrastructural observations on polyspermic penetration of zona pellucida-free human oocytes inseminated in vitro, *Fert. Steril.*, 26, 523, 1975.
77. **Morgenstern, L. L. and Soupart, P.**, Oocyte recovery from the human ovary, *Fert. Steril.*, 23, 751, 1972.
78. **Lopata, A., Johnston, I. W. H., Leeton, J. F., Muchnicki, D., Talbot, J. M., and Wood, C.**, Collection of human oocytes at laparoscopy and laparotomy, *Fert. Steril.*, 25, 1030, 1974.
79. **Berger, M. J., Smith, D. M., Taymor, M. L., and Thompson, R. S.**, Laparoscopic recovery of mature human oocytes, *Fert. Steril.*, 26, 513, 1975.
80. **Renou, P., Trounson, A. O., Wood, C., and Leeton, J. F.**, The collection of human oocytes for in vitro fertilization. I. An instrument for maximizing oocyte recovery rate, *Fert. Steril.*, 35, 409, 1981.
81. **Blandau, R. J.**, In vitro fertilization and embryo transfer, *Fert. Steril.*, 33, 3, 1980.
82. **Rock, J. and Menkin, M. F.**, In vitro fertilization and cleavage of human ovarian eggs, *Science*, 100, 107, 1944.
83. **Menkin, M. F. and Rock, J.**, In vitro fertilization and cleavage of human ovarian eggs, *Am. J. Obstet. Gynecol.*, 55, 440, 1948.
84. **Hayashi, M.**, Fertilization in vitro using human ova, *Int. Conf. Planned Parenthood*, 7, 505, 1963.
85. **Edwards, R. G., Donahue, R. P., Baramki, T. A., and Jones, H. W.**, Preliminary attempts to fertilize human oocytes matured in vitro, *Am. J. Obstet. Gynecol.*, 96, 192, 1966.
86. **Dukelow, W. R. and Fujimoto, S.**, Capacitation of sperm, in *Progress in Infertility*, Behrman, S. J. and Kistner, R. W., Eds., Little, Brown, Boston, 1975, 745.
87. **Edwards, R. G.**, Stages of the development of the human egg, in *Advances in the Biosciences, 4*, Raspé, G., Ed., Pergamon Press, Vieweg, 1970, 235.
88. **Bavister, B. D., Edwards, R. G., and Steptoe, P. C.**, Identification of the midpiece and tail of the spermatozoan during fertilization of human eggs in vitro, *J. Reprod. Fert.*, 20, 159, 1969.
89. **Seitz, H. M., Rocha, G., Brackett, B. G., and Mastroianni, L.**, Cleavage of human ova in vitro, *Fert. Steril.*, 22, 255, 1971.
90. **McMaster, R., Yanagimachi, R., and Lopata, A.**, Penetration of human eggs by human spermatozoa in vitro, *Biol. Reprod.*, 19, 212, 1978.
91. **Shutt, D. A. and Lopata, A.**, The secretion of hormones during the culture of human preimplantation embryos with corona cells, *Fert. Steril.*, 35, 413, 1981.
92. **Croxatto, H. B., Diaz, S., Fuentealba, B., Croxatto, H. D., Carrillo, D., and Fabres, C.**, Studies on the duration of egg transport in the human oviduct. I. The time interval between ovulation and egg recovery from the uterus in normal women, *Fert. Steril.*, 23, 447, 1972.
93. **Avendano, S., Croxatto, H. D., Pereda, J., and Croxatto, H. B.**, A seven cell human egg recovered from the oviduct, *Fert. Steril.*, 26, 1167, 1975.

94. **Pereda, J. and Croxatto, H. B.**, Ultrastructure of a seven-cell human embryo, *Biol. Reprod.*, 18, 481, 1978.
95. **Steptoe, P. C. and Edwards, R. G.**, Reimplantation of a human embryo with subsequent tubal pregnancy, *Lancet*, 1, 880, 1976.
96. **Steptoe, P. C., Edwards, R. G., and Purdy, J. M.**, Clinical aspects of pregnancies established with cleaving embryos grown in vitro, *Br. J. Obstet. Gynaecol.*, 87, 757, 1980.
97. **Steptoe, P. C. and Edwards, R. G.**, Birth after the reimplantation of a human embryo, *Lancet*, 2, 366, 1978.
98. **deKretser, D., Hudson, D. P., Leeton, J., Lopata, A., Outch, K., Talbot, J., and Wood, C.**, Transfer of a human zygote, *Lancet*, 2, 728, 1973.
99. **Lopata, A., Johnson, I. W. H., Hoult, I. J., and Speirs, A. I.**, Pregnancy following intrauterine implantation of an embryo obtained by in vitro fertilization of a preovulatory egg, *Fert. Steril.*, 33, 117, 1980.
100. **Lopata, A.**, Successes and failures in human in vitro fertilization, *Nature (London)*, 288, 642, 1980.
101. **Trounson, A. and Wood, C.**, Extracorporeal fertilization and embryo transfer, in Clinics in Obstetrics and Gynaecology, Vol. 8, No. 3. *Developments in Infertility Practice*, Hull, M. G. R., Ed., W. B. Saunders Co., London, 1981, 681.

Chapter 10

# FACTORS AFFECTING THE SUCCESS OF EGG TRANSFER

**C. E. Adams**

## TABLE OF CONTENTS

## I. INTRODUCTION

Recognition of the various factors that may affect the success of egg transfer has grown in parallel with the development of the subject, and most notably over the last 25 years, as the more important principles, e.g., the need for synchronization, became established.

Among experimental investigations concerned specifically with factors affecting survival and development of transferred eggs are those of McLaren and Michie[1] in the mouse, Moore and Rowson[2] and Moore and Shelton[3] in sheep, and Rowson, Moor, and Lawson[4] and Sreenan, Beehan, and Mulvehill[5] in cattle. Other articles containing reviews are available.[6-9]

Presently, the aim is to give a general view of the subject rather than an exhaustive list of factors affecting the success of egg transfer. It is perhaps pertinent to note that the list is constantly being added to as new facets of egg transfer are explored. Moreover, some factors are only recognized *post hoc.*

Broadly speaking, the various factors fall into three main categories, associated with (1) the donor, (2) the recipient, and (3) technical aspects.

## II. THE DONOR

### A. Selection, Management, and Preparation

The selection, management, and preparation of the donor can have a considerable influence on performance, as reflected in maximizing the yield of fertilized eggs. Gates[10] observed that the use of mice in good physical condition is of great importance to the success of superovulation and subsequent mating. Apparently some inbred strains of mice consistently produce a better ovarian response than others; the use of F1 hybrids from a cross between BALB/C and a second inbred strain was recommended.

It is generally agreed that the viability of eggs from superovulated females compares favorably with that of eggs from untreated females. However, much of the evidence stems from indirect observations rather than from primary studies. In the sheep high levels of superovulation (>20 ovulations) have been associated with impaired fertilization,[11] but this is now believed to depend upon the endocrine milieu and defective sperm transport rather than inherently defective eggs. Another factor to be considered is that when deep frozen spermatozoa are inseminated into superovulated cattle, "between bull" differences may be exaggerated, resulting in a higher proportion of unfertilized eggs than under normal conditions.[12]

Gonadotrophin treatment permits the use of prepuberal animals as egg donors.[10,13] Gates[10] noted that stages greater than four cells do not develop well in the genital tract of the prepuberal mouse.

### B. Examination and Selection of Eggs

The selection of eggs for transfer has to be made largely on the basis of morphological criteria, though it must be admitted that such criteria do not in themselves afford an infallible guide. In the final analysis the ability of an egg to develop normally to term can only be tested by transfer, and this is particularly true under experimental conditions. Conversely, of course, failure to survive at whatever stage does not necessarily mean that the egg was defective.

The most useful criterion for assessing an egg's viability is "on time" cleavage or development. Experience shows that the prognosis for eggs with retarded cleavage — two or more divisions in arrears — is poor, and they should be discarded. A difference of one cleavage division, however, is not serious; in fact, it is quite common to find mixtures of 2- and 4-cell, 8- and 16-cell, or 16- and 32-cell eggs.

Considerable variation may exist between the eggs of different species in relation to the symmetry of cleavage and in the size and appearance of individual blastomeres. As a result, standards set, for example, for the rabbit cannot be applied directly to cattle; experience in assessing the eggs of a particular species is essential.

The use of a good quality microscope is also most important. With poor optics or at too low magnifications, some stages may present problems; for example, the late morula in the rabbit can be confused with a one-cell, unfertilized egg. In such cases, the presence of accessory spermatozoa, as revealed by high magnifications, provides supportive evidence of fertilization.

The way eggs are recovered may also affect their chances of survival. Recovery from slaughtered animals was apparently less satisfactory than recovery in vivo,[4] possibly due to changes *in utero* post mortem.

## III. THE RECIPIENT

Notwithstanding how well a transfer is actually performed, the inherent fertility of the recipient will have a considerable bearing on the outcome. Already in 1950, Venge[14] suspected that the efficiency of recipient rabbits was affected by their genetic constitution, since in his experience crossbred does were markedly superior to purebred does. McLaren and Michie[1] pointed out that transferred eggs are presumably as much exposed as are native eggs to the natural hazards causing loss between mating and parturition in unoperated females of the same stock. The existence of wide variations in performance is well exemplified by two strains of mice, C57BL and DBA, in which 16% and 42%, respectively, of eggs were lost between fertilization and birth.[15] Seidel[16] has suggested that in the future as much emphasis might be placed on selection of the recipient as the donor.

Apart from variations in fertility status, depending upon species, breed, strain, and individual, factors of a temporary nature — some defined and others as yet poorly understood — may also affect a recipient's performance. In sheep, it has been estimated that 20 to 30% of ewes are incapable of becoming pregnant at any one estrous period, so called "transient infertility".[2]

Women seeking the alleviation of sterility by means of in vitro fertilization and egg transfer obviously represent a special case; criteria for their selection have been recently described.[17]

The management of recipients either prior to or after egg transfer can affect their performance. McLaren and Michie[1] attributed what they regarded as a low pregnancy rate, 60%, in mice to the recipients having been kept prior to mating in a "relatively bad environment", i.e., large stock boxes each containing 20 to 30 mice. In a subsequent experiment,[18] they recorded significantly higher pregnancy rates in their controls, which they attributed to improved conditions prior to mating (not more than eight animals per cage) and the use of Fl hybrid females having a high reproductive performance.

In rodents it is essential to either mate recipients with a vasectomized (sterile) male or use a vibrator[19,20] to transform the cyclic corpora lutea into functional ones, thereby making the animal pseudopregnant. The transfer of fertilized eggs to unstimulated females does not lead to pregnancy.

Recipients can be divided into those which fail to become pregnant and those which become pregnant but lose some or all of the embryos between transfer and parturition. The underlying causes may be quite different. Early loss of the whole inoculum, particularly in polytocous species, is often symptomatic of inappropriate transfer conditions.

## IV. TECHNICAL CONSIDERATIONS

### A. Media and Operating Room Conditions

For the eggs of some species simple salt solutions are quite satisfactory for recovery and short-term storage pending transfer. However, the use of some organic component, such as homologous serum or bovine plasma albumin fraction V, is usually beneficial, particularly when the interval from collection to transfer is extended. But the presence of an ovicidal factor in the serum of certain species[21] must be taken into account. For example, homologous serum is an unsuitable medium for cow eggs, whereas TCM 199 is highly satisfactory.[4] Nevertheless, eggs stored for longer than 2 hr (121 to 480 min) in either TC 199 (B) (sodium bicarbonate buffer) or TC 199 (H) (Hepes buffer) survived less well ($p<0.01$) than those similarly kept but for a shorter time (40 to 120 min).[5] Certain stages in egg development may also be more sensitive than others to a particular medium.[22]

Whittingham[23] noted that wide and rapid temperature fluctuations are possibly detrimental to embryo survival, adding that in his own experience mouse and rat embryos maintained at a constant room temperature from collection to transfer survived much better than embryos moved in and out of the 37°C incubator, at least for holding times of not more than 3 hr.

According to Murray[24] a contaminated surgery room atmosphere can be extremely detrimental to embryo survival; he recommends the use of germicidal ultraviolet lamps to minimize microbial contamination. We have also noted that whenever building operations have been in progress on site the risk of airborne infections contaminating flushing media seems to become a problem.

The early luteal phase uterus is particularly susceptible to infection. In the rabbit infection via the transfer medium is probably the principal cause of failure, and the whole inoculum is affected. Therefore, it is most important that all media and glassware should be sterile.

### B. Synchronization of Donor and Recipient

The need for close synchronization of estrus between donor and recipient animals, as originally demonstrated in the rabbit,[25] has since been shown to hold in a wide variety of species, including the rat[26] (Figure 1), ferret,[27] sheep[3,28] (Figure 2), and cow[29] (Figure 3).

To provide optimal conditions for egg survival, the objective should be near exact synchrony, avoiding as far as possible differences exceeding ±24 hr. However, this is not meant to imply that wider limits of asychrony, e.g., ±48 hr, may not sometimes yield satisfactory results. The limits of tolerance are affected both by species variations and the stage of development of the egg. In cattle, synchronization requirements are "slightly more acute" than in sheep.[29] In sheep, using eggs aged 48 to 84 hr, the best results were obtained when synchronization was within ±12 hr,[3] whereas with 5-, 7-, and 9-day embryos a difference of ±48 hr still gave a satisfactory pregnancy rate.[28]

Some reports[1,26] indicate that the combination "donor in advance of recipient", e.g., 3½ → 2½ days, is more successful than the reciprocal, 2½ → 3½ days. This has been attributed to the eggs suffering a delay in development of about 12 hr due to manipulation;[30] as a result synchronization is either enhanced or worsened, depending upon the combination.

Except for species like the cat, rabbit, ferret, and mink, in which ovulation is non-spontaneous, synchronization in spontaneously ovulating species is commonly achieved by artificial means. Certain treatments, however, while effective in synchronizing estrus, may impair fertility at the first heat period following treatment.[31]

In recipient mice synchronized with low doses of gonadotrophins the proportion

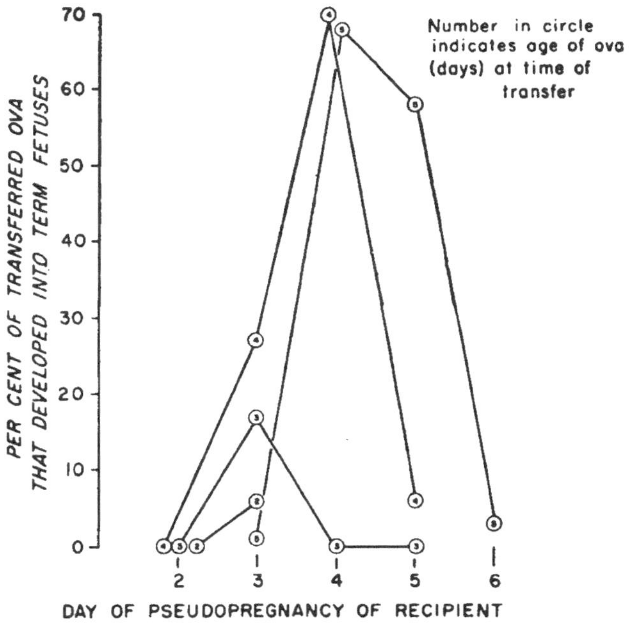


FIGURE 1. The proportion of transferred ova that developed into term fetuses when the ova were of various ages in relation to the stage of uterine development of the recipient rats. (From Noyes, R. W. and Dickmann, Z., *J. Reprod. Fert.*, 1, 186, 1960. With permission.)

becoming pregnant with transferred embryos was reduced (61% compared with 91% in naturally mated recipients).[32]

### C. Site of Egg Transfer

The choice of site of transfer, whether to the oviduct or uterus, is largely governed by the stage of egg development (donor) and interval elapsing from ovulation to transfer (recipient). In general, if eggs are transferred to the uterus prematurely, i.e., >1 day before they normally enter the uterus, the proportion developing is reduced.[25,33,34] Of day-2 rabbit eggs transferred synchronously to the uterus, only about 25% survived to term,[25] and the bulk of the loss occurs before implantation.[34] Eggs are "lost" either through destruction in utero and/or expulsion into the vagina.[34-36]

In sheep no two-cell and only 16% of four-cell eggs developed into lambs following uterine transfer.[37] Some workers have found that the transfer of eight- or more cell eggs to the uterine horns proved more successful than that of eggs having less than eight-cells to the oviducts.[2,38] However, others[3] have reported the converse, and suggested that in the earlier work site of transfer and age of eggs were confounded.

The synchronous transfer of day-3 cow eggs resulted in a significantly lower conception rate than that of day 4, 5, 6, or 7 embryos to synchronized uteri.[33] But significantly more 4-day eggs survived in a day-3 uterus than did 3-day eggs (47% vs. 12%).

Another factor to be considered is the site of transfer relative to the location of the corpus luteum. In cattle, embryo survival was higher in uteri ipsilateral to the corpus luteum (69%) than in the contralateral uteri.[39] However, the incidence of twin pregnancy was higher if one egg was transferred to each uterine horn (73%) than if both eggs were transferred to the horn ipsilateral to the corpus luteum (45%).[5,40] In the rabbit the success of implantation was not improved by transfer to both uterine horns as opposed to a single horn.[41]

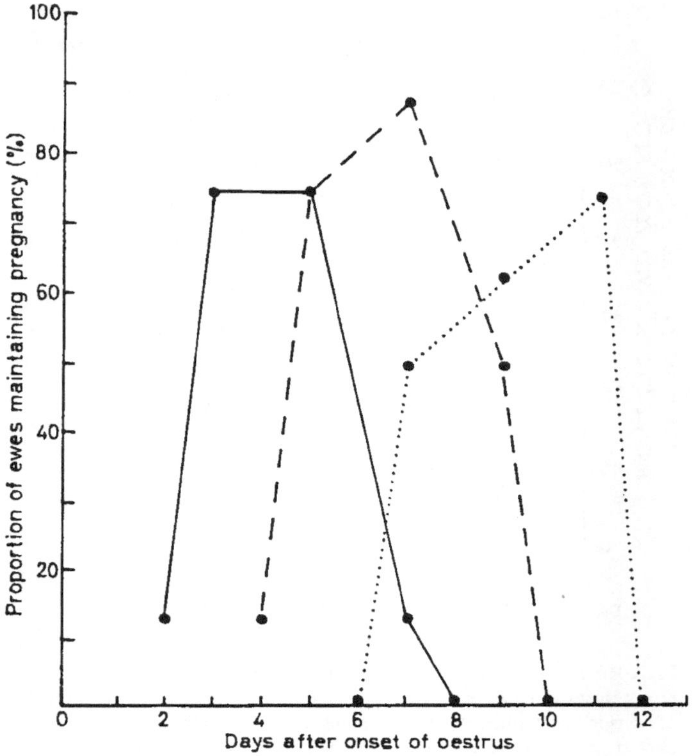


FIGURE 2. The proportion of recipient ewes becoming pregnant following the transfer of 5-day morulae (—), 7-day blastocysts (---), and 9-day blastocysts (....) at different stages of the estrous cycle. (From Rowson, L. E. A. and Moor, R. M., *J. Reprod. Fert.*, 11, 207, 1966. With permission.)

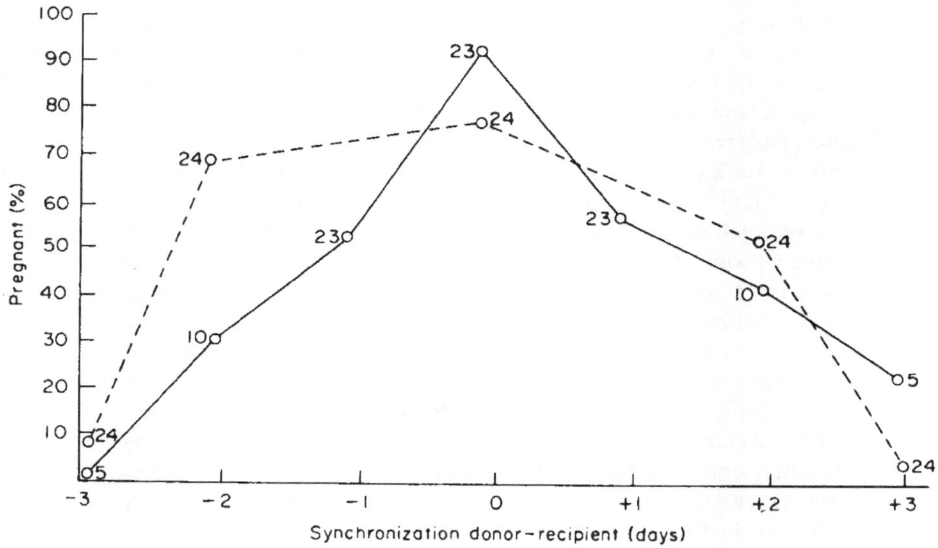


FIGURE 3. The effect of the degree of synchronization of donor and recipient on pregnancy rate: — cattle, —— sheep. The number of animals is given against each symbol. (From Rowson, L. E. A., Lawson, R. A. S., Moor, R. M., and Baker, A. A., *J. Reprod. Fert.*, 28, 427, 1972. With permission.)

### D. Number of Eggs Transferred

In polytocous species, the number of eggs transferred can have a marked effect on their prospects of survival. In some species a certain basic number of eggs is required to establish and maintain pregnancy. In the pig, for example, it appears that more than four embryos are necessary for the consistent maintenance of luteal function during early pregnancy.[42] In 25 gilts, recipients of four or fewer eggs on days 3 to 5, only six were pregnant on day 28 compared with seven out of eight which had received eight eggs. In the rabbit, the pregnancy rate (day 10) was higher in animals receiving two eggs than in those receiving one egg (70% vs. 53%), and still higher (86%) when five eggs were transferred. Moreover, only 20% of does carrying one fetus maintained pregnancy to term compared with 87% of those with two implantations.[43]

To maximize the yield of fetuses or young per egg transferred, the number of eggs should not be too high. Within the range 7 to 17 eggs, a higher proportion of eggs survived to 90 to 100 days in pigs receiving the smaller numbers, as follows: 7 to 9 eggs, 58%; 10 to 12 eggs, 49%; 14 to 17 eggs, 36%.[44] In sheep receiving either two or five eggs the pregnancy rate was similar but the proportion of eggs surviving was significantly lower after the transfer of five eggs.[2]

### E. Stage of Recipient's Estrous Cycle or Pseudopregnancy

In rodents and lagomorphs the period during which eggs can be successfully transferred is limited because attachment and implantation take place relatively early, on or before day 7 p.c. In farm animals where the preattachment phase is quite long, limitations are ultimately imposed by the need for the presence of embryos to prevent regression of the corpus luteum. In sheep[45,46] and cattle[47] embryos can be successfully transferred up to the 12th and 16th days, respectively, of the cycle. However, in the unmated pig transfers must be performed not later than days 7 or 8[48,49] to obtain success. In other species information on this aspect is still lacking.

### F. Time of Day of Transfer

It has been suggested[17] that a diurnal rhythm may play a role in establishing pregnancy with cleaving embryos in women, since each of four pregnancies occurred following transfers in the late evening. Possible explanations include the general quiescent state of the patients in the late evening, or the absence of stress and other factors associated with hospital life, rather than to their own endogenous rhythms. Another possibility is the period of time spent by oocytes in the ovary after the LH surge began up to their aspiration; the interval was 24 hr or more in the four patients who became pregnant.[17]

There is no evidence from experimental animals that the time of day when transfers are carried out influences the results.

### G. Skill

The outcome of egg transfers will be influenced to a greater or lesser degree by the manual dexterity, coordination, and experience of the operator. Obviously, more skill is required to transfer eggs successfully to the oviduct of the mouse or rat than to that of the rabbit. Experience has shown that transfers done at the beginning of a series are often associated with poorer results; this can depend not only upon being out of practice but also upon failing to arrange the best conditions or to assemble the right equipment for the job.

In an experiment lasting 18 weeks, McLaren and Michie[18] noted that their performance, as measured in yield of embryos per egg transferred, showed marked improvement during the course of the experiment, rising from 7 to 36%, or about 1.7% per week; this they attributed to increase in manual dexterity.

### H. Other Factors

Under certain conditions, it appears that a double series of manipulations, involving egg recovery from a temporary recipient and retransfer, can have a deleterious effect on the egg.[50,51] The method of transfer, whether surgical or nonsurgical, can also affect the level of success.

## REFERENCES


1. **McLaren, A. and Michie, D.**, Studies on the transfer of fertilized mouse eggs to uterine foster-mothers. I. Factors affecting the implantation and survival of native and transferred eggs, *J. Exp. Biol.*, 33, 394, 1956.
2. **Moore, N. W. and Rowson, L. E. A.**, Egg transfer in sheep. Factors affecting the survival and development of transferred eggs, *J. Reprod. Fert.*, 1, 332, 1960.
3. **Moore, N. W. and Shelton, J. N.**, Egg transfer in sheep: effect of degree of synchronization between donor and recipient, age of egg and site of transfer on the survival of transferred eggs, *J. Reprod. Fert.*, 7, 145, 1964.
4. **Rowson, L. E. A., Moor, R. M., and Lawson, R. A. S.**, Fertility following egg transfer in the cow; effect of method, medium and synchronization of oestrus, *J. Reprod. Fert.*, 18, 517, 1969.
5. **Sreenan, J. M., Beehan, D., and Mulvehill, P.**, Egg transfer in the cow: factors affecting pregnancy and twinning rates following bilateral transfers, *J. Reprod. Fert.*, 44, 77, 1975.
6. **Adams, C. E.**, Egg transfer and fertility in the rabbit, *Proc. 3rd Int. Congr. Anim. Reprod. (Cambridge)*, 2, 5, 1957.
7. **Newcomb, R.**, Fundamental aspects of ovum transfer in cattle, *Vet. Rec.*, 99, 40, 1976.
8. **Betteridge, K. J., Ed.**, A review of techniques and applications, in Embryo Transfer in Farm Animals, Monograph 16, Canada Department of Agriculture, 1977.
9. **El Badrawi, H. and Hafez, E. S. E.**, Physiological factors influencing the success of embryo transfer, *Arch. Androl.*, 5 (Abstr. 94), 97, 1980.
10. **Gates, A. H.**, Maximizing yield and developmental uniformity of eggs, in *Methods in Mammalian Embryology*, Daniel, J. C., Jr., Ed., W. H. Freeman, San Francisco, 1971, 64.
11. **Moore, N. W.**, Ph.D. thesis, University of Cambridge, 1959.
12. **Newcomb, R.**, Investigation of factors affecting superovulation and non-surgical embryo recovery from lactating British Friesian cows, *Vet. Rec.*, 106, 48, 1980.
13. **Adams, C. E.**, Some aspects of ovulation, recovery and transplantation of ova in the immature rabbit, in *Mammalian Germ Cells*, Wolstenholme, G. E. W., Ed., Churchill Livingstone, London, 1953, 198.
14. **Venge, O.**, Studies on the maternal influence on the birth weight in rabbits, *Acta Zool.*, 31, 1, 1950.
15. **Fekete, E.**, Differences in the effect of uterine environment in the dba and C 57 Black strains of mice, *Anat. Rec.*, 98, 409, 1947.
16. **Seidel, G. E., Jr.**, Applications of embryo preservation and transfer, *Beltsville Symp. Agric. Res.*, Hawk, H. W., Ed., Allanheld, Osmun, Montclair, N.J., 1979, 195.
17. **Edwards, R. G., Steptoe, P. C., and Purdy, J. M.**, Establishing full-term human pregnancies using cleaving embryos grown in vitro, *Br. J. Obstet. Gynaecol.*, 87 (new series), 737, 1980.
18. **McLaren, A. and Michie, D.**, Studies on the transfer of fertilized mouse eggs to uterine foster-mothers. II. The effect of transferring large numbers of eggs, *J. Exp. Biol.*, 36, 40, 1959.
19. **De Feo, V. J.**, Vaginal-cervical vibration: a simple and effective method for the induction of pseudopregnancy in the rat, *Endocrinology*, 79, 440, 1966.
20. **Diamond, M. and Yanagimachi, R.**, Induction of pseudopregnancy in the golden hamster, *J. Reprod. Fert.*, 17, 165, 1968.
21. **Chang, M. C.**, Effects of heterologous sera on fertilized rabbit ova, *J. Gen. Physiol.*, 32, 291, 1949.
22. **Whittingham, D. G. and Bavister, B. D.**, Development of hamster eggs fertilized in vitro and in vivo, *J. Reprod. Fert.*, 38, 489, 1974.
23. **Whittingham, D. G.**, In vitro fertilization, embryo transfer and storage, *Br. Med. Bull.*, 35, 105, 1979.
24. **Murray, F. A.**, Embryo transfer in large domestic animals, in *Methods in Mammalian Reproduction*, Daniel, J. C., Jr., Ed., Academic Press, New York, 1978, 285.
25. **Chang, M. C.**, Development and fate of transferred rabbit ova or blastocyst in relation to the ovulation time of recipients, *J. Exp. Zool.*, 114, 197, 1950.

26. **Noyes, R. W. and Dickmann, Z.**, Relationship of ovular age to endometrial development, *J. Reprod. Fert.*, 1, 186, 1960.
27. **Chang, M. C.**, Development of transferred ferret eggs in relation to the age of corpora lutea, *J. Exp. Zool.*, 171, 459, 1969.
28. **Rowson, L. E. A. and Moor, R. M.**, Embryo transfer in the sheep: the significance of synchronizing oestrus in the donor and recipient animal, *J. Reprod. Fert.*, 11, 207, 1966.
29. **Rowson, L. E. A., Lawson, R. A. S., Moor, R. M., and Baker, A. A.**, Egg transfer in the cow: synchronization requirements, *J. Reprod. Fert.*, 28, 427, 1972.
30. **Tarkowski, A.**, Experiments on the transplantation of ova in mice, *Acta Theriologica*, 2, 251, 1959.
31. **Polge, C.**, Embryo transfer and embryo preservation, in *Artificial Breeding of Non-Domestic Animals*, Watson, P. F., Ed., Academic Press, London, 1978, 303.
32. **Whittingham, D. G.**, Low temperature storage of mammalian embryos, in *Basic Aspects of Freeze Preservation of Mouse Strains*, Muhlbock, O., Ed., Gustav Fischer, Stuttgart, 1976, 45.
33. **Newcomb, R. and Rowson, L. E. A.**, Conception rate after uterine transfer of cow eggs, in relation to synchronization of oestrus and age of eggs, *J. Reprod. Fert.*, 43, 539, 1975.
34. **Adams, C. E.**, Consequencies of accelerated ovum transport, including a re-evaluation of Este's operation, *J. Reprod. Fert.*, 55, 239, 1979.
35. **Adams, C. E.**, Egg survival relative to maternal endocrine status, in *Ovum Transport and Fertility Regulation*, Harper, M. J. K., Pauerstein, C. J., Adams, C. E., Coutinho, E. M., Croxatto, H. B., and Paton, D. M., Eds., Scriptor, Copenhagen, 1976, 425.
36. **Adams, C. E.**, Retention and development of eggs transferred to the uterus at various times after ovulation in the rabbit, *J. Reprod. Fert.*, 60, 309, 1980.
37. **Averill, R. L. W. and Rowson, L. E. A.**, Ovum transfer in the sheep, *J. Endocrinol.*, 16, 326, 1958.
38. **Averill, R. L. W.**, Studies in the Fertility of the Ewe, Ph.D. thesis, University of Cambridge, 1956.
39. **Newcomb, R. and Rowson, L. E. A.**, unpublished data, 1976.
40. **Rowson, L. E. A., Lawson, R. A. S., and Moor, R. M.**, Production of twins in cattle by egg transfer, *J. Reprod. Fert.*, 25, 261, 1971.
41. **Adams, C. E.**, Studies on prenatal mortality in the rabbit, Oryctolagus cuniculus; the effect of transferring varying numbers of eggs, *J. Endocrinol.*, 13, 325, 1962.
42. **Day, B. M., Polge, C., Moor, R. M., Rowson, L. E. A., and Booth, W. D.**, Embryo numbers and luteal maintenance during early pregnancy in swine, *J. Anim. Sci.*, 28, 1499, 1967.
43. **Adams, C. E.**, Maintenance of pregnancy relative to the presence of few embryos in the rabbit, *J. Endocrinol.*, 48, 243, 1970.
44. **Dziuk, P. J., Polge, C., and Rowson, L. E. A.**, Intrauterine migration and mixing of embryos in swine following egg transfer, *J. Anim. Sci.*, 23, 37, 1964.
45. **Moor, R. M. and Rowson, L. E. A.**, Influence of the embryo and uterus on luteal function in the sheep, *Nature (London)*, 201, 522, 1964.
46. **Moor, R. M. and Rowson, L. E. A.**, The corups luteum of the sheep: functional relationship between the embryo and the corpus luteum, *J. Endocrinol.*, 34, 233, 1966.
47. **Betteridge, K. J., Eaglesome, M. D., Randall, G. C. B., and Mitchell, D.**, Collection, description and transfer of embryos from cattle 10 to 16 days after oestrus, *J. Reprod. Fert.*, 59, 205, 1980.
48. **Hunter, R. H. F., Polge, C., and Rowson, L. E. A.**, The recovery, transfer and survival of blastocysts in pigs, *J. Reprod. Fert.*, 14, 501, 1967.
49. **Webel, S. K., Peters, J. B., and Anderson, L. L.**, Synchronous and asynchronous transfer of embryos in the pig, *J. Anim. Sci.*, 30, 565, 1970.
50. **Dickmann, Z. and De Feo, V. J.**, The rat blastocyst during normal pregnancy and during delayed implantation, including an observation on the shedding of the zona pellucida, *J. Reprod. Fert.*, 13, 3, 1967.
51. **Adams, C. E.**, The development of rabbit eggs in the ligated oviduct and their viability after retransfer to recipient rabbits, *J. Embryol. Exp. Morphol.*, 29, 133, 1973.

Chapter 11

# MICROMANIPULATION OF EMBRYOS OF THE LARGE DOMESTIC SPECIES

S. M. Willadsen

## TABLE OF CONTENTS

# I. INTRODUCTION

## A. Early Experiments Involving Micromanipulation of Mammalian Embryos

For almost a century micromanipulation of embryos has been used in invertebrates and lower vertebrates to elucidate patterns of early development, and it is largely through such investigations that cellular interactions and the role played by individual blastomeres have been established. The potential value of a similar approach in the study of early development in mammals has long been recognized, but it was only after methods of embryo collection, handling, and transfer had reached a high level of efficiency that this became technically feasible.

Among the earliest experiments involving micromanipulation of mammalian embryos were those of Nicholas (1933),[1] who transplanted isolated blastomeres from two-cell rat embryos to the kidney capsule and observed varying degrees of development. Pincus (1936)[2] transplanted single blastomeres from two-cell rabbit embryos to the oviducts of pseudopregnant does and obtained normally differentiating but small-sized blastocysts, and in a simiar experiment Nicholas and Hall (1942)[3] showed that rat blastomeres isolated at the two-cell stage were capable of forming complete embryos. These results were confirmed and extended when Seidel (1952)[4] and Tarkowski (1959)[5] obtained viable young in the rabbit and in the mouse, respectively, from two-cell embryos in which one of the blastomeres had been killed by puncturing with a fine needle.

The technique of selective destruction of individual blastomeres suffers from the obvious limitation that only some of the original blastomeres are available for study. Investigations aimed at studying the developmental potential of each individual blastomere of a cleaving embryo necessitate the removal of the zona pellucida, so that the blastomeres may be completely separated. This also applies to investigations concerning the ability of blastomeres or whole embryos to regulate their development when aggregated with foreign blastomeres or embryos. Such experiments have depended upon effective methods for culture of embryos in vitro, since early cleavage stages in mammals are generally unable to survive in the female genital tract if the zona pellucida has been extensively damaged or removed,[6-9] whereas the survival of embryos in vitro does not depend on the presence of the zona. Methods for embryo culture in vitro have only been developed relatively recently[10-13] and are still inadequate for the embryos of the majority of species.

### B. Micromanipulation of Mouse Embryos

The mouse embryo, which will undergo complete preimplantation development in vitro, has lent itself readily to experiments involving radical micromanipulations. Over the last 20 years a number of refined microsurgical and related techniques have been developed by which early embryonic development in this species has been studied. These include methods for the removal of the zona pellucida,[14,15] dissociation of blastomeres,[16] aggregation of denuded embryos and blastomeres,[14,17] injection of single or aggregated cells into the cavity of early blastocysts,[18] and insertion of isolated blastomeres into foreign zonae pellucidae.[19] By means of these and an ever increasing variety of biochemical and immunological techniques, the biology of the mouse embryo has been analyzed in greater detail than that of any other mammalian embryo.[20-23] As a consequence most of our knowledge of early development and differentiation in mammals is based on studies of mouse embryos.

#### *1. Normal Cleavage and Early Differentiation*

In cleaving mouse embryos all blastomeres belonging to the same cell generation are not only morphologically similar but also possess essentially the same developmental potential.[20-22] The first overtly differentiated stage of development is the 3½-day blastocyst, in which the characteristics of the trophectodermal cells differ from those of the cells belonging to the inner cell mass.[24,25] The contribution made by the progeny of individual cells in younger embryos to inner cell mass and trophectoderm, respectively, is influenced by the presence of fellow blastomeres and governed by positional relationships within the late morula. Those cells which are completely surrounded by others at the time of blastulation give rise to the inner cell mass while the rest form the trophectoderm.[26] A slight asynchrony in cell cycle between early blastomeres appears to be perpetuated in their daughter cells, and this results in a disproportionately high representation in the inner cell mass of cells deriving from the one of the first two blastomeres which entered mitosis first.[27,28]

#### *2. The Developmental Potential of Isolated Blastomeres*

Normally the presence of the zona pellucida ensures that all the blastomeres arising from one zygote — and only these — participate in the formation of the blastocyst and that only a single blastocyst develops. However, if the blastomeres are separated at an early stage, each of them continues to develop according to the original schedule and to differentiate according to the usual principles, eventually giving rise to a diminutive blastocyst or a trophectodermal vesicle.[26] The latter form, which contains no inner cell mass and is therefore not viable, results if the embryo consists of so few cells at the time of blastulation that none is completely enveloped by fellow blastomeres. In the mouse single blastomeres from two-cell embryos and pairs of blastomeres from four-cell embryos are capable of developing into normal conceptuses,[5,19] whereas single blastomeres from four-cell embryos are not.[19] The separation of blastomeres has served as the basis for the production of monozygotic twins, but the efficiency of this approach in the mouse appears to be relatively low.[29]

*3. The Regulatory Capacity of Aggregated Whole Embryos and Dissociated Blastomeres*

If two or more denuded embryos of nearly the same stage of development are brought into close contact, they are able to regulate their development in such a way that they give rise to a single large blastocyst.[14,15] Such composite blastocysts are viable and have served to produce chimaeric animals.[14,17,30] The use of dissociated blastomeres for the production of chimaeric embryos allows cell number, origin, and position to be varied within wider limits and thus offers an even greater multitude of experimental possibilities. The blastocyst injection technique of Gardner[31] has been particularly valuable for the study of cell determination and fate.[21-23]

*4. Conclusions*

The general conclusions which may be drawn from the accumulated results of manipulations of cell number and position in mouse embryos are that early blastomeres have the ability to regulate their development following experimental interference so that a normally organized blastocyst is produced, and that this only fails when the cell number is so drastically reduced that too few cells occupy an inside position at the time of cavitation for the formation of a functional inner cell mass. Furthermore, it has been demonstrated that this regulatory capacity at least in some instances transgresses species boundaries, since composite but normally organized blastocysts have been produced from mouse embryos aggregated with rat embryos,[25] and also when rat inner cell masses were injected into mouse blastocysts.[32]

There is no experimental evidence to prove that the basic principles and patterns of early embryonic development in other eutherian mammals differ from those observed in the mouse. However, the precise limits to the regulatory capacity of blastomeres must be established experimentally for each species. In the rabbit, for instance, it has been shown that sometimes single blastomeres from four- and even eight-cell embryos can form viable blastocysts,[33] whereas this is not the case in the mouse.[19] This is probably related to the different cell number reached by the embryos of these two species at the time of blastulation. In the mouse blastulation normally occurs after about five cleavage divisions when the embryo consists of approximately 32 cells,[34,35] whereas in the rabbit blastulation normally takes place after about seven cleavage divisions when the embryo contains approximately 128 cells.[36] The time of blastulation is correlated with the cell generation rather than cell number.[34] Accordingly blastulation in a mouse embryo produced from a single blastomere of a four-cell embryo will occur when the embryo consists of about eight cells, hardly any of which are in an inside position. Blastulation therefore tends to result in the formation of a trophectodermal vesicle.[19,26] By contrast, an embryo produced from a single cell of a four-cell rabbit embryo will blastulate when it contains about 32 cells, several of which occupy an inside position. Hence, in this instance a normally organized blastocyst is formed. The disproportionate reduction in the number of inside cells in embryos consisting of a lower than normal number of cells at blastulation has its counterpart in the disproportionate increase in the number of inside cells in embryos consisting of a higher than normal number of cells. Thus, chimaeric blastocysts, formed by aggregation of several embryos, have larger than normal inner cell masses.[37] Since the young produced from micromanipulated embryos, irrespective of the initial number of cells, have in most instances been of normal size, regulation of cell number must have taken place at a subsequent stage of development. In the mouse this appears to occur soon after implantation.[5,37] Interestingly, the highest number of genotypes recorded in chimaeric mice is three,[38] suggesting that the actual fetus is produced from a very small number of inner cells, irrespective of the bulk of the inner cell mass.

### C. Micromanipulation of Embryos of the Large Domestic Species

The micromanipulation of cleaving embryos of the large domestic species is of interest both with a view to confirming or establishing basic mechanisms of development and differentiation and with a view to exploiting for practical purposes the developmental potential and regulatory capacity of blastomeres. However, the lack of satisfactory methods of embryo culture in vitro has until recently virtually ruled out the application of the microsurgical techniques used for detailed analyses of embryonic development in the mouse. Consequently, there are very few reports in the literature of experiments involving micromanipulation of embryos of the large domestic species.

#### *1. The Developmental Potential of Isolated Blastomeres*

Moore, Polge, and Rowson[9] studied the development of single blastomeres of four- and six-cell pig embryos. Thirty-five per cent of single blastomeres contained in their own zonae pellucidae and isolated by puncturing of the fellow blastomeres developed into blastocysts after transfer to recipient gilts. However, the viability of such blastocysts was not tested, and none of 30 single blastomeres inserted into evacuated zonae pellucidae underwent further development in parallel experiments. Their study was aimed at the artifical production of monozygotic twins. This was also the case in subsequent experiments in the sheep by Trounson and Moore,[39] who bisected 8- to 20-cell embryos and morulae of more than 20 cells. None of the "half" embryos of the first category developed normally during subsequent culture in vitro, but 30% of those belonging to the second category developed into apparently normal blastocysts in vitro, and 7 of the original 37 embryos each gave rise to two blastocysts. However, only two lambs were obtained from the 14 blastocysts transferred to synchronous recipient ewes, and these lambs were not monozygotic twins. When nine embryos collected on day 7 were similarly bisected and the "half" embryos thus produced transferred directly without culture in vitro, none of the recipient ewes became pregnant.

#### *2. The Regulatory Capacity of Aggregated Whole Embryos and Dissociated Blastomeres*

Attempts to produce chimaeric sheep by micromanipulation of cleaving embryos have fared almost equally badly. Pighills, Hancock, and Hall[40] attempted to adopt Tarkowski's method[14] for fusion of two genetically unrelated embryos. However, no pregnancy resulted when eight supposedly chimaeric embryos fused at the 16- to 32-cell stage were transferred to recipient ewes after culture in vitro. When 14 embryos fused in the same way were transferred on the day of collection, three lambs were produced. One (stillborn) lamb was subsequently shown to be chimaeric. In a much larger series of experiments in the sheep, Tucker, Moor, and Rowson[41] injected one to four blastomeres from day 3 to day 7 embryos into genetically unrelated day 3 embryos. Ninety-two embryos produced in this way were transferred directly to recipient ewes, and 33 lambs were subsequently born. Of the latter, three were demonstrated to be chimaeric and a fourth lamb appeared to have developed solely from the injected blastomeres. The chimaeric lambs all derived from embryos produced in a single trial.

#### *3. Conclusions*

In discussing their results the authors of all the above reports recognized that the limited success achieved in their respective experiments was to a considerable extent attributable to technical difficulties and imperfections. In the light of more recent experiments in this laboratory, there can be little doubt that this was indeed the case. It

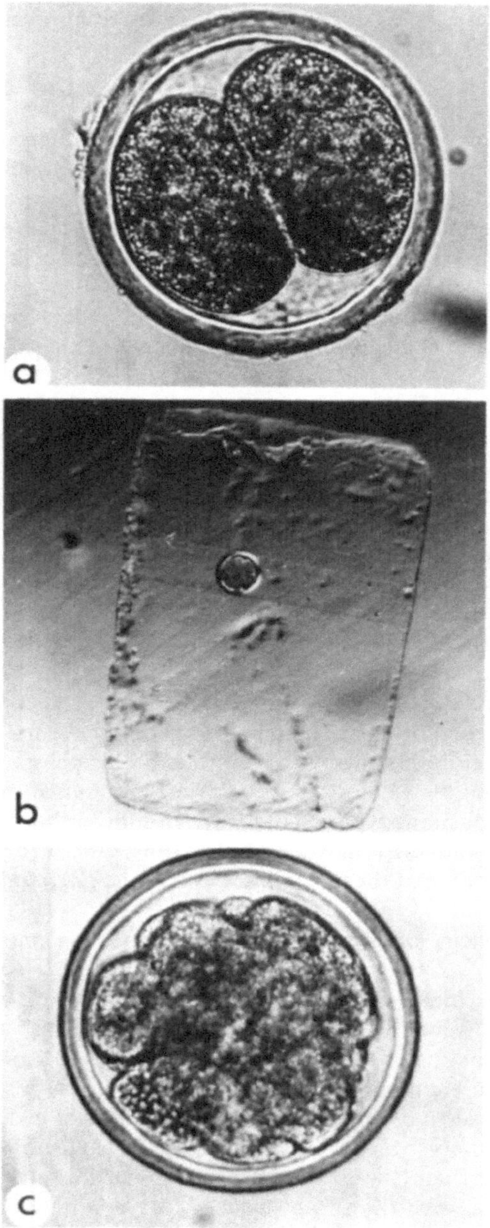

FIGURE 1. (a) Normal two-cell sheep embryo (magnification × ∼380). (b) Agar cube, containing sheep embryo, recovered after 3 days in a ligated rabbit oviduct (magnification × 50). (c) The embryo in Figure 1b at higher magnification. This embryo, which contains about 32 cells, was embedded in agar and transferred to a rabbit when it was at the two-cell stage (magnification × ∼380).

may also be suggested with near certainty that the central problem in all instances arose from the lack of adequate methods for culture of the micromanipulated embryos in vitro combined with the inability of the embryonic cells to survive in vivo after the zona pellucida had been ruptured or removed. On the other hand, the results do indicate that micromanipulation of cleaving embryos of the large domestic species would prove equally informative and useful as similar manipulations of mouse embryos, provided this purely technical problem could be overcome. However, although considerable progress has been made with respect to the culture in vitro of embryos from the large domestic animals,[12,13] none of the methods currently used is entirely satisfactory for the culture of early embryos, i.e., the cleavage stages which are most dependent on the zona pellucida for their survival in vivo.[8,9]

### D. Agar Embedding and Culture in Vivo as an Alternative to Culture in Vitro of Micromanipulated Embryos

#### *1. The Role of the Zona Pellucida in Early Development*

The function of the zona pellucida during embryo cleavage seems to be mainly of a physical nature: to prevent dispersal of the blastomeres,[7] to prevent direct contact between the embryo and alien cells whether these be the epithelial lining of the oviduct or the uterus,[6] leucocytes,[33] spermatozoa, or the cells of other embryos, and to facilitate the passage of the embryo through the oviduct.[6] Except in some species, notably the rabbit, in which the zona plays a role during the expansion of the blastocyst,[42] it no longer appears to be essential for the continued normal development in vivo once the fully compacted late morula stage has been reached.[6-8] Indeed, in the majority of mammals the blastocyst hatches from the zona in the course of expansion (in the sheep and the cow 7 to 8 days after fertilization, in the pig 6 to 7 days after fertilization). But for less advanced embryos to develop normally in vivo, not only must the zona be present, but it must also be relatively intact.[8,9]

#### *2. The Use of Agar for the Protection in Vivo of Embryos with Defective Zonae Pellucidae*

In the procedure which is described below for micromanipulation of early cleavage stages of the large domestic species, the embedding of the micromanipulated embryos in agar is the central technique. The agar coating, which is practically insoluble in the female genital tract, seals any holes in the zona pellucida, thus allowing the embryonic cells to survive and develop in vivo. In preliminary experiments it was found that small cubes of agar generally remained intact when left for several days in ligated rabbit oviducts. Subsequently, it was observed that sheep embryos embedded in similar cubes and placed in ligated rabbit oviducts developed at an apparently normal rate (Figure 1a,b,c). Finally, it was observed that single blastomeres from two-cell sheep embryos inserted into evacuated zonae pellucidae and subsequently embedded in agar and transferred to ligated rabbit oviducts also underwent further cleavage, provided the agar coating remained intact (Figure 2a,b). These observations strongly suggested that agar embedding could be used to protect micromanipulated embryos in vivo without significantly interfering with their development. Since then the technical problems related to the application of the general principle have been largely solved; consequently agar embedding followed by transfer to the oviduct of a temporary recipient can now be regarded as an efficient if somewhat laborious method for the culture of micromanipulated embryos.

## II. TECHNIQUES OF MICROMANIPULATION, AGAR EMBEDDING, AND CULTURE IN VIVO OF EMBRYOS OF THE LARGE DOMESTIC SPECIES

### A. General Principles

#### *1. Outline of Approach*

The successive steps in the present procedure for micromanipulation of embryos in

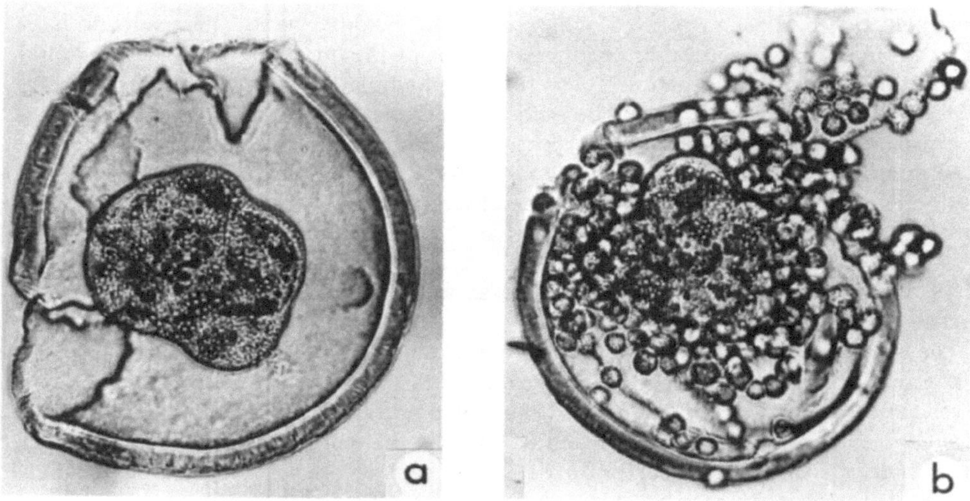


FIGURE 2. (a) Morula which developed from a single blastomere of a two-cell sheep embryo during culture in a ligated rabbit oviduct after being embedded in agar (magnification × ∼420). (b) Embryo invaded by leucocytes during culture in the rabbit due to defects in the agar coating (magnification × ∼420). In this instance some embryonic cells remain, but more often this is not the case.

the large domestic species are briefly as follows: the zona pellucida is removed, and the blastomeres are separated into cell groups or individual cells as required. The separated blastomeres are injected into evacuated foreign zonae pellucidae. After the perivitelline space has been filled with sheep serum, the embryos are embedded in a tiny cylinder of agar, which in turn is embedded in a larger, more robust agar cylinder. This agar 'chip' is transferred to the ligated oviduct of a ewe. It is allowed to remain there for as long as it normally takes an embryo of the stage of development from which the cells were derived to reach the early blastocyst stage. The chip is then recovered, the embryos released from the agar and transferred to definitive hosts.

*2. Micromanipulations*

The techniques used for removal of the zona pellucida, separation of blastomeres, and injection of blastomeres into evacuated zonae have much in common with micromanipulative procedures employed by other workers,[9,14,19,33] although they may differ in detail. The main consideration has been to minimize the risk of injury to the embryonic cells so as to enable the subsequent development of all blastomeres from a particular embryo to be studied. The procedures are carried out under conditions which do not appear to reduce the viability of normal embryos, and the techniques depend on mechanical means only. The use of enzymes and chemical agents, whose influence on the embryonic cells is not entirely clear, has deliberately been avoided. Pronase, which is used routinely for the removal of the zona pellucida in the mouse,[15,17] in most instances fails to dissolve the zonae of sheep, pig, and cow embryos collected from the oviduct, i.e., the stages of development which so far have been of particular interest for micromanipulation.

*3. Agar Embedding*

The technique of embedding the micromanipulated embryos in agar was developed by the author. Several embedding techniques were tried with the following considerations in mind: (1) the agar coating should completely cover the embryo, but not interfere with cell movements within the embryo during its development nor with the interaction between cells in embryos composed of several separated blastomeres; (2) it should be of sufficiently small dimensions to allow easy transfer to and recovery from the sheep oviduct while at the same time being sufficiently strong to avoid breaking during handling and while residing in the sheep oviduct; (3) it should allow individual embryos to be readily identified after culture; (4) it should be easy to produce and remove without injury to the embryo.

Results obtained with four different embedding techniques, none of which was found to be entirely satisfactory, and with the one which was finally chosen are summarized in Table 1. The last technique, which is described in detail below, is advantageous in several respects. It is easy to ensure that the embryos are completely covered by agar. The short period during which the embryos are in direct contact with fluid agar and the presence of relatively viscous sheep serum in the perivitelline space combine to prevent agar from penetrating to the blastomeres. Although the agar chip is of considerable bulk, this is to some extent offset by the possibility of embedding two or more of the small cylinders, each containing several embryos, within the same large cylinder. The identification of individual embryos after culture presents no difficulty provided their respective positions have been recorded in a sketch of the chip before transfer (see Section II,[3] of this chapter).

Single blastomeres may be embedded without a zona pellucida, but this is not advisable, because it makes it very difficult to release the embryos at a later stage. On the other hand, the micromanipulated embryos develop equally well whatever the source of the zonae, and a supply of foreign zonae may conveniently be stored by deep-freezing. Storage at −196°C (in liquid nitrogen) results in less mechanical damage to the zonae than storage at −20°C.

For the production of chimaeric embryos containing two or three times the normal number of cells, zonae of fully expanded pig blastocysts may be used. If the cell number exceeds that of three normal embryos an agar cylinder with a preformed spheric cavity of appropriate size may be used as the first embedding.[43]

*4. Temporary Recipients*

Initially, the oviducts of follicular phase rabbits were used for the culture of agar-embedded embryos. However, the viability of nonmanipulated sheep and cow embryos is reduced considerably if the culture period in the rabbit is extended beyond 3 days,[44] and that of pig embryos after an even shorter period.[45] The ligated sheep oviduct, on the other hand, offers an adequate environment for the development of not only sheep embryos, but also of cow and pig embryos up to the expanded blastocyst stage.[46] Since sheep are more convenient to operate on than pigs or cows, the ligated sheep oviduct has been adopted as the standard culture system for micromanipulated embryos of all the large domestic animals. In the first experiments which concerned the developmental potential of single blastomeres from two-cell sheep embryos, ewes in which the onset of estrus had occurred synchronously with that of the respective donors were used as temporary recipients.[47] However, it was subsequently found that synchrony is not essential[48] (although it may be inadvisable to use as a temporary recipient a ewe which is expected to come into estrus during the culture period). Indeed, for the culture of micromanipulated cow, pig, and horse embryos, the oviduct of an anestrous ewe has been found to be quite satisfactory.[49-51]

When the embryos have reached the expanded blastocyst stage, they must be released from the agar if they are to develop any further. This is the major practical disadvantage of the present embedding technique since it necessitates the recovery of the agar chip. However, 20 or more micromanipulated embryos may be cultured in a single temporary recipient, and the deferment of selection of embryos for transfer to defini-

Table 1
COMPARISON OF DIFFERENT TYPES OF AGAR EMBEDDING FOR THE CULTURE IN LIGATED RABBIT AND SHEEP OVIDUCTS OF SINGLE BLASTOMERES FROM TWO-CELL SHEEP EMBRYOS INSERTED INTO FOREIGN ZONAE PELUCIDAE

| Temporary recipient | Design of agar embedding | No. of embryos transferred | No. of embryos recovered[a] with intact embedding (%) | No. of recovered embryos which had continued to develop (%) |
|---|---|---|---|---|
| Rabbit (culture period 3 to 3½ days) | Cube (1.5 × 1.5 × 1.5 mm) containing one embryo | 10 | 5 (50) | 5 (100) |
| | Sphere (diameter 1.5 mm) containing several embryos | 29 | 16 (55) | 12 (75) |
| | Single cylinder (0.5 × 2 mm) containing several embryos | 16 | 10 (63) | 7 (70) |
| Sheep (culture period 3½ to 5 days) | Single cylinder (0.5 × 2 mm) containing several embryos | 22 | 13 (59) | 12 (92) |
| | Single cylinder (0.15 × 0.5 mm) containing one embryo | 45 | 9 (20) | 9 (100) |
| | Double cylinder (0.8 × 2 mm) containing several embryos | 48 | 47 (98) | 46 (98) |

[a] Of the remaining 70 embryos transferred, 7 were recovered which had continued to develop, a further 24 were represented by empty zonae pellucidae, while 39 were lost.

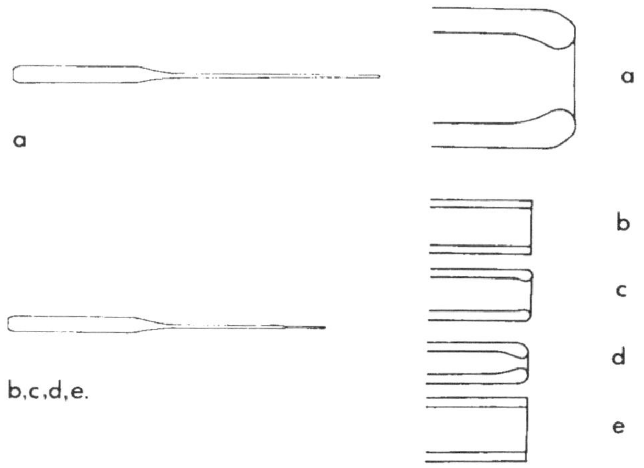


FIGURE 3. Pasteur pipettes prepared from 6-mm Pyrex® glass tubing (cf. text). (a) Standard pipette for handling and transfer of embryos; (b) pipette for cleaning the zona pellucida; (c) pipette for handling naked blastomeres and embryos; (d) pipette for separation of blastomeres; (e) pipette for first agar embedding.

tive recipients until they have developed into blastocysts results in a more economical use of recipients.

## B. Preparation of Instruments

A range of specially designed glass instruments is required for the various steps in the procedure. Most of these are made with the aid of an ordinary Bunsen burner and a diamond-tipped pencil, but for making the micro-instruments to be used with the micromanipulator, a microforge (De Fonbrune) and an electrode pipette puller are needed. In the preparation of glass instruments attention to detail is required in order to avoid frustration during the manipulation of embryos.

### *1. Small Bore Pipettes*

#### *a. Standard Transfer and Handling Pipettes (Figure 3a)*

A 10- to 15-mm middle section of a circa 10-cm piece of 6-mm Pyrex® glass tubing is heated by rotating over a Bunsen burner until pliant, and the tube is pulled to a length of 40 to 45 cm from which two circa 12-cm long pipettes are produced. The tip and base of each pipette are flame polished. Apart from being used for routine handling and transfer of embryos, this type of pipette is also used for the preparation of pipettes with sharp edged tips for cleaning and agar embedding of embryos and fine bore pipettes with polished tips for separation and handling of naked blastomeres. The latter four types of pipettes are made by pulling a segment circa 3 cm from the tip of the standard pipette into a capillary of the appropriate diameter which is cut 1 to 1.5 cm from its base.

#### *b. Cleaning Pipette (Figure 3b)*

This pipette has an inner tip diameter a fraction smaller than the diameter of the intact zona pellucida of the embryo to be manipulated. The tip is left with unpolished edges.

#### *c. Pipette for Handling of Naked Blastomeres and Embryos*

The capillary portion of this pipette is of about the same bore as that of the cleaning pipette, but the tip is flame polished.

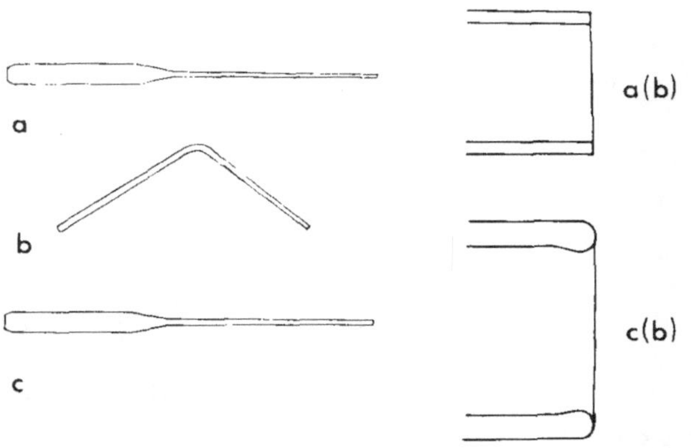


FIGURE 4. Pasteur pipettes and flushing tube prepared from 230 mm soda glass Pasteur pipettes (cf. text). (a) Pipette for second agar embedding; (b) tube for recovery of agar chips from the sheep oviduct; (c) pipette for transfer of agar chips to the sheep oviduct.

***d. Blastomere Separation Pipette (Figure 3d)***

The diameter of the capillary portion of this pipette is 100 μ or less. The opening of the tip is narrowed and polished with a Bunsen burner, or, more conveniently, with the microforge, so that it allows the passage of only one blastomere of the stage of development for which it is to be used.

***e. Small Bore Agar Embedding Pipette (Figure 3e)***

The capillary portion of this pipette has a diameter of circa 150 μ. The tip is left with unpolished edges.

*2. Large Bore Pipettes and Flushing Tube*

Soda glass Pasteur pipettes (230 mm, Bilbate Ltd. Daventry, England) are used for the preparation of large bore agar embedding pipettes, flushing tubes, and large bore transfer pipettes.

***a. Large Bore Agar Embedding Pipette (Figure 4a)***

The capillary part of a soda glass pipette is cut off. A 10- to 15-mm middle section of the remaining tube is heated over a Bunsen burner and the tube pulled to a length of 50 to 60 cm. From this two circa 12-cm pipettes are produced. The base of each pipette is flame polished, but the tip, which should have an inner diameter of 0.5 to 0.8 mm, is left unpolished.

***b. Flushing Tube (Figure 4b)***

The cut off capillary portions of soda glass pipettes are used for the preparation of flushing tubes. The wide end of the tube, which is to be inserted into the oviduct, should have an inner diameter of 1.5 to 2 mm. The edges of this end are flame polished but without narrowing of the opening. The tube is cut to a length of circa 11 cm, and the middle section is bent circa 60° over a Bunsen burner. The narrow end, which should have an inner diameter of 1.0 to 1.5 mm, is left unpolished.

***c. Large Bore Transfer Pipette (Figure 4c)***

The capillary part of a soda glass pipette is cut circa 9 cm from its base. The edges of the tip, which should have an inner diameter of 1 to 1.5 mm, are flame polished

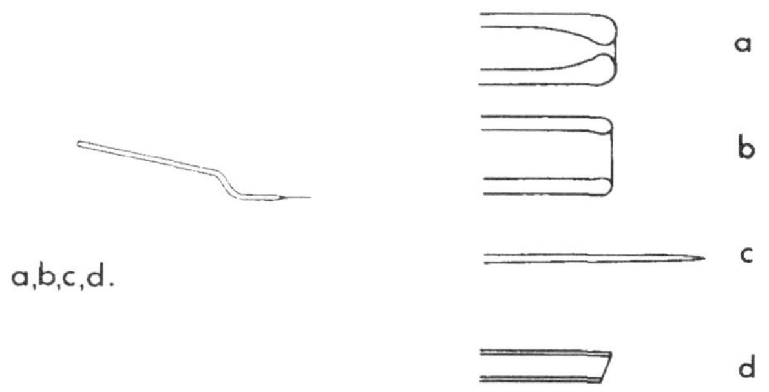


FIGURE 5. Microinstruments produced from 1.5-mm glass tubes (cf. text). (a) Small bore holding pipette; (b) large bore holding pipette; (c) needle; (d) blastomere injection pipette.

without narrowing of the opening. The wide portion of the pipette is cut and flame polished so that the total length of the instrument is circa 12 cm.

*3. Microinstruments*

For the preparation of microinstruments special capillary tubes may be used, but the capillary ends of soda glass Pasteur pipettes are also quite satisfactory for this purpose. For blastomere injection pipettes, those tubes which have the smallest ratio between wall thickness and diameter are selected.

A short section of the middle of a circa 15 cm × 1.5 mm capillary tube is heated over a Bunsen burner and pulled to a length of circa 3 cm and an outer diameter of 120 to 150 $\mu$. Two capillary pipettes are produced from this, each having a thin portion 8 to 10 mm long. The shafts of the pipettes are bent circa 60° about 2 cm from the tip, and again circa 45° backwards in the same plane circa 1 cm further back. The shaft is cut and flame polished circa 4 cm from the second bend.

***a. Small Bore Holding Pipette (Figure 5a)***

A curved capillary pipette is inserted in the holder of the microforge with its tip close to the hot platinum wire so that the opening is reduced to circa 20 $\mu$. The tip should be cut at a right angle with its axis, and the opening should be exactly at the center of the tip so that the zona pellucida is well supported during positioning and microsurgery.

***b. Large Bore Holding Pipette (Figure 5b)***

This is prepared in the same way as the small bore holding pipette except that the sharp edges of the tip are only blunted and polished without reduction of the diameter of the opening. Again it is important to cut the tip at a right angle.

***c. Needle (Figure 5c)***

A curved pipette is inserted in the holder of the microforge, and the tip is brought into contact with the hot platinum wire and pulled into a thin needle circa 1.5 mm long with a diameter of about 30 $\mu$ at its base. The axis of the needle should be the same as that of the adjacent capillary section.

***d. Blastomere Injection Pipettes (Figure 5d)***

These are prepared with the aid of the electrode puller and the microforge. The middle section of a capillary tube is prepulled by hand to a diameter of circa 300 $\mu$ before being drawn with the pipette puller. The two pipettes thus produced are curved as described for the other microinstruments. The curved pipette is then placed with its tip horizontal in the holder of the microforge. The tip is moved forward to be fused superficially with the hot filament at a point where the diameter of the pipette is appropriate. The heat is turned off so that the tip of the pipette sticks to the filament. The contraction of the filament during cooling usually suffices to break the pipette tip cleanly just behind the point of fusion. It is an advantage if the break occurs in a plane which makes the upper edge protrude slightly. The edges of the tip are polished with the aid of the microforge without reduction of the diameter of the opening.

After being prepared all glass instruments are cleaned and rinsed in distilled water and then siliconized. The microinstruments are stored with their shafts embedded in modelling clay in 9-cm Petri dishes. Before and after use they are rinsed thoroughly with boiling distilled water while fitted into an instrument holder. The other instruments are stored in Pyrex® glass boxes lined with gauze and sterilized in a hot air oven.

If properly handled, cleaned, and stored, all the glass instruments may be reused many times.

## C. Technical Procedures

### *1. Arrangement of Instruments*

The main working area (Figure 6a) is enclosed under a Perspex hood. It contains a stereo microscope (Wild M5) fitted with a Perspex platform which allows several Petri dishes to be placed in the immediate vicinity of the microscope so that they may be moved in and out of the field of vision without having to be lifted off the stage. The hood also contains a polystyrene pad on which flasks with agar solution are placed immediately before the embedding of embryos.

The micromanipulation system (Figure 6b) is placed next to the main working area. It comprises a micromanipulator (Leitz, left and right micromanipulator units plus base), a stereo microscope (Wild M7), and two microsyringes (De Fonbrune). Ordinary 10- or 20-mℓ plastic syringes may be used instead of the microsyringes. The microscope is placed on a solid block so that its stage is within horizontal working distance of the microtools.

The microsyringes are placed on the left hand side of the micromanipulator. One is connected to a plastic tube which leads to a glass adaptor taped to the top of the left micromanipulator unit. The other is similarly connected via a longer plastic tube to a glass adaptor placed on top of the right micromanipulator unit. A modified version of the standard Leitz instrument holder is used for the microtools. It consists of a stainless steel tube with a captive nut (Figure 7). This relatively short holder allows great versatility in the positioning of the instrument heads of the micromanipulator. The hole in the captive nut is slightly larger than that of the standard holder in order to accommodate tools made from wider capillary tubes than those intended for use with the standard holder. For fitting of micropipettes into the holder, the captive nut is slid onto the pipette which is then inserted into a silicone rubber tube threaded through the shaft of the holder. The pipette is gently wedged into the shaft, and the captive nut tightened. The holder is placed in the instrument head and the silicone rubber tube connected with the glass adaptor on top of the micromanipulator unit. The glass needle is fitted into the instrument holder in a similar way by means of a circa 1-cm piece of silicone rubber tubing.

It is useful to have a Pyrex® beaker with boiling distilled water placed near the working area for sterilizing instruments when micromanipulations are in progress. Embedding pipettes should not be rinsed in boiling water during use since this tends to make the agar cylinder stick in the pipette.

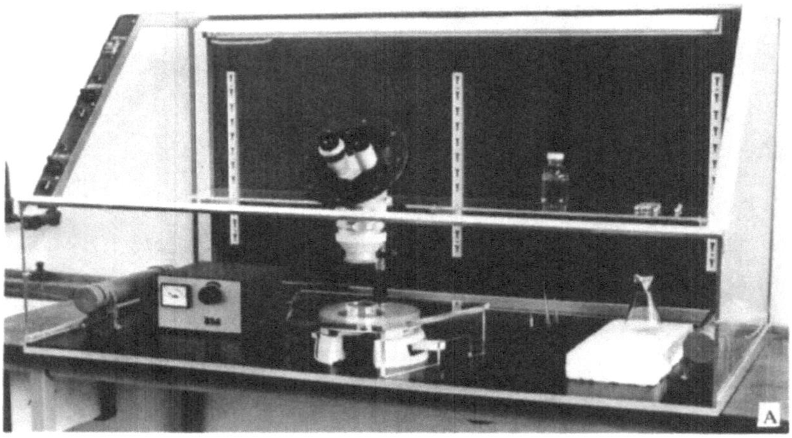


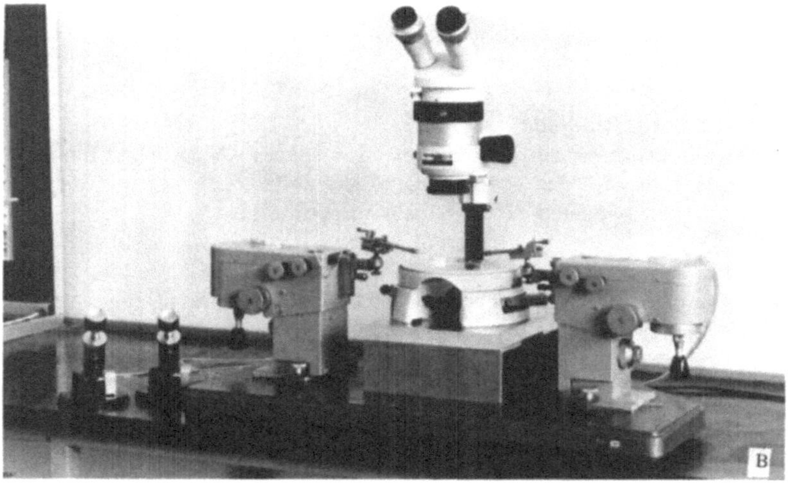


FIGURE 6. Arrangement of instruments. (a) Main working area; (b) micromanipulation system.

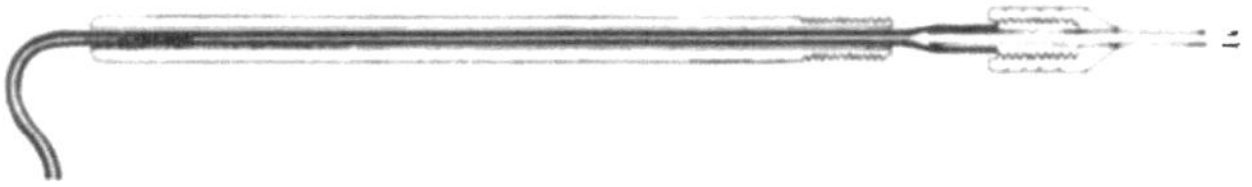

FIGURE 7. Stainless steel holder for microinstruments showing fitting of microinstrument. The shaft is 86 mm long with an outer diameter of 4.7 mm and an inner diameter of 2.5 mm. The captive nut is 18 mm long with an outer diameter of 6.2 mm. The inner diameter of its conical part is 1.5 mm. The 10-mm threads of shaft and captive nut are 3/16 Whitworth. A 2-mm silicone rubber tube connects the micropipettes with the microsyringe.

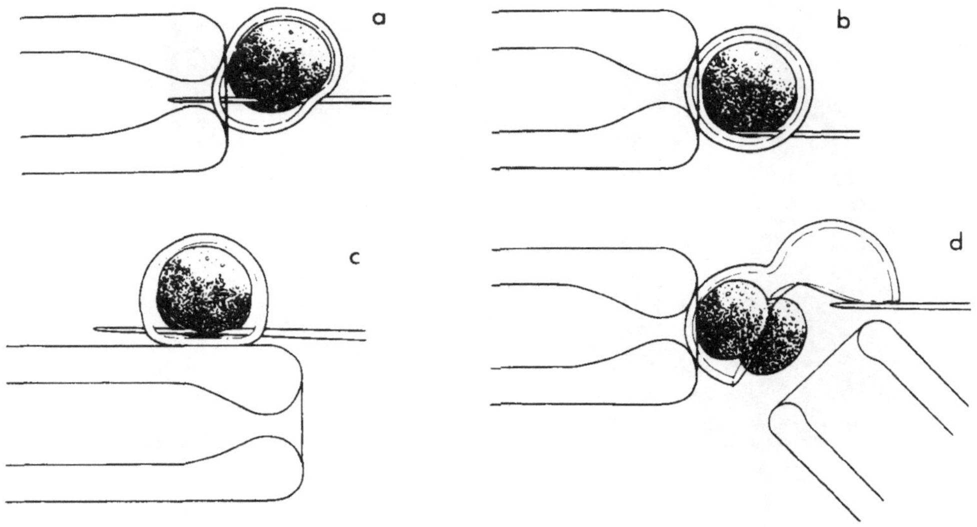


FIGURE 8. Removal of the zona pellucida (cf. text).

*2. Micromanipulations*

***a. Cleaning of the Zona Pellucida***

Before micromanipulation all oviducal cells and debris which might block the holding pipette or interfere with the positioning of the embryo during the removal of the zona pellucida, are gently peeled off with a cleaning pipette.

***b. Removal of the Zona Pellucida***

The lid of a 9-cm Petri dish is placed on the micromanipulation stage and filled with phosphate buffered saline (PBS).[52] An instrument holder, fitted with a small bore holding pipette, is placed in the instrument head of the left micromanipulator unit with the bends of the pipette shaft in a vertical plane. Another instrument holder, fitted with a glass needle, is placed in the head of the right unit. The working parts of the microtools are lowered into the PBS and positioned with their axes on a straight line and parallel with the bottom of the dish and with their tips almost touching at the center of the field of vision. The embryo is then transferred to the dish and picked up with the holding pipette. While minimal suction is applied to the zona, the embryo is rotated on the tip of the pipette with the point of the needle until it presents an equatorial segment where no blastomeres will be in the way of the penetrating needle. The embryo is then held firmly while the point of the needle pierces the zona and traverses the perivitelline space, gently pushing the blastomeres aside, to penetrate the zona again where it is held by the holding pipette (Figure 8a,b). Suction is discontinued while the needle point enters well into the lumen of the pipette. The needle is then withdrawn with the embryo still on it, and rubbed against the wall of the pipette, so that the section of the zona it spans (90 to 120°) is torn open (Figure 8c). In most instances the embryo remains attached to the needle after the tear in the zona has been made and may therefore be returned directly to the tip of the pipette. After the zona has been opened extreme care is necessary during the repositioning of the embryo in order to avoid damaging blastomeres if suction is accidentally applied over the tear. After repositioning of the embryo another tear is made in continuation of the first one, and if necessary yet another. When the zona has been opened two thirds to three quarters round its equator, it is positioned with the slit in a vertical plane and held firmly by one of its poles. The lips of the slit are held widely apart with the point of the needle and the blastomeres flushed out with a handling pipette (Figure 8d). If the blastomeres stick to the inside of the zona it may be necessary to gently loosen them with the needle before attempting to flush them out. The denuded blastomeres are immediately transferred to another dish.

### *c. Separation of the Blastomeres*

First the denuded cells are repeatedly pipetted with a blastomere separation pipette whose tip diameter allows the whole embryo to pass in and out under some distortion of shape. When the cohesion between the blastomeres has been gradually loosened, a pipette with a smaller tip diameter is used to dissociate them into groups. The final separation into single blastomeres is carried out with a pipette which allows only one, slightly compressed blastomere of that particular stage of development to pass at a time (Figure 9). The mechanical separation of blastomeres is the part of the whole procedure which is most likely to destroy blastomeres. As one would expect, the more advanced the stage of development, the greater the cohesive forces between the blastomeres. However, there are also marked species differences, with sheep blastomeres being extremely difficult and cow blastomeres being particularly easy to separate without injury. The separated blastomeres are left close together in the Petri dish.

### *d. Preparation of Evacuated Zonae Pellucidae*

Pig, sheep, and cow ovaries collected at slaughter are convenient sources of zonae pellucidae. The oocytes are aspirated from ovarian follicles with a 23-g needle and incubated in PBS at 37°C for 24 hr. During the brief culture period the corona radiata cells become less firmly attached to the zona so that they are easy to remove with a cleaning pipette. Also, the microvilli of the oocyte retract from the zona, thus facilitating its complete evacuation before use. After the corona cells have been peeled off, the oocytes are transferred in a small amount of PBS to 1-mℓ glass ampoules, which are sealed and stored in liquid nitrogen until they are required for use in an experiment. An ampoule containing the appropriate number of frozen oocytes is then rewarmed in water at room temperature. The oocytes are transferred to the micromanipulator stage. The instruments and their arrangements are the same as for removal of the zona pellucida. Only oocytes with intact zonae are used. The oocyte is held firmly with the small holding pipette and pierced from pole to pole with the needle, the point entering well into the lumen of the pipette. The oocyte is rubbed against the pipette wall so that a 180° tear is made in the zona. The cytoplasm is sucked out with a blastomere separation pipette, and the size of the tear is checked to ensure that it is large enough to accommodate the tip of the blastomere injection pipette to be used. The evacuated zona is transferred to the dish containing the separated blastomeres.

### *e. Injection of Blastomeres into Evacuated Zonae Pellucidae*

The Petri dish containing PBS on the micromanipulation stage is renewed. The small holding pipette is placed in the left micromanipulator unit, and a blastomere injection pipette of appropriate tip diameter is placed in the right unit. The capillary part of the blastomere injection pipette is positioned with its tip near the center of the field of vision and a fraction closer to the bottom of the dish than its base. Its axis should be at an angle of circa 30° with that of the capillary portion of the holding pipette.

An evacuated zona is placed near the tip of the holding pipette, picked up, and positioned with the tear in a vertical plane in such a way that its lips may be parted easily with the tip of the injection pipette. The blastomere to be injected is placed close to the tip of the injection pipette, picked up, and held near the opening. While the zona is held firmly, the tip of the injection pipette is inserted into the tear and the blastomere deposited within the zona (Figure 10).

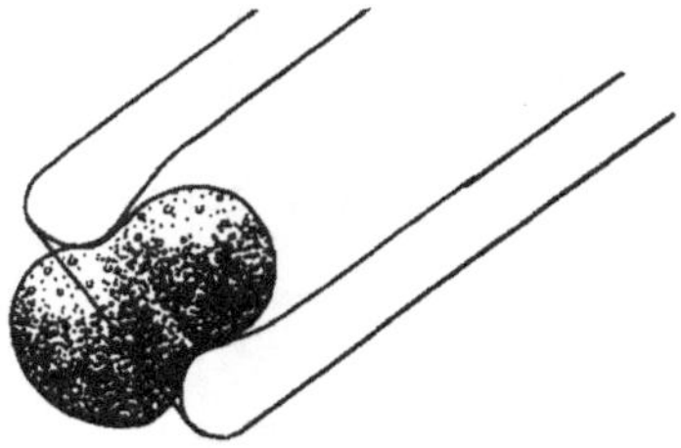

FIGURE 9. Separation of blastomeres (cf. text).

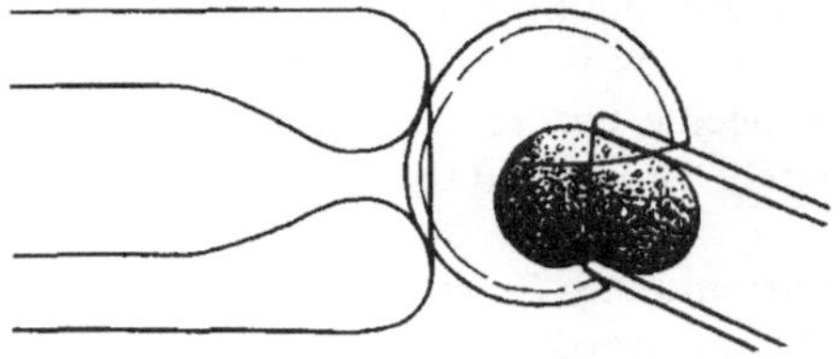

FIGURE 10. Insertion of single blastomeres into evacuated zona pellucida (cf. text).

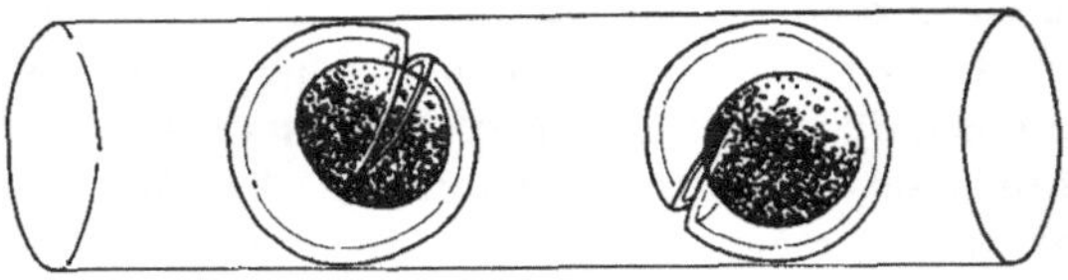

FIGURE 11. First agar embedding (cf. text).

*3. Embedding of Micromanipulated Embryos in Agar*

***a. First Embedding***

A 1.0% solution of agar (Difco) in 0.9% NaCl in distilled water is prepared by boiling. The flask containing the solution is left to cool at room temperature, and when its temperature has dropped to circa 50°C, it is placed on a polystyrene pad. Meanwhile, a 5-cm Petri dish is placed on the micromanipulation stage and half filled with PBS. Five to 6 mℓ of heat treated (60°C for 30 min), sterile filtered sheep serum is gently added to form a layer at the bottom of the dish. The embryos to be embedded are transferred to the serum layer with a small bore agar embedding pipette. An empty 5-cm Petri dish is placed on the stage of the other microscope with another 5-cm Petri dish filled with PBS next to it.

When the agar solution has cooled to 38°C, or slightly below, the embryos are picked up in a minimal amount of sheep serum with a small bore agar embedding pipette. Five to 10 mℓ of agar solution are poured into the empty Petri dish. The embryos are transferred to the agar, and the capillary portion of the pipette is rinsed with the solution. After a few seconds the embryos are picked up with the pipette and held near its tip in a small amount of agar for about 10 sec after withdrawal from the solution, and then expelled in the small cylinder of agar into the Petri dish containing PBS. The section of the cylinder containing the embryos is cut as short as possible, with the aid of two 24-g hypodermic needles fitted on 1-mℓ syringes, without squeezing the embryos (Figure 11).

***b. Second Embedding***

A 1.2% solution of agar in 0.9% NaCl is prepared. When the solution has cooled to 38°C, the small cylinder containing the embryos is embedded in a larger one with the aid of a large bore embedding pipette, using the same procedure as described for the first embedding. It is important to ensure that the small cylinder is completely covered by the second embedding. Finally the section of the agar cylinder containing the embryos is cut to form a chip not longer than 2.5 to 3 mm (Figure 12).

*4. Transfer and Recovery of Agar Chips*

***a. Transfer of the Agar Chip to the Temporary Recipient***

For a general description of the surgical procedures used in embryo transplantation in the sheep see Willadsen.[53]

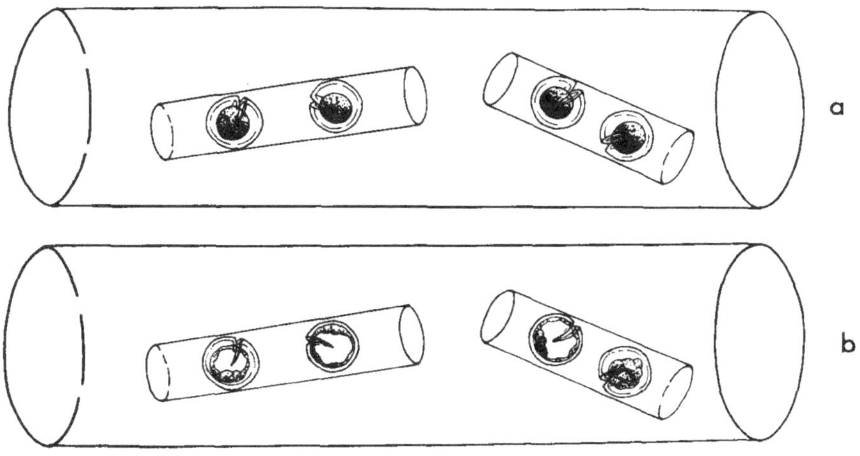


FIGURE 12. Agar chip. (a) Before transfer to the sheep oviduct; (b) after recovery from the sheep. The position of the individual embryo within the chip allows it to be readily identified after culture.

The genital tract of the recipient ewe is exteriorized via a midventral laparotomy, and the oviduct is ligated with braided silk a few millimeters from the utero-tubal junction. The vessels which run along the oviduct are not included in the ligature. The agar chip is picked up with a large bore transfer pipette in a minimal amount of PBS and held near the tip with 1 or 2 mm of air between it and the opening. The tip of the pipette is gently eased circa 3 cm into the oviduct via the infundibulum with the aid of a pair of small plain forceps. The chip is expelled with as little air as possible while the pipette is slowly withdrawn.

*b. Recovery of the Agar Chip from the Temporary Recipient*

At a second operation the genital tract is exteriorized and the oviduct ligature removed. The wide end of a flushing tube is inserted 2 to 3 cm into the oviduct via the infundibulum and held in position by one operator, while the oviduct is flushed with circa 10 m$\ell$ of PBS via a blunted 21-g hypodermic needle inserted into the uterine tip by a second operator. The flushing fluid is collected into a glass cup from which the chip is transferred to a fresh lot of PBS.

*5. Removal of the Agar Embedding*

The lid of a 9-cm Petri dish filled with PBS is placed on the micromanipulation stage. A large bore holding pipette is placed in the instrument head of the left micromanipulator unit and positioned with its capillary portion parallel to the bottom of the Petri dish and its tip near the center of the field of vision. A 24-g hypodermic needle is screwed onto the distal end of an instrument holder shaft, which is placed in the right micromanipulator unit, with the cutting edges of the needle in a vertical plane and the lower edge parallel to the bottom of the dish.

The agar chip is transferred to the dish and while still resting on the bottom, held firmly at a point near the embedded embryo. The cavity containing the embryo is opened by transsection of the chip with the needle. The section containing the embryo is then held firmly while the point of the needle is inserted between the zona pellucida and the surrounding agar to dislodge the embryo. Immediately after its release the embryo is transferred to another dish. With practice it is possible to release the embryo without the micromanipulator using two 24-g hypodermic needles fitted on the barrels of 1-m$\ell$ syringes.

FIGURE 13. Two pairs of monozygotic sheep twins produced from "half" embryos. The parent embryos were at the four-cell and eight-cell stage, respectively, when blastomere separation was performed.

## III. RECENT INVESTIGATIONS OF THE DEVELOPMENTAL POTENTIAL AND REGULATORY CAPACITY OF BLASTOMERES FROM CLEAVING EMBRYOS OF THE LARGE DOMESTIC SPECIES

### A. Sheep

The methods described above were first used to investigate the developmental potential of single blastomeres from two-cell sheep embryos with a view to the artificial production of monozygotic twins.[47] These experiments demonstrated that at the two-cell stage each blastomere has the capacity to develop into a normal full-term conceptus. Subsequent experiments have shown that monozygotic twins may also be produced from four- and eight-cell embryos by separation of the blastomeres into two groups of two and two groups of four, respectively[48] (Figure 13). This may be more convenient if the aim is simply to produce monozygotic twins, both because it is almost impossible to ensure that all embryos are at the two-cell stage at the time of collection, and because two-cell sheep embryos are particularly difficult to micromanipulate without damaging one or both of the blastomeres. "Half" embryos (Figure 14b), whether they have been produced from two-, four-, or eight-cell embryos, appear to be equally viable as normal sheep embryos (Figure 14a). They may be stored by deep-freezing, and this has been used to produce pairs of monozygotic twins in which the two animals were of different age.[48] The viability of "quarter" embryos (Figure 14c), produced by separating the blastomeres of four-cell embryos into four single cells or those of eight-cell embryos into four pairs, is somewhat lower than normal. Nevertheless, one group of monozygotic quadruplet lambs and several pairs of monozygotic twins have been produced by this approach.[54] The viability of embryos produced from single blastomeres of eight-cell embryos is almost negligible.[54] In most instances this is no doubt due to the low cell number attained by such embryos at the time of blastulation, which tends to result in the formation of trophectodermal vesicles rather than proper blastocysts (Figure 14d).

The same general methods have been used to produce chimaeric sheep (Figure 15).[55] This area of investigation promises to be particularly interesting due to the great versatility of the methods with respect to choice of stage of development, number, and even species of the component blastomeres.

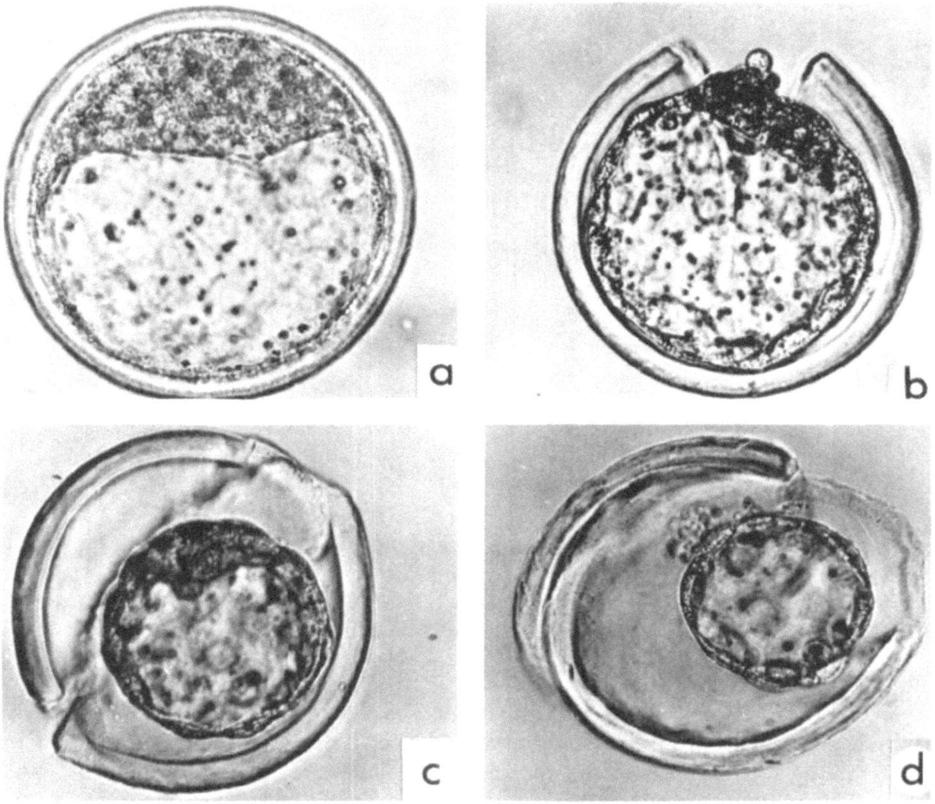


FIGURE 14. Day-7 sheep embryos (magnification × 400). (a) Normal early blastocyst; (b) "half" embryo:early blastocyst produced from four blastomeres of an eight-cell embryo; (c) "quarter" embryo:early blastocyst produced from two blastomeres of an eight-cell embryo; (d) "eight" embryo:trophectodermal vesicle produced from a single blastomere of an eight-cell embryo. Note the relatively small size of the inner cell mass in (b) and (c) compared to (a) and the apparent absence of inner cells in (d).

FIGURE 15. Chimaeric sheep produced by aggregation of blastomeres from purebred Welsh Mountain embryos and Welsh Mountain × Suffolk embryos.

FIGURE 16. Monozygotic triplet heifers produced from "quarter" embryos. The parent embryo was at the eight-cell stage at blastomere separation. The fourth embryo did not survive to term.

## B. Other Large Domestic Species

As mentioned earlier the sheep oviduct offers an adequate environment for the development of cleaving embryos from the other large domestic species. The agar-embedding technique has, therefore, opened the way for studies involving micromanipulations of cow, pig, and horse embryos. So far, such studies have been on a more limited scale than those with sheep embryos. However, they have confirmed both the general usefulness of the technique and the basic similarity between blastomeres from embryos of different species with respect to their ability to regulate their development.

### *1. Cattle*

Cow embryos are morphologically very similar to sheep embryos, but the cohesive forces between blastomeres are weaker and, therefore, the cells may be separated mechanically much more easily without injury. The first experiments with cow embryos were aimed at the production of monozygotic quadruplets from eight-cell embryos. The initial survival rates obtained after transfer of such "quarter" embryos were quite high but were reduced considerably as the pregnancies advanced.[49] Nevertheless, these experiments resulted in the birth of one group of monozygotic triplets (Figure 16) and two pairs of monozygotic twins. Subsequent experiments have been aimed at the use of nonsurgically recovered day-5 and day-6 morulae (Figure 17a) for the production of monozygotic twins. The results indicate that this approach is also feasible (Figure 17b).[56] As in the sheep, such "half" embryos can survive deep-freezing.[57]

### *2. Pigs*

A comprehensive study of the developmental potential of blastomeres from pig embryos has not yet been made, but "half" embryos produced from two- and four-cell embryos have been demonstrated to be viable, and several pairs of monozygotic twins have been produced.[50] A degree of intrauterine competition between embryos and the problems associated with identifying monozygotic offspring make the pig the least attractive of the domestic species for this type of research, although monozygotic twins, triplets, or quadruplets would no doubt be of very considerable practical interest.

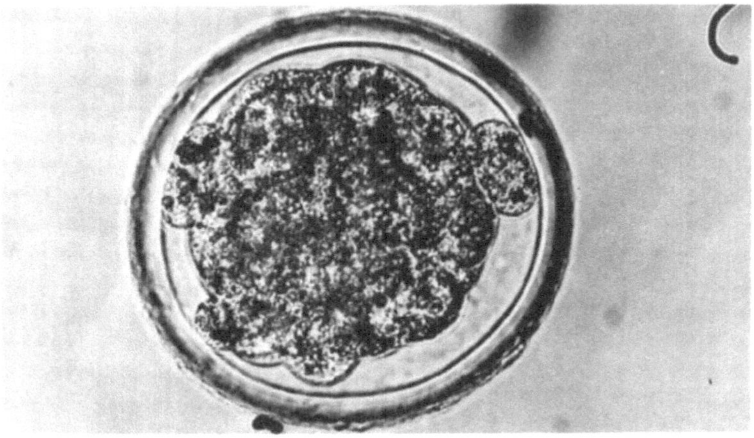

A

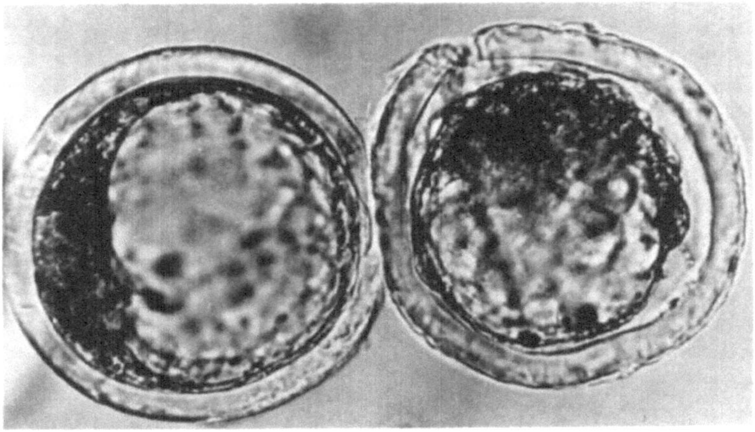

B

FIGURE 17. (a) Cow embryo collected nonsurgically on day 6. The embryo is a morula containing about 48 cells (magnification × ~450). (b) Monozygotic "half" embryos produced from day 6 cow morula (magnification × ~400).

*3. Horses*

Micromanipulations of horse embryos have of necessity been on a very limited scale. "Half" embryos have been found to be viable, but interestingly, the two pairs of monozygotic twin foals (Figure 18) obtained so far have resulted from the transfer of "quarter" embryos.[51]

In summary it may be stated that the techniques described above have proved useful in the study of certain aspects of early embryonic development in the large domestic animals and have, at the same time, offered unique opportunities for the production of several monozygotic or chimaeric animals of preselected genetic background. The possibilities of transferring two monozygotic embryos to two different recipients, and of storing embryos with a view to the production of monozygotic animals of differing

FIGURE 18. Monozygotic pony twins produced from "quarter" embryos. The parent embryo was at the four-cell stage at blastomere separation.

age, further add to the practical value of the techniques. It seems reasonable to suggest that the potential usefulness of agar embedding combined with culture in vivo extends both to other species and to problems other than those dealt with hitherto. At the same time it is recognized that the present techniques of micromanipulation and agar embedding leave considerable room for improvement, and that adequate methods for culture in vitro of embryos of the large domestic species are still urgently required.

## REFERENCES


1. **Nicholas, J. S.**, The development of rat embryonic tissues after transplantation of the egg to the kidney, *Anat. Rec.*, 55 (Abstr.), 31, 1933.
2. **Pincus, G.**, *The Eggs of Mammals*, Macmillan, New York, 1936, 96.
3. **Nicholas, J. S. and Hall, B. V.**, Experiments on developing rats. II. The development of isolated blastomeres and fused eggs, *J. Exp. Zool.*, 90, 441, 1942.
4. **Seidel, F.**, Die Entwicklungspotenzen einer isolierten Blastomere des Zweizellenstadiums in Säugetierei, *Naturwissenschaften*, 39, 306, 1952.
5. **Tarkowski, A. K.**, Experiments on the development of isolated blastomeres of mouse eggs, *Nature (London)*, 184, 1286, 1959.
6. **Modlinski, J. A.**, The role of the zona pellucida in the development of mouse eggs in vivo, *J. Embryol. Exp. Morphol.*, 23 (3), 539, 1970.
7. **Bronson, R. A. and McLaren, A.**, Transfer to the mouse oviduct of eggs with and without the zona pellucida, *J. Reprod. Fert.*, 22, 129, 1970.
8. **Trounson, A. O. and Moore, N. W.**, The survival and development of sheep eggs following complete or partial removal of the zona pellucida, *J. Reprod. Fert.*, 41, 97, 1974.
9. **Moore, N. W., Polge, C., and Rowson, L. E. A.**, The survival of single blastomeres of pig eggs transferred to recipient gilts, *Aust. J. Biol. Sci.*, 22, 979, 1969.
10. **Brinster, R. L.**, In vitro culture of mammalian embryos, *J. Anim. Sci.*, 27 (Suppl.), 1, 1969.

11. **Biggers, J. D., Whitten, W. K., and Whittingham, D. G.,** The culture of mouse embryos in vitro, in *Methods in Mammalian Embryology,* Daniel, J. C., Ed., W. H. Freeman, San Francisco, 1971, 86.
12. **Betteridge, K. J., Ed.,** Embryo Transfer in Farm Animals, Canada Department of Agriculture, Ottawa, 1977.
13. **Anderson, G. B.,** Advances in large animal embryo culture, in *Methods in Mammalian Reproduction,* Daniel, J. C., Ed., Academic Press, New York, 1978, 273.
14. **Tarkowski, A. K.,** Mouse chimaeras developed from fused eggs, *Nature (London),* 190, 857, 1961.
15. **Mintz, B.,** Experimental study of the developing mammalian egg: removal of the zona pellucida, *Science,* 138, 594, 1962.
16. **Mulnard, J. G.,** Studies of regulation of mouse ova in vitro, in *Preimplantation Stages of Pregnancy,* Wolstenholme, G. E. W. and O'Connor, M., Eds., Churchill Livingstone, London, 1965, 123.
17. **Mintz, B.,** Allophenic mice of multi-embryo origin, in *Methods in Mammalian Embryology,* Daniel, J. C., Ed., W. H. Freeman, San Francisco, 1971, 186.
18. **Gardner, R. L.,** Production of chimaeras by injecting cells or tissues into the blastocyst, in *Methods in Mammalian Reproduction,* Daniel, J. C., Ed., Academic Press, New York, 1978, 137.
19. **Rossant, J.,** Postimplantation development of blastomeres isolated from 4- and 8-cell mouse eggs, *J. Embryol. Exp. Morphol.,* 36 (2), 283, 1976.
20. **Wilson, J. B. and Stern, M. S.,** Organization in the preimplantation embryo, in *The Early Development of Mammals,* Balls, M. and Wild, A. E., Eds., Cambridge University Press, Cambridge, 1975, 81.
21. **Gardner, R. L. and Papaioannou, V. E.,** Differentiation in the trophectoderm and inner cell mass, in *The Early Development of Mammals,* Balls, M. and Wild, A. E., Eds., Cambridge University Press, Cambridge, 1975, 107.
22. **Gardner, R. L. and Rossant, J.,** Determination during embryogenesis, in *Embryogenesis in Mammals,* Ciba Foundation Symp. 40 (new series), Elsevier, Amsterdam, 1976, 5.
23. **Adamson, E. D. and Gardner, R. L.,** Control of early development, in *Reproduction,* Short, R. V., Ed., British Council, London, 1978, 113.
24. **Gardner, R. L.,** An investigation of inner cell mass and trophoblast tissue following their isolation from the mouse blastocyst, *J. Embryol. Exp. Morphol.,* 28, 279, 1972.
25. **Rossant, J.,** Investigation of inner cell mass determination by aggregation of isolated rat inner cell masses with mouse morulae, *J. Embryol. Exp. Morphol.,* 36 (1), 163, 1976.
26. **Tarkowski, A. K. and Wroblewska, J.,** Development of blastomeres of mouse eggs isolated at the 4- and 8-cell stage, *J. Embryol. Exp. Morphol.,* 18 (1), 155, 1967.
27. **Kelly, S. J., Mulnard J. G., and Graham, C. F.,** Cell division and cell allocation in early mouse development, *J. Embryol. Exp. Morphol.,* 48, 37, 1978.
28. **Graham, C. F. and Lehtonen, E.,** Formation and consequences of cell patterns in pre-implantation mouse development, *J. Embryol. Exp. Morphol.,* 49, 277, 1979.
29. **Mullen, R. J., Whitten, W. K., and Carter, S. C.,** in *Annual Report of the Jackson Laboratory,* Bar Harbor, Maine, 1970, 67.
30. **McLaren, A.,** *Mammalian Chimaeras,* Cambridge University Press, Cambridge, 1976.
31. **Gardner, R. L.,** Mouse chimaeras obtained by the injection of cells into the blastocyst, *Nature (London),* 220, 596, 1968.
32. **Gardner, R. L. and Johnson, M. H.,** Investigation of early mammalian development using interspecific chimaeras between rat and mouse, *Nature (London),* 246, 86, 1973.
33. **Moore, N. W., Adams, C. E., and Rowson, L. E. A.,** Developmental potential of single blastomeres of the rabbit egg, *J. Reprod. Fert.,* 17, 527, 1968.
34. **Tarkowski, A. K.,** Preimplantation Stages of Pregnancy, *Wolstenholme, G. E. W. and O'Connor, M., Eds., Churchill Livingstone, London, 1965, 139.*
35. **Ducibella, T.,** Surface changes of the developing trophoblast cell, in *Development in Mammals,* Johnson, M., Ed., North-Holland, Amsterdam, 1977, 5.
36. **Adams, C. E.,** personal communication, 1978.
37. **Buehr, M. and McLaren, A.,** Size regulation in chimaeric mouse embryos, *J. Embryol. Exp. Morphol.,* 31 (1), 229, 1974.
38. **Markert, C. L. and Petters, R. M.,** Manufactured hexaparental mice show that adults are derived from three embryonic cells, *Science,* 202, 56, 1978.
39. **Trounson, A. O. and Moore, N. W.,** Attempts to produce identical offspring in the sheep by mechanical division of the ovum, *Aust. J. Biol. Sci.,* 27, 505, 1974.
40. **Pighills, E., Hancock, J. L., and Hall, J. G.,** Attempted induction of chimaerism in sheep, *J. Reprod. Fert.,* 17, 543, 1968.
41. **Tucker, E. M., Moor, R. M., and Rowson, L. E. A.,** Tetraparental sheep chimaeras induced by blastomere transplantation, *Immunology,* 26, 613, 1974.
42. **Adams, C. E.,** personal communication, 1979.

43. **Fehilly, C. B. and Willadsen, S. M.**, unpublished data, 1979.
44. **Lawson, R. A. S., Adams, C. E., and Rowson, L. E. A.**, The development of sheep eggs in the rabbit oviduct and their viability after re-transfer to ewes, *J. Reprod. Fert.*, 29, 105, 1972.
45. **Polge, C., Adams, C. E., and Baker, R. D.**, Development and survival of pig embryos in the rabbit oviduct, in *Proc. 7th Int. Congr. Anim. Reprod. A. I.*, Deutsche Gesellschaft für Züchtungskunde, Bonn, 1972, 513.
46. **Willadsen, S. M. and Polge, C.**, unpublished data, 1978.
47. **Willadsen, S. M.**, A method for culture of micromanipulated sheep embryos and its use to produce monozygotic twins, *Nature (London)*, 277, 298, 1979.
48. **Willadsen, S. M.**, The viability of early cleavage stages containing half the normal number of blastomeres in the sheep, *J. Reprod. Fert.*, 59, 357, 1980.
49. **Willadsen, S. M. and Polge, C.**, Attempts to produce monozygotic quadruplets in cattle by blastomere separation, *Vet. Rec.*, 108, 211, 1981.
50. **Willadsen, S. M. and Polge, C.**, unpublished data, 1979.
51. **Willadsen, S. M., Pashen, R., and Allen, W. R.**, unpublished data, 1980.
52. **Whittingham, D. G.**, Survival of mouse embryos after freezing and thawing, *Nature (London)*, 233, 125, 1971.
53. **Willadsen, S. M.**, Embryo transplantation in sheep, in *The Management and Diseases of Sheep*, British Council, Ed., Commonwealth Agricultural Bureaux, Slough, England, 1979, 69.
54. **Willadsen, S. M.**, The developmental capacity of blastomeres from 4- and 8-cell sheep embryos, *J. Embryol. Exp. Morph.*, 65, 165, 1981.
55. **Fehilly, C. B., Willadsen, S. M., and Tucker, E. M.**, unpublished data, 1979.
56. **Willadsen, S. M., Lehn-Jensen, H., Fehilly, C. B., and Newcomb, R.**, The production of monozygotic twins of preselected parentage by micromanipulaion of non-surgically collected cow embryos, *Theriogenology*, 15, 23, 1981.
57. **Lehn-Jensen, H., Willadsen, S., and Fehilly, C. B.**, unpublished data, 1980.

Chapter 12

# APPLICATIONS OF EGG TRANSFER TO ANIMAL PRODUCTION, BREEDING, AND RESEARCH

I. Wilmut

## TABLE OF CONTENTS

# I. INTRODUCTION

Many exciting opportunities have been created by the development of techniques associated with embryo transfer, for example, storage for prolonged periods by freezing, the production of identical offspring, and "chimaeras". The possibility of obtaining increasing numbers of offspring from selected farm animals has attracted many breeders, and the possibility of being able to specify the sex of the offspring has excited attention even further. Similarly, the research worker welcomes the chance to manipulate reproduction and so investigate subjects as varied as factors which influence fertility, maternal effects on fetal growth and birth weight, sexual differentiation, and susceptibility to tumor growth.

The objective of this discussion is to present an assessment of the commercial value of the techniques before considering their research applications.

# II. COMMERCIAL APPLICATIONS

## A. Introduction

Two basic principles are helpful in understanding the commercial value of embryo transfer. First, application of the technique can only be justified if the value of the product is greater than the cost of the operation, and as the value of agricultural products varies between countries and with time, any one economic analysis has only limited relevance. In order to make the considerations of more general value, the biological effects will be shown in order to allow the reader to complete the analysis by use of current estimates of costs and benefits. Use of embryo transfer in cattle can, for example, increase the number of twin pregnancies, and so increase the amount of meat produced from a given input of animal feed, and in this way improve the biological efficiency, but it does not necessarily follow that the cost of the operation would be recovered from the birth of additional calves.

Second, each additional manipulation will cause the loss of some embryos. It may be convenient to freeze embryos or to discover their sex by cytogenetic analysis, but fewer embryos will survive to term after application of these techniques. It is preferable to use as simple a system as possible in order to maximize the yield of offspring.

### *1. Exceptionally Valuable Donors*

Embryo transfer has been employed commercially in sheep and pigs as well as cattle, but the majority of the work has been with cattle because of the greater value of each animal. The initial transfers multiplied the numbers of a breed or type of animal which was scarce in a particular country. In Britain, North America, and Australia, continental European breeds of cattle commanded extraordinary prices up to 20 times that of a comparable animal of a native breed, which enabled breeders to recover transfer costs of around U.S. $2000 per pregnancy. Similar situations have encouraged the multiplication of polled Merino ewes and Wiltshire Horn ewes in Australia[1,2] and of Texel ewes in Great Britain.[3] Embryo transfer has been developed in Australia to enable breeders to multiply Angora goats which have commanded exceptional prices.[4] Clearly such applications can only be justified as long as the scarcity of a particular breed causes it to realize exceptional prices.

The second application, which will continue, increases the number of young from valuable pedigree families. Heifers from well-known herds can command a price some five times that of typical commercial heifers of the same breed, and outstanding bulls realize very much higher prices. Embryo transfer can increase the number of young from chosen mothers, and they can either all be sold, or the owner may apply rigorous selection to produce the one animal which will command the highest price. Similarly a breeder, having purchased one elite animal, may use embryo transfer to increase the number of animals of that type in his herd. Such an application may be very attractive when each pregnancy costs around U.S. $600.

It may also seem reasonable to invest such sums of money in an attempt to obtain further young from a proven female which has apparently become "infertile" or which, because of lameness or illness, would be unable to carry a pregnancy to term. Although the number of calves was lower than with normal animals, some have been obtained from "infertile" donors.[5]

### B. Breed Improvement

By increasing the number of young from selected animals, embryo transfer could also be exploited to increase the rate of genetic improvement of farm livestock. The impact of such investment is likely to be greatest in cattle because of their low reproductive rate, and least in pigs with the highest reproductive rate among farm livestock. Furthermore, the greater value of individual cattle will justify more costly husbandry procedures than would be appropriate for sheep or pigs. The value of embryo transfer will be considered for improvement of beef, dairy, and dual-purpose cattle. Selection schemes may be applied by individual farmers or by groups of breeders in artificial insemination (A.I.) organizations or breed societies. The potential value of the technique will be assessed both at the level of the individual herd and in larger improvement schemes.

#### *1. Dairy Herds*

Individuals may choose to produce more calves from superior cows in their herds, by using the best cows as donors and the other inferior animals as recipients.[6] The value of the additional milk produced by the superior offspring is small in comparison to the cost of transfer.[7,8] Although, for example, an animal may yield 2700 *ℓ* per lactation more than the herd average, and after additional feed costs, this may be valued in Britain at around U.S. $420 per lactation, her offspring are expected to yield only 290 *ℓ* per lactation more than the offspring born after conventional selection schemes. After additional feed costs this extra milk was valued at U.S. $42 in Britain in 1978.[8] This relatively small economic return reflects the facts that milk yield is a poorly inherited characteristic (around 25% compared with 40% or more for many of the beef characteristics), that a calf obtains only half its genes from its mother, and that conventional schemes improve yield. The relative value of embryo transfer to the breeder is emphasized by the fact that A.I. with semen from a progeny tested bull may improve yield by 600 *ℓ* and cost around $10, providing a greater improvement than that obtained by embryo transfer for a much smaller cost.

#### *2. Beef Herds*

Similarly a breeder may wish to improve his herd for a growth characteristic (e.g., 400-day weight, efficiency of food conversion). The improvement resulting from the use of only the top 10% of animals as mothers of replacements is around 7.9 kg at 400 days, for each generation of selection.[8] This is clearly a small gain in comparison to the cost of transfer if the improvement is being carried out merely to obtain more meat from the herd. However, improvement of this kind might enable a breeder to obtain very high prices for animals sold for breeding.

#### *3. Dairy Schemes*

Several workers have estimated the increase in the rate of improvement which can be obtained by use of embryo transfer to produce bulls for A.I. centers from a smaller number of mothers.[7,9-11] The increase is determined by the efficiency with which conventional schemes are being applied, by the number of calves from each mother, and by the extent to which inbreeding can be tolerated.[10,11] If an existing scheme is efficient and increased inbreeding cannot be tolerated, the improvement may be 2½ to 4%, whereas it may reach 6 to 8% if inbreeding can be tolerated (e.g., if two or more bulls may be used from each donor). The economic return for the improvement that may be obtained has been contrasted with the additional costs of transfer.[10]

*4. Beef-Improvement Schemes*

Some of the more important performance characteristics of beef animals can be accurately estimated in animals of both sexes by 300 to 400 days of age (e.g., weight at 400 days, efficiency of food conversion). In this way, it is possible to select both sire and dams of the replacements by normal breeding age. The rate of improvement of the herd can be increased by 50 to 80% over conventional breeding, depending upon the degree of inbreeding considered to be acceptable, the number of offspring obtained from each donor, and the number of mates for each bull.[12] This improvement could amount to an additional 14 kg weight at 400 days of age for each generation of selection. If in turn bulls from such schemes were used by A.I., each calf would have an additional 7 kg at 400 days which was valued at £3 in Britain in 1978.[7] As a bull may be used for tens of thousands of inseminations, it may be possible to recover a considerable investment.

*5. Dual-Purpose Cattle*

It is possible to use embryo transfer to produce several bull calves from bull-dams that were chosen on the basis of their own performance, and then to combine information on the bull's own growth with a conventional progeny test of milk yield of his daughters.[13] In this way, it is possible to maintain selection for milk yield while introducing selection for a beef character. This approach seems attractive, particularly in countries which obtain much of their beef from the dairy herd.

*6. Implications for the Future*

At the present time it seems that the techniques can only be justified if the breeder is to sell some animals as breeding stock at a premium and if he can command high prices. It cannot be justified simply to increase the yield of meat or milk. In breed improvement schemes, the increase in the rate of response over that which could be obtained at present will not be greatly affected by an increase in the number of calves which can be obtained from each donor. In beef improvement schemes, the additional increase in response is very low when more than eight calves are produced from each donor.[12,14] Only if very many more embryos are available cheaply will it be possible to consider direct improvement of a large part of the national herd in a manner which is comparable to that obtained by use of A.I. This cannot be envisaged at present. As the cost of embryo transfer decreases, it will be possible to employ the technique more widely; however, the costs of transfer are only one part of the cost of improvement schemes.

Part of the effect of embryo transfer on breed improvement depends upon a reduction of the generation interval, and the greatest effect would be obtained by treating prepubertal animals to induce them to ovulate. While ovulation has been induced and fertilization of the oocyte achieved, subsequent development of the embryo has only rarely led to birth of young.[15-17] It may be that this approach will be more successful if longer more complex hormone treatments are applied. In order to gain full advantage from such opportunities it will be necessary to develop physiological methods of predicting potential future performance, perhaps by measurements of metabolites or hormones. In this way embryos would only be recovered from selected young females.

### C. Increased Production of Beef Calves

Beef production may be relatively inefficient for two reasons whose effect can be reduced by use of embryo transfer. As a cow will rear less than one calf each year, even in the best circumstances, the cost of maintaining the cow contributes greatly to the cost of the calf. Embryo transfer can reduce this disadvantage by creation of twin pregnancies, and while it is possible to establish a high proportion of twin pregnancies by transfer of two embryos into an unmated recipient or one embryo into a mated recipient (see Chapter 5), present costs are high in relation to the value of calves.[7] Secondly, some dairy breeds such as the Jersey produce poor calves for beef production, even after mating to a bull of one of the beef breeds. Egg transfer can be used to produce a calf from two beef breeds, but except in a few circumstances (e.g., perhaps with Jerseys), the value of a calf that is a cross-bred from two beef breeds is not so much more than the value of a cross-bred calf from a dairy cow as to cover the costs of egg transfer.

### D. The Value of Methods of Embryo Freezing

The ability to freeze and thaw embryos with very little detrimental effect on their survival greatly facilitates routine embryo transfer, and also creates new opportunities for importation of lines or breeds with the minimum of risk of introducing disease. The extraordinary variability of the response to treatments to induce ovulation and the requirement for synchronous transfer create problems at embryo-transfer centers, because the number of recipients required cannot be estimated accurately (see Chapter 5). By contrast, storage techniques allow transfer into recipients as they become available. The loss of embryos due to freezing is small, but nevertheless, it is preferable to synchronize a group of recipients to take some embryos on the day of recovery, while intending to freeze any excess.

Particular farms or countries may seek to exclude certain diseases, and just as A.I. permits introduction of new genes without a health risk, transfer of frozen embryos is even more effective. It is essential to screen donors for the particular diseases, while holding the embryos in quarantine for a period longer than the incubation period of the disease, and after transfer to monitor the recipients for symptoms of the disease.[18] It has also been argued that embryo transfer offers a useful way to introduce an animal to a different climate.[19] It is suggested that the young animals will acquire from their mothers immunity to some of the diseases associated with the different climate and that as a result the new breed may adapt to its new environment more quickly.

At the present time embryo transfer has only been carried out in a relatively small number of laboratory and domestic species, but it seems certain that other species will be studied. In time, the ability to control reproduction, to recover embryos, or fertilize oocytes in vitro and then to freeze them before transfer will facilitate the preservation and multiplication of endangered species.[20]

## III. RESEARCH APPLICATIONS

The research applications of embryo transfer are concerned either directly with reproduction or with genetic and maternal effects.

### A. Reproduction

#### *1. Introduction*

Embryo transfer has been used to study pregnancy from the earliest cleavage stages of embryo development to parturition. At any time when either the embryo or the mother influences the other, it has been possible to analyze the relationship by using embryo transfer to create unusual situations. In this way the time when an embryo first acts on the mother to establish pregnancy has been defined, and the role of the embryo at parturition analysed. The events of conception, embryonic and fetal development and parturition will be considered in turn.

*2. Early Embryonic Development*

Embryonic development has been studied by embryo transfer and associated procedures of micromanipulation, immunological surgery, and protein analysis. Most research has been carried out in laboratory mammals, but some observations have been made with sheep, cattle, and pigs.

Studies in mice show that the cells of four- and eight-cell embryos are undifferentiated in that they have the ability to differentiate into all types of tissue.[21] By contrast, individual blastomeres within a zona do not have the ability to develop to term, apparently because they lack sufficient "mass".[22,23] Differentiation begins as the embryo develops into a morula and then a blastocyst. The cells undergo a process of compaction in which the borders of adjoining cells come into very close contact and "gap junctions" are formed.[24,25] Differentiation of an "inner cell mass" is believed to follow isolation of cells within the morula.[26,27] It is not known when cells lose the ability to develop into other tissues.[28-33]

The control of embryonic development rests partly within the embryo and partly with the environment within the reproductive tract. The proteins that are synthesized by the embryo change during the early cleavage stages.[34,35] The first RNA controlling this synthesis is in the oocyte before fertilization, but transcription of RNA from the embryonic genome occurs even at the two-cell stage and contributes progressively more RNA during early cleavage.[36,37] The influence of the maternal environment is discussed in Section III.A.3.

Growth of embryos is subject to a remarkable control mechanism which is not understood. When chimaeric mouse embryos are formed by fusion of two entire embryos, they obviously contain twice the normal number of cells for that stage of development, but by 6 days post-coitum the egg cylinder is normal in size. The regulation seems to involve all parts of the cylinder but is not understood.[39]

The trophectoderm cells of mouse blastocysts are in two regions, the polar trophectoderm adjacent to the inner cell mass and the remaining mural trophectoderm. The polar trophectoderm forms the ectoplacental cone under the influence of the inner cell mass,[39] whereas the mural trophectoderm gives rise to "giant cells". In mice, the giant cells contain up to a thousand times the diploid DNA content, and could be formed by fusion of cells or by chromosome replication without cell division (endoreduplication). The formation of these and other multinucleate cells has been studied in chimaeric mice formed from strains which had different forms of an enzyme. Giant cells in chimaeric embryos produced the form of the enzyme associated with each parent, but no other form,[40] whereas skeletal muscle fibers contained hybrid forms of the enzyme as well as the parental forms.[41] These observations demonstrate that muscle myotubes form by cell fusion, and imply that trophoblast giant cells are formed by endoreduplication.

*3. Establishment of Pregnancy*

It was soon established that if embryos were transferred from one female to another, maximum conception was only obtained if their estrous cycles were at appropriate stages.[42] In this regard there are two groups of animals: those such as the sheep in which donor and recipient animals should be at the same stage of the cycle, and those such as the mouse in which conception rates are at least as high and may be higher if the donor is at a more advanced stage than the recipient. In mice in two of three studies, maximum embryo survival was obtained with embryos from day 3 donors transferred into day 2 recipients;[43,44] in the other study, embryo survival in these circumstances was equal to that obtained after synchronous transfer.[45] By contrast, embryos from day 2 donors were rarely able to survive in day 3 recipients.[43] Similar results were obtained in rats, when embryos 1 day younger than the recipient were rarely able to survive, but older embryos (day 3 or 4) were able to survive as synchronous embryos (day 2 or 3).[46] In both species, the young which develop from "advanced" embryos are heavier than those from synchronous embryos.[44,46,47]

These species are among those which may have a delay to implantation induced by lactation or by treatment with progesterone after ovariectomy,[48,49] and this has facilitated studies of the endocrine control of implantation. It seems that a perinidatory estrogen secretion which begins on day 3 and lasts for several days in rats[50] causes release into the uterine lumen on day 4 of proteins which bind to embyos,[51,52] and which in turn activate the embryo.[49,53] When younger embryos are transferred in rats, their development continues normally until day 4 of pseudopregnancy, near the time of normal implantation, after which they degenerate and begin to lose their ability to implant if transferred to a recipient synchronous with the original donor.[54] By contrast, older embryos implant at the same time as synchronous embryos after continuing to develop, so that if day 5 embryos are transferred into a day 4 recipient and recovered 1 day later they are similar to normal day 6 embryos.[54] It seems probable that in rats the proteins released on day 4 stimulate embryos of that age or older to implant, whereas younger embryos are damaged by the change in the environment. The greater weight of young from advanced embryos has been attributed to their ability to form a larger trophoblast and so a larger implantation site.[44,47]

The importance of synchrony within mouse embryos has also been studied by injection of cells of differing stages into blastocysts.[55] If cells 0, 2, and 4 days older than the blastocyst were injected, chimaeras formed in 15, 13, and 0% of young, indicating that a difference of 4 days prevents participation in the development of the embryo.

In sheep, the conception rate after transfer between donor and recipients whose stage of the estrous cycles differ by 2 days is very little less than that obtained if they are perfectly synchronized.[56,57] By contrast, a difference of 3 days between the onset of estrus in donor and recipient results in very few pregnancies.[57] There is evidence that the precise requirement differs at different stages of the cycle, as day 9 embryos were more able to tolerate an advanced environment than day 5 or day 7 embryos,[57] however, this effect has never been confirmed. In cattle the effect of asynchrony is the same whether the recipient be "advanced" or "retarded", although estimates of the effects of 1 day's asynchrony differ.[58-61] The only information available in pigs suggest that the response in this species is the same as that in the cow.[62] The effects of asynchrony in the rabbit are also essentially symmetrical.[42,63]

In these species, delayed implantation does not occur under any circumstances, and there is no evidence of a systemic perinidatory surge of estrogen. In rabbits, sheep, and cattle, pregnancy can be maintained in animals ovariectomized soon after mating by treatment with progesterone alone.[64-67]

The details of the hormonal control of early pregnancy in the ewe have been investigated by transfer of embryos into ewes which had been ovariectomized and then treated with estrogen and progesterone. Three phases have been demonstrated: one dominated by progesterone equivalent to the luteal phase of a first cycle, the second dominated by estrogen equivalent to the next estrus at which mating would have occurred, and the third dominated by progesterone equivalent to the luteal phase which would lead into pregnancy.[68] Treatment of ewes with estradiol equivalent to that present during the early part of the cycle does not influence embryo development.[69,70]

Furthermore, two experiments have demonstrated that the time when progesterone concentration increases has a major role in determining the time of changes in uterine functions. After injecting ewes with progesterone daily from the onset of estrus, day 10 embryos were able to develop after transfer to treated ewes on the 6th day afer estrus, but not after transfer on the 10th day.[71] Apparently, the development of uterine function had been advanced by 4 days. As ewes which received this hormone treatment had estrous cycles which were 4 days shorter than normal, it seems that the same mechanisms control both uterine secretions and the uterine luteolytic factors.

A similar effect was observed in two groups of ovariectomized ewes which received treatments with progesterone which differed only in the time after estrus when the amount of progesterone injected began to increase.[68] When day 4 embryos were transferred to these ewes on day 2½, 4, or 5½, in one group maximum survival was observed on day 4, whereas in the second group in which the progesterone concentration increased 1½ days earlier, maximum embryo survival was observed after transfer to ewes on day 2½. Clearly uterine function had been advanced by the same 1½-day interval by which the increase in progesterone concentration had been advanced. It is also known that variation in concentration of progesterone alters uterine function as assessed by effects on cycle length[72] and development of embryos.[73,74] It is not known whether natural variation in concentration of estrogen at the time of estrus alters uterine function.

A similar effect of premature exposure to progesterone on uterine function and embryo development has been noted in the rabbit, and in this species changes in uterine secretions have been shown.[75] However, in rabbits there is also a delaying effect on uterine secretions of estrogen given post-coitally.[76] Transfer of embryos has shown that day 2 embryos survive best in estrogen-treated rabbits on day 5, and day 4 embryos on day 8.[63,77] In contrast to sheep, it seems that in rabbits both estrogen and progesterone are important in determining the rate of change of the uterine environment.

Detailed information on the hormonal control of early pregnancy is not available for pigs or cattle. However, a comparison of the observations for sheep and rabbits shows clear differences between these species. The time of changes in uterine function in sheep is apparently determined mainly by progesterone, although earlier exposure to estrogen and progesterone is essential for normal function. In rabbits, the time of changes is apparently influenced by both estrogen and progesterone.

Although it has been shown for many years that embryos transferred "asynchronously" have a reduced opportunity to develop, the cause of their death has not been adequately investigated. In rats, embryos in an advanced uterus are unaffected until the recipient reaches day 4 of pseudopregnancy.[54] In rabbits morulae exposed to advanced uteri are damaged more rapidly as the degree of asynchrony is widened, so that morulae were adversely affected after only 24 hr in a day 9 or 11 recipient.[78] In both species, the effect of the adverse environment is to cause degeneration of the embryos.

In sheep the effect depends upon the stage of development of the embryo.[79] Embryos first recovered on day 3 were unaffected by spending 3 days in a recipient which had been estrous 3 days before the donor. By contrast, embryos first recovered as morulae on day 6 were larger than normal after 3 days in a recipient in estrus 3 days before the donor, but had a much reduced ability to implant when transferred to a second recipient synchronous with the original donor. Embryos transferred to a less advanced recipient grew less quickly than normal,[2] but their ability to implant has not been assessed.

As sheep embryos are able to develop in the sheep or rabbit oviduct until the blastocyst stage,[73,80] it seems possible that their development is relatively independent until this stage, but that beyond this point the uterine environment is critically important. As the uterus is greatly influenced by progesterone, any spontaneous anomaly in its secretion or any change induced by stresses imposed on the ewe such as undernutrition[81] may cause embryonic death by affecting progesterone secretion and so uterine function.

In polytocous species, the uterus has the additional function of distributing embryos along its length. In pigs, which have a bicornuate uterus, transfer of embryos from breeds with different skin colors into different uterine horns demonstrated that movement occurs randomly throughout the two horns.[82] It is known that migration occurs between day 8 and 13,[83,84] but the mechanism and endocrine control are not known. Effective distribution of the embryos is essential for maintenance of the corpora lutea, as areas of uterus not occupied by embryos may act to induce luteal regression.[85]

While embryo transfer has demonstrated all of these effects of the mother on embryo development, it has also shown the role of the embryo in establishing pregnancy. In sheep and cattle, two species in which implantation does not occur until after the time when a nonpregnant animal returns to estrus, it has been established that an embryo must be present by day 12 and 16, respectively, in order to prevent luteolysis.[86,87] Conversely in sheep, removal of embryos on day 12 is followed by a normal return to heat.[88] In pigs, embryo transfer is only successful up to day 5 and transfer beyond this stage yields few pregnancies.[89] This failure is not understood. It is known that at these later stages, more embryos survive if transferred to mated recipients than if transferred to unmated recipients.[90]

Infusion of homogenates of day 14 to 15 embryos into the uterine lumen of unmated ewes from day 12 onwards prolonged the cycle until day 23 to 24,[91] and a protein extracted at pH 9.6 from sheep embryos has since been identified and named trophoblastin.[92] It seems that in sheep, embryos act to overcome the luteolytic effect on the uterus, and it has been suggested that the embryo may induce the uterus to secrete $PGE_2$, which is known to inhibit luteolysis induced by $PGF_2$.[93] Preliminary observations support this hypothesis.[94] In pigs estrogen synthesized by the embryos[95] may modify the manner of release of $PGF_2$ from the uterus.[96]

The effect of the embryo on corpus luteum function varies from the antiluteolytic effect shown in sheep and pigs when the size and secretory activity of the corpora lutea are not increased during pregnancy, to the stimulatory effect in guinea pigs and rats when both size and secretory activity are increased during pregnancy.[97] Much remains to be learned of the complex hormonal mechanisms concerned with the establishment and maintenance of pregnancy.

*4. Maternal Influences on the Establishment of Pregnancy*

A decline in reproductive performance occurs with advancing age, and in polytocous species such as pigs, mice, and rabbits has been measured by a decline in litter size. In aged mice and rabbits some degeneration of embryos occurs within the first 4 days,[98,99] however, in both species embryos from old females which appeared to be normal survived as well as embryos from young donors when both types of embryo were transferred to young recipients.[98,100,101] In these species it seems that the death of embryos early in pregnancy in aged females is caused by an unfavorable uterine environment, which is due to age as such rather than to the effect of successive pregnancies.[102] In sheep a poor conception rate is commonly observed in ewe lambs. When embryos were transferred from either lambs or mature ewes into mature ewes, only 33% of those from lambs survived to term, whereas 73% of those from adult donors survived.[103] The cause of this difference is not known. Similarly, fertility is low in ewes which are nursing lambs at the time of mating, and embryo survival is higher if embryos are transferred from such ewes to dry ewes. The lower recovery of eggs on day 2 or day 7 indicates that the maternal environment is inappropriate before the eggs enter the uterus.[104] Conception rate increased from 7% to 37% and then 60% when embryos were transferred to lactating ewes induced to ovulate 20, 36, or 65 to 90 days after parturition.[105] In ewes subjected to a heat stress embryos have been shown to be damaged within 3 days of conception, although transfer of embryos from normal ewes to heat stressed ewes showed that after this time the maternal environment continues to have a deleterious effect on embryo survival.[106]

As the relationship between mother and embryo becomes more clearly understood, it is probable that embryo transfer will be used again to investigate maternal and other environmental effects on the establishment of pregnancy.

### *5. Embryonic and Fetal Differentiation*

The embryonic differentiation of lower animals is straightforward to define and map as the two halves of such an animal develop from the two sides of the embryo.[107] By contrast, in mammals analysis of chimaeric young shows that cell migration is extensive. In 1976 McLaren concluded that in no instance has an organ or tissue of a chimaeric animal been derived from only one of the embryos.[38] The particular pattern found is influenced by the rate of cell division, by differential death rate of particular cell types, by migration, and by "association" or "repulsion" of cells. At present, conclusions can be drawn on the number of cells which initially contributed to particular structures, but no general model to describe events or analyze control mechanisms is possible.[38] Chimaeras carrying genetic markers will be invaluable experimental material for future analysis.

### *6. Embryonic and Fetal Growth*

Embryo transfer has been used to investigate the effect of an increased number of embryos in mice, rats, pigs, and sheep, and in all species embryo death follows "crowding of embryos", although in some species the number of young born may be increased. In pigs, litter size is not increased by transfer of additional embryos to mated recipients, but it is not clear when the number of embryos is reduced to within the normal range (cf. Bazer et al.[108-111]). In mice, litter size may be increased if appropriate numbers of embryos are added,[43] and in sheep and cattle the number of young can be increased by transfer of more than the normal number of embryos (e.g., Rowson et al.[112,113]). In sheep comparisons have been made by transfer within and between breeds.[114-117] All of the breeds had the ability to produce more young than their typical "litter size", but some breeds such as the Finnish Landrace tended to produce more young than others.[115-117]

In two studies in sheep, embryo survival has been higher when embryos were transferred between strains or breeds than after transfer within the strain or breed. The effect achieved statistical significance in the larger study[118] but not in the other.[115] The reason for the advantage of the alien embryo is not known.

The death of embryos or fetuses after transfer of large number of embryos is presumed to follow increased competition for essential nutrients. It occurs before or after implantation in mice,[43] before implantation in sheep,[114,119] and probably after implantation in pigs.[111]

### *7. Sexual Differentiation*

The fusion of a male and a female embryo in a chimaera has created fascinating opportunities for the study of sexual differentiation. In general terms such young are more likely to develop as males, but the proportion varies from being all males in some combinations to an equal number of males and females in other combinations.[38] The proportion is believed to be determined by the ratio of XX to XY cells in the primordial gonad. Dominance of the XX component has never been recorded. A few animals show at least rudimentary organs of both sexes, and in some cases male organs on one side may be associated with a testis while female organs on the other are associated with a rudimentary ovary. A local rather than systemic influence of the gonad is implied by such observations. Furthermore XX cells may be morphologically normal in secondary sexual glands and capable of secreting proteins specific to such organs under the influence of male environment.[120] Conversely, XY cells are apparently capable of developing within mammary glands.[38]

The progeny of XX/XY males are all derived from the XY component,[38] and it seems extremely unlikely that XX germ cells are able to differentiate into sperm even in a testis. The extreme rarity of XX/XY females makes it less certain whether or not XY cells can differentiate in the ovary.[38] The fate of germ cells is probably influenced by the normal pattern of meiosis of germ cells and differentiation of somatic cells. Do germ cells begin meiosis spontaneously or under the influence of somatic cells? Synchronous maturation of germ cells and somatic cells is presumed to be essential for normal development, and so in a combination of tissues with different schedules of development one part of the combination would be expected to degenerate.

By contrast to the situation in mixed sex chimaeras, when both components of the animal are of the same chromosomal sex, germ cells from both component embryos can form gametes.[38,121] However, the germ cells of one component are quite frequently excluded, particularly in some strain combinations.[38] Such exclusion need not necessarily be reflected in similar domination in somatic cells of the testes or any other organ, and may reflect an incomplete mixing of the two cell types when germ cells first segregate.

The phenotypes of sperm in XY/XY chimaeras are apparently determined by the sperm themselves.[122] In chimaeras formed from strains with very different sperm types, under no circumstances were different sperm types formed.

It is certain that studies of chimaeras will continue to add greatly to our knowledge of sexual differentiation and its control.

### *8. Placental and Fetal Hormones*

Increasing use will be made of embryo transfer to study the hormones secreted by the placenta. It is known, for example, that in equids concentrations of PMSG are greatly influenced by fetal genotype.[123] Is the level of placental lactogen or estrogen determined by fetal or maternal genotype, or do both have an influence?

### *9. Parturition*

Transfer of embryos between breeds has shown that the fetus has a major role in determining gestation length.[114,124] Finnish Landrace lambs were carried for around 143 days whether in ewes of their own breed or Border Leicester or Welsh Mountain breeds, although the typical gestation of these other breeds was around 146 days. By contrast, Tasmanian Merino lambs were carried for 150 days. The effect was analyzed further by reciprocal crossing when one third of the variation was found to arise from the breed of the mother and two thirds from the breed of the lamb. These observations are compatible with the finding that fetal hormones have a major role in initiating parturition.[125]

### *10. Chimaeras, Cancer, and Tolerance*

The very existence of chimaeras raises the obvious question of how two different populations of cells can exist in the same animal. While it has been established that two populations of cells do exist throughout the immune system, it is not known how mutual tolerance is established. Two models have been suggested: one postulating the elimination of any cells which react against "self", the other proposing that protective mechanisms prevent damage (see McLaren[38] for a discussion). Evidence for protective molecules has been sought by in vitro incubation of lymph cells from chimaeras with cells from the two component strains. In some studies cytotoxicity was prevented specifically by serum from chimaeras, suggesting the presence of protective molecules.

However, such protection was not observed consistently. When in vivo effects were sought, chimaera lymphoid cells were found not to induce a response in $F_1$ parental crosses as judged by lymph node weight gain.[133] Clearly a great deal remains to be learned about tolerance, rejection, and autoimmunity by use of chimaeras.

When chimaeras are formed from strains with very different incidence of specific tumors, the frequency of formation of tumors is primarily determined by the susceptible component (see McLaren[38]). This conclusion applies to the mammary tumors, hepatomas, and leukemias which have been studied. However, in an exceptional case a resistant line determined the susceptibility to tumor. The question is therefore raised whether such animals provide new opportunities of understanding tumor rejection in resistant lines and tumor formation in susceptible animals.

**B. Maternal Effects**

Embryo transfer can be used to determine whether the mother or the young or both are defining a particular characteristic. In sheep the technique has been used to study maternal influences on birth weight and litter size, on fleece type, and on copper metabolism.

Birth weight of lambs is influenced by maternal breed, but fetal genotype exerts a far greater influence.[116,126,127] In general, the uterus is capable of responding to the demands of larger fetuses or to a larger number of fetuses, but according to the law of diminishing returns.[127] The lamb's birth weight appeared to be maximum if it was born as a single lamb to a ewe of a large breed. A significant maternal effect on birth weight has also been demonstrated in rabbits[128] and mice.[129] The study in mice partitioned the maternal effect on growth into pre- and post-natal phases and established the marked importance of the prenatal phase.[129] Comparable studies have not been performed in farm animals.

A number of studies have shown maternal effects on fleece type which persist and some of which probably involve mechanisms other than nutritional deprivation following transfer of embryos from a large to a small breed.[130-132] There have been shown to be maternal effects, embryo effects, and maternal-fetal interactions. The maternal environment affected follicle density, primary/secondary follicle ratio, fiber medullation, fiber diameter, and fiber length.[130]

A marked maternal effect on copper in the liver of lambs was shown by reciprocal transfer of embryos between Welsh Mountain ewes, with a high level, and Scottish Blackface ewes, which have a low copper level. Copper concentration in lambs born to Blackface ewes was less than half that of lambs born to Welsh Mountain ewes.[134]

It seems probable that embryo transfer will be used increasingly to analyze the relationship between mother and young in farm animals.

## IV. FUTURE DEVELOPMENTS

**A. A General Perspective**

By far the most exciting part of research in this area is to imagine what may be achieved in the future, and while breakthroughs carry obvious rewards, the long hard task of applying new methods on a field scale is the most difficult of all. As a result, over a period of several decades, scientific and popular journals have been full of unrealized predictions about "test-tube babies", about choosing the sex of the offspring, and using culture techniques to allow maturation of the thousands of follicular oocytes before in vitro fertilization and culture to a stage that can be transferred to the uterus. More recently "cloning" in its various forms has received attention.

In the foreseeable future techniques may be developed to achieve in vitro maturation and fertilization of oocytes, discovery of the sex of the embryo, and the several forms of cloning.

### B. Oocyte Maturation and Fertilization

As the mammalian ovary contains thousands of primordial follicles, it is clearly an important objective to seek ways of promoting oocyte maturation and fertilization in vitro in order to obtain a large number of young from selected females.

Normal maturation of the liberated oocyte in vitro to yield an oocyte capable of fertilization and development is not yet possible with any species. Instead, it has been found that success depends upon culture within a follicle so that the whole complex can mature. Numerous studies have characterized changes in the hormones being synthesized in the follicle, in intercellular junctions, and in proteins being synthesized by the oocyte.[135-139] However, the mechanisms involved in oocyte maturation remain obscure. Live young have been born after fertilization of rabbit and sheep oocytes which matured during culture.[140,141] The follicles cultured were quite large antral follicles which are available in only relatively limited numbers. As it takes many weeks in farm animals for the growth from a primordial follicle to a large antral follicle,[142] it is evident that several new techniques must be developed before the ultimate reserve of follicles can be exploited.

Fertilization in vitro has been achieved so far in several species (hamster, mouse, human, Chinese hamster, guinea pig, Mongolian gerbil, monkey, rat, rabbit, dog, and cat, see Rogers[143] for review of methods and criteria used). In those species in which it is successful, fertilization depends upon incubation of the spermatozoa in salt solutions which usually contain macromolecules which are believed to play a crucial role in inducing capacitation of the spermatozoa. The search in other species is for suitable salt solutions and macromolecules.

Finally, before commercial exploitation the embryos should be cultured to a stage capable of being transferred to the uterus. This is impracticable in vitro with embryos of the farm animals, but can be achieved in the rabbit oviduct.[80,144]

Quite clearly an image of a ready supply of embryos generated by in vitro maturation and fertilization is an illusion, for even when the techniques have been developed, the costs of the sophisticated procedures will be relatively high. Furthermore, the total yield of embryos may be smaller than expected unless the success of each step is extraordinarily high. If four steps are employed each with a 60% success, only 13% ($0.6^4$) of oocytes become embryos fit for transfer if the success of each step is completely independent.

### C. Sexing of Embryos

As separation of X and Y bearing sperm seems a formidable task, the only possible method of choosing the sex of an embryo is to discover the sex and select those which are of the desired sex. This has been achieved by detection of the Barr body and by chromosomal analysis. In trophoblast cells from rabbit embryos on day 5, the female embryos can be identified by the sex chromatin which is present against the nuclear membrane. This method was used successfully to identify female embryos before transfer.[145] Trophoblast from cattle and sheep embryos has been subjected to successful chromosome analysis before transfer of the remainder of the embryo.[146] While these methods have research applications, they are too cumbersome for commercial applications.

An exciting new possibility was created by the demonstration that half of a group of eight-cell mouse embryos were lysed by antiserum against the male antigen — H-Y antigen.[147] These observations have now been extended and confirmed by karyotyping treated embryos and establishing that 92% of unaffected embryos were female.[148] It remains first to be discovered whether or not the H-Y antigen is present on embryos of other species, and second to devise a way of revealing the male embryos without lysis of the cells.

While such a procedure may be extremely convenient it does not increase the number of young of a particular sex, and if the test system is harmful it may actually decrease the number. The commercial value of the procedure would clearly be greatest in situations where the breeder requires only young of one sex, as would be the case for an A.I. center trying to breed bulls from exceptional cows.

### D. Cloning

The word is derived from the Greek "klon" meaning twig or cutting and has been used to describe a number of forms of vegetative reproduction in animals. Several different procedures all have the end result of producing a number of identical young which are known as clones.

The simplest procedure is to bisect an embryo and obtain two "identical offspring". This has recently been achieved with sheep,[149] some 50 years after human baby hair was used to create the first vertebrate clones by dividing the eggs of the salamander.[150] A strict limit to the number of young that can be obtained is imposed by the requirement for a certain cytoplasmic mass (see Section III.A.2).

Clones could also be produced by transplantation of nuclei removed from embryonic cells into enucleated oocytes. This procedure can be carried out with amphibian eggs but not yet with mammalian eggs.[151,152] The major difficulty with mammalian eggs is their small size, being approximately one tenth the diameter of amphibian eggs. Nevertheless, it seems probable that successful procedures will be devised at some time in the near future. It has been established with two species of amphibian that the technique is increasingly less successful as nuclei are removed from older embryos (see McKinnell[151]). This change has been presumed to reflect a greater differentiation of the nuclei, which we are unable to reverse, but recent evidence suggests that the change may reflect differences in the ability of nuclei to integrate with cytoplasmic structures.[153] This ability was partly restored by incubation of the differentiated nuclei in oocytes at diplotene. Clearly a great deal remains to be learned about these interrelationships in oocytes and somatic cells. The fact that cells of mouse embryos are only totipotent at the earliest stages of development suggest that similar patterns will be found in mammals.

The ultimate procedure, with an almost endless opportunity to create clones, would involve taking nuclei from an adult animal and transferring them into oocytes. Although some activation of the oocyte has been achieved in amphibia and man, no adult animals developed after use of this procedure.[151]

The potential value of these procedures is probably greatest in research. As well as contributing to investigations into events in reproduction, in a manner similar to transfer itself, a number of groups of clones are uniquely suited for research into diseases and their treatment. It would be extremely useful to be able to apply a variety of treatments to clones which had all been given the same infection. Similarly, clones offer exciting opportunities for research into cancers and their treatment, and into the changes which occur during aging.

In animal breeding, the simpler forms of cloning do offer the possibility of a few extra young from selected donors, and it may be that methods of incorporating nuclei into cells with the aid of viruses will be developed to simplify the procedure. The ability to reproduce selected adult animals that had proved themselves would clearly be very useful, but it is not certain that the costs could be recovered. Furthermore, the low heritability of many commercial traits may lead to disappointment when clones fail to live up to the reputation of their "parent".

### E. Genetic Engineering

Greater understanding of the structure and function of the DNA in chromosomes has recently permitted the introduction of a specific gene into bone marrow cells which were then returned to a mouse.[154] While this technique itself has great potential value for the treatment of cancer and some genetic diseases, improvement of animal performance might well depend upon modification of a greater number of cell types. It might be possible to do this by manipulation of embryonic cells; however, before this could be achieved, a great deal must be learned about specific genes in farm animals in order to define those that are limiting and the more valuable alleles. While the abstract concept can be considered, its practical application cannot yet be imagined.

## V. A BRAVE NEW WORLD OF ANIMAL BREEDING?

Embryo transfer and its ancillary techniques have already contributed to animal breeding and more particularly to research, and their impact will undoubtedly increase. The commercial advantage in being able to produce the next generation from fewer dams is obvious, as is the potential hazard arising from excessive inbreeding. At present it seems that the time when inbreeding may become a severe risk is so far away, that detailed calculations of the dangers and prescription of satisfactory remedies can await the development of the more effective procedures. Furthermore, there are many novel ideas that remain to be investigated, and their impact cannot yet be imagined (see for example a stimulating essay by Seidel[155]).

Animal breeding is only just entering a new era and a great deal will be achieved.

## ACKNOWLEDGMENTS

I am grateful to many people for constructive criticism and, in particular, thank Drs. R. B. Land, R. H. F. Hunter, and C. J. M. Hinks.

## REFERENCES

1. **Moore, N. W. and Shelton, J. N.**, The application of the technique of egg transfer to sheep breeding, *Aust. J. Agric. Res.*, 13, 718, 1962.
2. **Lawson, R. A. S.**, Research applications of embryo transfer in sheep and goats, in *Embryo Transfer in Farm Animals*, Betteridge, K. J., Ed., Monogr. 16, Canada Department of Agriculture, Ottawa, 1977, 72.
3. **Wilmut, I.**, unpublished data, 1980.
4. **Moore, N. W.**, Multiple ovulation and ovum transfer in the goat, *Proc. Aust. Soc. Anim. Prod.*, 10, 246, 1974.
5. **Bowen, R. A., Elsden, R. P., and Seidel, G. E.**, Embryo transfer for cows with reproductive problems, *J. Am. Vet. Med. Assn.*, 172, 1303, 1978.
6. **Nicholas, F. W.**, The potential effect of multiple ovulation and embryo transfer on response to selection in dairy cattle, 3rd Int. Congr. Soc. Adv. Breed. Res. Asia Oceania, Anim. Breed. Paper 9, 1977.
7. **Miller, P.**, Implications of Ova Transplantation for Genetic Improvement of Cattle, duplicated report Am. Breed. Serv., De Forest, Wis., 1975.
8. **Wilmut, I. and Hume, A.**, The value of embryo transfer to cattle breeding in Britain, *Vet. Rec.*, 103, 107, 1978.
9. **Skjervold, H.**, Breeding aspects in case of practical application of egg transplant, in *Symp. Egg. Transplantation*, Assoc. Swed. Livestock Breed. Prod., Hållsta 1974, 87.
10. **Cunningham, E. P.**, The use of egg transfer techniques in genetic improvement, in *Egg Transfer in Cattle*, Rowson, L. E. A., Ed., Commission of the European Communities, Luxembourg, 1976, 345.
11. **Hill, W. G. and Land, R. B.**, Superovulation and ovum transplantation in genetic improvement programmes, in *Egg Transfer in Cattle*, Rowson, L. E. A., Ed., Commission of the European Communities, Luxembourg, 1976, 355.

12. **Land, R. B. and Hill, W. G.**, The possible use of superovulation and embryo transfer in cattle to increase response to selection, *Anim. Prod.*, 21, 1, 1975.
13. **Krausslich, H.**, Application of superovulation and egg transplantation in A.I. breeding programmes for dual purpose cattle, in *Egg Transfer in Cattle*, Rowson, L. E. A., Ed., Commission of the European Communities, Luxembourg, 1976, 333.
14. **Elsen, J. M. and Mocquat, J. C.**, Intérêt de la transplantation d'oeufs pour accroitre l'efficacité de la selection de bovins à viande, *Ann. Genet. Sel. Anim.*, 8, 331, 1976.
15. **Casida, L. E., Meyer, R. K., McShan, W. H., and Wisnicky, W.**, Effects of pituitary gonadotropins on the ovaries and the induction of superfecundity in cattle, *Am. J. Vet. Res.*, 4, 76, 1943.
16. **Foote, R. H.**, Induced superovulation and fertility in prepuberal cattle, sheep and swine, *J. Anim. Sci.*, Suppl. 49, 1971.
17. **Seidel, G. E., Larson, L. L., Spilman, C. H., Hahn, J., and Foote, R. H.**, Cutlure and transfer of calf ova, *J. Dairy Sci.*, 54, 923, 1971.
18. **Wrathall, A. E., Done, J. T., Stuart, P., Mitchell, D., Betteridge, K. J., and Randall, G. C. B.**, Successful intercontinental pig conceptus transfer, *Vet. Rec.*, 87, 226, 1970.
19. **Wilmut, I. and Rowson, L. E. A.**, Experiments on the low temperature preservation of cow embryos, *Vet. Rec.*, 92, 686, 1973.
20. **Polge, C.**, Embryo transfer and embryo preservation, in *Artificial Breeding of Non-Domestic Animals*, Watson, P. F., Ed., Academic Press, London, 1978, 303.
21. **Kelly, S. J.**, Studies of the developmental potential of 4- and 8-cell stage mouse blastomeres, *J. Exp. Zool.*, 200, 365, 1977.
22. **Fiser, P. S. and Macpherson, J. W.**, Development of embryonic structures from isolated mouse blastomeres, *Can. J. Anim. Sci.*, 56, 33, 1976.
23. **Rossant, J.**, Post-implantation development of blastomeres isolated from 4- and 8-cell mouse eggs, *J. Embryol. Exp. Morphol.*, 36, 283, 1976.
24. **Enders, A. C. and Schlafke, S. J.**, The fine structure of the blastocyst: some comparative studies, in *Preimplantation Stages of Pregnancy*, Wolstenholme, G. E. W. and O'Connor, M., Eds., Little, Brown, Boston, 1965, 29.
25. **Calarco, P. G. and McLaren, A.**, Ultrastructural observations of pre-implantation stages of the sheep, *J. Embryol. Exp. Morphol.*, 36, 609, 1976.
26. **Tarkowski, A. K. and Wroblewska, J.**, Development of blastomeres of mouse eggs isolated at the 4- and 8-cell stage, *J. Embryol. Exp. Morphol.*, 18, 155, 1967.
27. **Hillman, N., Sherman, M. I., and Graham, C.**, The effect of spatial arrangements on cell determination during mouse development, *J. Embryol. Exp. Morphol.*, 28, 263, 1972.
28. **Stern, S. M. and Wilson, I. B.**, Experimental studies on the organisation of the pre-implantation mouse embryos. I. Fusion of asynchronously cleaving eggs, *J. Embryol. Exp. Morphol.*, 28, 247, 1972.
29. **Rossant, J.**, Investigation of the determinative state of the mouse inner cell mass. I. Aggregation of isolated inner cell masses with morulae, *J. Embryol. Exp. Morphol.*, 33, 979, 1975.
30. **Rossant, J.**, Investigation of the determinative state of the mouse inner cell mass. II. The fate of isolated inner cell masses transferred to the oviduct, *J. Embryol. Exp. Morphol.*, 33, 991, 1975.
31. **Handyside, A. H.**, Time of commitment of inside cells isolated from pre-implantation mouse embryos, *J. Embryol. Exp. Morphol.*, 45, 37, 1978.
32. **Hogan, B. and Tilly, R.**, In vitro development of inner cell masses isolated immunosurgically from mouse blastocysts. I. Inner cell masses from 3 to 5 day old p.c. blastocysts incubated for 24 hrs before immunosurgery, *J. Embryol. Exp. Morphol.*, 45, 93, 1978.
33. **Hogan, B. and Tilly, R.**, In vitro development of inner cell masses isolated immunosurgically from mouse blastocysts. II. Inner cell masses from 3.5- to 4.0-day p.c. blastocysts, *J. Embryol. Exp. Morphol.*, 45, 107, 1978.
34. **Van Blerkom, J., Barton, S. C., and Johnson, M. H.**, Molecular differentiation in the pre-implantation mouse embryo, *Nature (London)*, 259, 319, 1976.
35. **Van Blerkom, J. and McGaughey, R. W.**, Molecular differentiation of the rabbit ovum. II. During the pre-implantation development of in vivo and in vitro matured oocytes, *Dev. Biol.*, 63, 151, 1978.
36. **Levey, I. L., Stull, G. B., and Brinster, R. L.**, Poly (A) and synthesis of polyadenylated RNA in the pre-implantation mouse embryo, *Dev. Biol.*, 64, 140, 1978.
37. **Johnson, M. H.**, Intrinsic and extrinsic factors in pre-implantation development, *J. Reprod. Fertil.*, 55, 255, 1979.
38. **McLaren, A.**, *Mammalian Chimaeras*, Cambridge University Press, Cambridge, 1976.
39. **Gardner, R. L., Papaioannou, V. E., and Barton, S. C.**, Origin of the ectoplacental cone and secondary giant cells in mouse blastocysts reconstituted from isolated trophoblast and inner cell mass, *J. Embryol. Exp. Morphol.*, 30, 561, 1973.

40. **Chapman, V. M., Ansell, J. D., and McLaren, A.,** Trophoblast giant cell differentiation in the mouse: expression of glucose phosphate isomerase (GPI-1) electrophoretic variants in transferred and chimaeric embryos, *Dev. Biol.*, 29, 48, 1972.
41. **Mintz, B. and Baker, W. W.,** Normal mammalian muscle differentiation and gene control of isocitrate dehydrogenase synthesis, *Proc. Natl. Acad. Sci. U.S.A.*, 58, 592, 1967.
42. **Chang, M. C.,** Development and fate of transferred rabbit ova or blastocysts in relation to the ovulation time of recipients, *J. Exp. Zool.*, 114, 197, 1950.
43. **McLaren, A. and Michie, D.,** Studies on the transfer of fertilized mouse eggs to uterine foster-mothers. I. Factors affecting the implantation and survival of native and transferred eggs, *J. Exp. Biol.*, 33, 394, 1956.
44. **Aitken, R. J., Bowman, P., and Gauld, I.,** The effect of synchronous and asynchronous egg transfer on foetal weight in mice selected for large and small body size, *J. Embryol. Exp. Morphol.*, 37, 59, 1977.
45. **Noyes, R. W., Doyle, L. L., Gates, A. H., and Bentley, D. L.,** Ovular maturation and fetal development, *Fert. Steril.*, 12, 405, 1961.
46. **Noyes, R. V. and Dickmann, Z.,** Relationship of ovular age to endometrial development, *J. Reprod. Fertil.*, 1, 186, 1960.
47. **Noyes, R. W., Doyle, L. L., and Bentley, D. L.,** Effect of pre-implantation development on foetal weight in the rat, *J. Reprod. Fert.*, 2, 238, 1961.
48. **Psychoyos, A.,** Hormonal requirements for egg-implantation, *Adv. Biosci.*, 4, 275, 1969.
49. **McLaren, A.,** Endocrinology of implantation, *J. Reprod. Fert.*, Suppl. 18, 159, 1973.
50. **Nimrod, A., Ladony, S., and Lindner, H. R.,** Peri-nidatory ovarian oestrogen secretion in the pregnant rat determined by gas chromotography with electron capture detection, *J. Endocrinol.*, 53, 240, 1972.
51. **Surani, M. A. H.,** Hormonal regulation of protein in the uterine secretion of ovariectomised rats and the implications for implantation and embryonic diapause, *J. Reprod. Fert.*, 43, 411, 1975.
52. **Tzartos, S. J. and Surani, M. A. H.,** Affinity of uterine luminal proteins for rat blastocysts, *J. Reprod. Fert.*, 56, 579, 1979.
53. **Fishel, S. B. and Surani, M. A. H.,** Changes in responsiveness in preimplantation mouse embryos to serum, *J. Embryol. Exp. Morphol.*, 45, 301, 1978.
54. **Dickmann, Z. and Noyes, R. W.,** The fate of ova transferred into the uterus of the rat, *J. Reprod. Fert.*, 1, 197, 1960.
55. **Moustafa, L. A. and Brinster, R. L.,** Induced chimerism by transplanting embryonic cells into mouse blastocysts, *J. Exp. Zool.*, 181, 1972.
56. **Moore, N. W. and Shelton, J. N.,** Egg transfer in sheep. Effect of degree of synchronization between donor and recipient, age of eggs and site of transfer on the survival of transferred eggs, *J. Reprod. Fert.*, 7, 145, 1964.
57. **Rowson, L. E. A. and Moor, R. M.,** Embryo transfer in the sheep: the significance of synchronizing oestrus in the donor and recipient animal, *J. Reprod. Fert.*, 11, 207, 1966.
58. **Rowson, L. E. A., Moor, R. M., and Lawson, R. A. S.,** Fertility following egg transfer in the cow, effect of method, medium, and synchronization of oestrus, *J. Reprod. Fert.*, 18, 517, 1969.
59. **Rowson, L. E. A., Lawson, R. A. S., Moor, R. M., and Baker, A. A.,** Egg transfer in the cow. Synchronization requirement, *J. Reprod. Fert.*, 28, 427, 1972.
60. **Sreenan, J. M. and Beehan, D.,** Egg transfer in the cow; pregnancy rate and egg survival, *J. Reprod. Fert.*, 41, 497, 1974.
61. **Sreenan, J. M. and Beehan, D.,** Egg transfer in the cow; factors affecting pregnancy and twinning rates following bilateral transfers, *J. Reprod. Fert.*, 44, 77, 1975.
62. **Webel, S. K., Peters, J. B., and Anderson, L. L.,** Synchronous and asynchronous transfer of embryos in the pig, *J. Anim. Sci.*, 30, 565, 1970.
63. **Adams, C. E.,** Asynchronous egg transfer in the rabbit, *J. Reprod. Fert.*, 35, 613, 1973.
64. **Kwun, J. K. and Emmens, C. W.,** Hormonal requirements for implantation and pregnancy in the ovariectomised rabbit, *Aust. J. Biol. Sci.*, 27, 275, 1974.
65. **Foote, W. D., Gooch, L. A., Pope, A. L., and Casida, L. E.,** The maintenance of early pregnancy in the ovariectomised ewe by injection of ovarian hormones, *J. Anim. Sci.*, 16, 986, 1957.
66. **Moore, N. W. and Rowson, L. E. A.,** Maintenance of pregnancy in spayed ewes treated with progesterone alone, *Nature (London)*, 184, 1410, 1959.
67. **Hawk, H. W., Brinsfield, T. H., Turner, G. D., Whitmore, G. T., and Norcross, M. A.,** Embryo survival in first service and repeat-breeder cattle after ovariectomy and hormone therapy, *J. Dairy Sci.*, 46, 1397, 1963.
68. **Miller, B. G. and Moore, N. W.,** Effects of progesterone and oestradiol on RNA and protein metabolism in the genital tract and on survival of embryos in the ovariectomised ewes, *Aust. J. Biol. Sci.*, 29, 565, 1976.

69. **Trounson, A. O. and Moore, N. W.**, Effect of progesterone and oestrogen on the survival and development of fertilised ova in the ovariectomised ewe, *Aust. J. Biol. Sci.*, 27, 511, 1974.
70. **Miller, B. G., Moore, N. W., Murphy, L., and Stone, G. M.**, Early pregnancy in the ewe: effects of oestradiol and progesterone on uterine metabolism and on embryo survival, *Aust. J. Biol. Sci.*, 30, 279, 1977.
71. **Lawson, R. A. S. and Cahill, L. P.**, personal communication, 1974.
72. **Thwaites, C. J.**, Exogenous progesterone and oestrous cycle length in the ewe, *J. Agric. Sci. (Camb.)*, 77, 147, 1971.
73. **Wintenberger-Torres, S.**, Étude Experimentale de la Migration Tubaire des Oeufs et de Leur Segmentation chez la Brebis, these de doctorat d'etat es Sciences Naturelles, Paris, 1967.
74. **Lawson, R. A. S. and Findlay, J. K.**, Embryo maternal interactions concerned with recognition of pregnancy, in *Reproduction and Evolution*, Calaby, J. H. and Tyndale-Biscoe, C. H., Eds., Australian Academy of Science, Canberra, 1977, 349.
75. **McCarthy, S. M., Foote, R. H., and Maurer, R. R.**, Embryo mortality and altered uterine luminal proteins in progesterone treated rabbits, *Fert. Steril.*, 28, 101, 1977.
76. **Beier, M. and Mootz, U.**, Significance of maternal uterine proteins in the establishment of pregnancy, in *Maternal Recognition of Pregnancy*, Whelan, J., Ed., Ciba Foundation, Excerpta Medica, N.Y., 1979, 111.
77. **Beier, H. M., Mootz, U., and Kühnel, W.**, Asyncrone eitransplantationen während der verzögerten uterussekretion beim kaninchen, *Proc. 7th Int. Congr. Anim. Reprod. A. I. (Munchen)*, 3, 1891, 1972.
78. **Adams, C. E.**, The fate of fertilized eggs transferred to the uterus or oviduct during advancing pseudopregnancy, *J. Reprod. Fert.*, 26, 99, 1971.
79. **Wilmut, I. and Sales, D. I.**, Effect of an asynchronous environment on embryonic development in sheep, *J. Reprod. Fert.*, 61, 179, 1981.
80. **Lawson, R. A. S., Adams, C. E., and Rowson, L. E. A.**, The development of sheep eggs in the rabbit oviduct and their viability after re-transfer to ewes, *J. Reprod. Fert.*, 29, 105, 1972.
81. **Cummings, I. A., Mole, B. J., Obst, J., Blockey, M. A., de B., Winfield, C. G., and Goding, J. R.**, Increase in plasma progesterone caused by under nutrition during early pregnancy in the ewe, *J. Reprod. Fert.*, 24, 146, 1971.
82. **Dziuk, P. J., Polge, C., and Rowson, L. E. A.**, Intrauterine migration and mixing of embryos in swine following egg transfer, *J. Anim. Sci.*, 23, 37, 1964.
83. **Dhindsa, D. S., Dziuk, P. J., and Norton, H. W.**, Time of transuterine migration and distribution of embryos in the pig, *Anat. Rec.*, 159, 325, 1967.
84. **Polge, C. and Dziuk, P. J.**, Time of cessation of intrauterine migration of pig embryos, *J. Anim. Sci.*, 31, 565, 1970.
85. **Dhindsa, D. S. and Dziuk, P. J.**, Influence of varying the proportion of uterus occupied by embryos on maintenance of pregnancy in the pig, *J. Anim. Sci.*, 27, 668, 1968.
86. **Moor, R. M. and Rowson, L. E. A.**, The corpus luteum of the sheep: functional relationship between the embryos and corpus luteum, *J. Endocrinol.*, 34, 233, 1966.
87. **Betteridge, K. J., Eaglesome, M. D., Randall, G. C. B., and Mitchell, D.**, Collection, description and transfer of embryos from cattle 10 to 16 days after oestrus, *J. Reprod. Fert.*, 59, 205, 1980.
88. **Moor, R. M. and Rowson, L. E. A.**, The corpus luteum of the sheep: effect of removal of embryos on luteal function, *J. Endocrinol.*, 34, 497, 1966.
89. **Webel, S. K., Peters, J. B., and Anderson, L. L.**, Synchronous and asynchronous transfer of embryos in the pig, *J. Anim. Sci.*, 30, 565, 1970.
90. **Fenton, F. R., Bazer, F. W., Robison, D. W., and Ulberg, L. C.**, Superinduction of gilts with 7-day pig embryos, *J. Anim. Sci.*, 28(Abstr.), 144, 1969.
91. **Rowson, L. E. A. and Moor, R. M.**, The influence of embryonic tissue homogenate infused into the uterus, on the life-span of the corpus luteum in the sheep, *J. Reprod. Fert.*, 13, 511, 1967.
92. **Martal, J., Lacroix, M. C., Loudes, C., Saunier, M., and Wintenberger-Torres, S.**, Trophoblastin, an antiluteolytic protein present in early pregnancy in sheep, *J. Reprod. Fert.*, 56, 63, 1979.
93. **Henderson, K., Scaramuzzi, R., and Baird, D. T.**, Simultaneous infusion of prostaglandin $E_2$ antagonises the luteolytic action of prostaglandin $F_{2}\alpha$ in vivo, *J. Endocrinol.*, 72, 379, 1977.
94. **Ellinwood, W. E., Nett, T. M., and Niswender, G. D.**, Maintenance of the corpus luteum of early pregnancy in the ewe. II. Prostaglandin secretion by the endometrium in vitro and in vivo, *Biol. Reprod.*, 21, 845, 1979.
95. **Perry, J. S., Heap, R. B., and Amoroso, E. C.**, Steroid hormone production by pig blastocysts, *Nature (London)*, 245, 45, 1973.
96. **Bazer, F. W. and Thatcher, W. W.**, Theory of maternal recognition of pregnancy in swine based on estrogen controlled endocrine versus exocrine secretion of prostaglandin $F_{2\alpha}$ by the uterine endometrium, *Prostaglandins*, 14, 397, 1977.

97. **Moor, R. M.**, Effect of embryo on corpus luteum function, *J. Anim. Sci.*, 27(Suppl. 1), 97, 1968.
98. **Talbert, G. B. and Krohn, P. L.**, Effect of maternal age on viability of ova and uterine support of pregnancy in mice, *J. Reprod. Fert.*, 11, 399, 1966.
99. **Adams, C. E.**, Ageing and reproduction in the female mammal with particular reference to the rabbit, *J. Reprod. Fert.*, Suppl. 12, 1, 1970.
100. **Adams, C. E.**, The influence of advanced maternal age on embryo survival in the rabbit, *5th Int. Congr. Anim. Reprod. A.I. (Trento)*, 2, 305, 1964.
101. **Maurer, R. R. and Foote, R. H.**, Maternal ageing and embryonic mortality in the rabbit. I. Repeated superovulation, embryo culture and transfer, *J. Reprod. Fert.*, 25, 329, 1971.
102. **Jones, E. C. and Krohn, P. L.**, The relationships between numbers of oocytes and fertility in virgin and multiparous mice, *J. Endocrinol.*, 21, 469, 1961.
103. **Quirke, J. F. and Hanrahan, J. P.**, Comparison of the survival in the uteri of adult ewes of cleaved ova from adult ewes and ewe lambs, *J. Reprod. Fert.*, 51, 487, 1977.
104. **Cognie, Y., Hermandez-Barreto, M., and Saumande, J.**, Low fertility in nursing ewes during the non-breeding season, *Ann. Biol. Anim. Biochem. Biophys.*, 15, 329, 1975.
105. **McDonald, M. F. and Rowson, L. E. A.**, Ovum transfer to lactating ewes, *J. Reprod. Fert.*, 4, 205, 1962.
106. **Alliston, C. W. and Ulberg, L. C.**, Early pregnancy loss in sheep at ambient temperatures of 70° and 90°F as determined by embryo transfer, *J. Anim. Sci.*, 20, 608, 1961.
107. **Raven, C. P.**, *An Outline of Developmental Physiology*, 3rd ed., Pergamon Press, Oxford, 1966.
108. **Bazer, F. W., Clawson, A. J., Robison, O. W., and Ulberg, L. C.**, Uterine capacity in gilts, *J. Reprod. Fert.*, 18, 121, 1969.
109. **Fenton, F. R., Bazer, F. W., Robison, O. W., and Ulberg, L. C.**, Effect of quantity of uterus on uterine capacity in gilts, *J. Anim. Sci.*, 31, 104, 1970.
110. **Monk, E. L. and Erb, R. E.**, Effect of unilateral ovariectomy and hysterectomy on reproductive parameters in the gilt during early pregnancy, *J. Anim. Sci.*, 39, 366, 1974.
111. **Webel, S. K. and Dziuk, P. J.**, Effect of stage of gestation and uterine space on prenatal survival in the pig, *J. Anim. Sci.*, 38, 960, 1974.
112. **Moore, N. W., Rowson, L. E. A., and Short, R. V.**, Egg transfer in sheep. Factors affecting the survival and development of transferred eggs, *J. Reprod. Fert.*, 1, 332, 1960.
113. **Rowson, L. E. A., Lawson, R. A. S., and Moor, R. M.**, Production of twins in cattle by egg transfer, *J. Reprod. Fert.*, 25, 261, 1971.
114. **Moore, N. W.**, The survival and development of fertilised eggs transferred between Border Leicester and Merino ewes, *Aust. J. Agric. Res.*, 19, 295, 1968.
115. **Lawson, R. A. S. and Rowson, L. E. A.**, The influence of breed of ewe and offspring on litter size after egg transfer in sheep, *J. Reprod. Fert.*, 28, 433, 1972.
116. **Bradford, G. E., Taylor, St. C. S., Quirke, J. F., and Hart, R.**, An egg-transfer study of litter size, birth weight and lamb survival, *Anim. Prod.*, 18, 249, 1974.
117. **Land, R. B. and Wilmut, I.**, The survival of embryos transferred in large groups to sheep of breeds with different ovulation rates, *Anim. Prod.*, 24, 183, 1977.
118. **Trouson, O. A. and Moore, N. W.**, Ovulation rate and survival of fertilised ova in Merino ewes selected for and against multiple births, *Aust. J. Agric. Res.*, 23, 851, 1972.
119. **Robinson, T. J.**, The control of fertility in sheep. II. The augmentation of fertility by gonadotrophin treatment of the ewe in the normal breeding season, *J. Agric. Sci. (Camb.)*, 41, 6, 1951.
120. **Mintz, B., Domon, M., Hungerford, D. A., and Morrow, J.**, Seminal vesicle formation and specific male protein secretion by female cells in allophenic mice, *Science*, 175, 657, 1972.
121. **Mintz, B.**, Hermaphroditism, sex chromosomal mosaicism and germ cell selection in allophenic mice, *J. Anim. Sci.*, 27(Suppl. 1), 51, 1968.
122. **Burgoyne, P. S.**, Sperm phenotype and its relationship to somatic and germ line genotype. A study using mouse aggregation chimaeras, *Dev. Biol.*, 44, 63, 1975.
123. **Allen, W. R. and Stewart, F.**, The biology of pregnant mare serum gonadotrophin (PMSG), in *Control of Reproduction in the Cow*, Sreenan, J. M., Ed., Martinus Niyhoff, The Hague, 1978, 50.
124. **Bradford, G. E., Hart, R., Quirke, J. F., and Land, R. B.**, Genetic control of the duration of gestation in sheep, *J. Reprod. Fert.*, 30, 459, 1972.
125. **Thorburn, G. A., Challis, J. R. C., and Currie, W. B**, Control of parturition in domestic animals, *Biol. Reprod.*, 16, 18, 1977.
126. **Hunter, G. L.**, The maternal influences on size in sheep, *J. Agric. Sci.*, 48, 36, 1956.
127. **Dickinson, A. G., Hancock, J. L., Hovell, G. J. R., Taylor, St. C. S., and Wiener, G.**, The size of lambs at birth — a study involving egg transfer, *Anim. Prod.*, 4, 64, 1962.
128. **Venge, O.**, Studies of maternal influence on birth weight in rabbits, *Acta Zool.*, 30, 1, 1950.
129. **Brunby, P. J.**, The influence of the maternal environment on growth in mice, *Heredity*, 14, 1, 1960.
130. **Wiener, G. and Slee, J.**, Maternal and genetic influences on follicle and fleece development in Lincoln and Welsh Mountain sheep: study involving egg transfer, *Anim. Prod.*, 7, 333, 1969.

131. **Burns, M.**, Effects of ovum transfer on the birthcoats of lambs, *J. Agric. Sci.*, 78, 1, 1972.
132. **Burns, M. and Ryder, M. L.**, Effects of ovum transfer on the skin follicles and birthcoats of Finnish Landrace and Soay lambs, *J. Agric. Sci.*, 82, 209, 1974.
133. **Barnes, R. D. and Graham, C. G.**, Tolerance in early embryo aggregation-derived mouse chimaeras, *Cell. Immunol.*, 21, 146, 1976.
134. **Wiener, G., Wilmut, I., and Field, A.**, Maternal and lamb breed interactions in the concentration of copper in the tissues and plasma of sheep, in *Proc. 3rd Int. Symp. Trace Element Metab. Man Anim.*, Kirchgessner, M., Ed., Inst. für Ernahrungsphysiologie, Freising-Weihenstephan, 1977, 469.
135. **Amsterdam, A., Koch, Y., Lieberman, M. E., and Lindner, H. R.**, Distribution of binding sites for human chorionic gonadotrophin in the preovulatory follicle of the rat, *J. Cell Biol.*, 67, 894, 1975.
136. **Moor, R. M.**, Role of steroids in the maturation of ovine oocytes, *Ann. Biol. Anim. Biochem. Biophys.*, 18, 477, 1978.
137. **Warnes, G. M., Moor, R. M., and Johnson, M. H.**, Changes in protein synthesis during maturation of sheep oocytes in vivo and in vitro, *J. Reprod. Fert.*, 49, 331, 1977.
138. **Coons, L. W. and Espey, L. L.**, Quantitation of nexus junctions in the granulosa cell layer of rabbit ovarian follicles during ovulation, *J. Cell Biol.*, 74, 321, 1977.
139. **Burghardt, R. C. and Anderson, E.**, Hormonal modulation of ovarian interstitial cells with particular reference to gap junctions, *J. Cell Biol.*, 81, 104, 1979.
140. **Thibault, C., Gerard, M., and Menezo, Y.**, Preovulatory and ovulatory mechanisms in oocyte maturation, *J. Reprod. Fert.*, 45, 605, 1975.
141. **Moor, R. M. and Trounson, A. O.**, Hormonal and follicular factors affecting maturation of sheep oocytes in vitro and their subsequent developmental capacity, *J. Reprod. Fert.*, 49, 101, 1977.
142. **Turnbull, K. E., Braden, A. W. H., and Mattner, P. E.**, The pattern of follicular growth and atresia in the ovine ovary, *Aust. J. Biol. Sci.*, 30, 229, 1977.
143. **Rogers, B. J.**, Mammalian sperm capacitation and fertilisation in vitro (a critique of methodology), *Gamete Res.*, 1, 165, 1978.
144. **Lawson, R. A. S., Rowson, L. E. A., and Adams, C. E.**, The development of cow eggs in the rabbit oviduct and their viability after re-transfer to heifers, *J. Reprod. Fert.*, 28, 313, 1972.
145. **Gardner, R. L. and Edwards, R. G.**, Control of the sex ratio at full term in the rabbit by transferring sexed blastocysts, *Nature (London)*, 218, 346, 1968.
146. **Hare, W. C. D., Singh, E. L., Betteridge, K. J., Eaglesome, M. D., Randall, G. C. B., and Mitchell, D.**, Embryo sexing with particular reference to cattle, in *Control of Reproduction in the Cow*, Sreenan, J. M., Ed., Martinus Nijhoff, The Hague, 1978, 441.
147. **Krco, C. J. and Goldberg, E. H.**, H-Y (male) antigen: detection on 8-cell mouse embryos, *Science*, 193, 1134, 1976.
148. **Epstein, C. J., Smith, S., and Travis, B.**, Expression of H-Y antigen on pre-implantation mouse embryos, *Tissue Antigens*, 15, 63, 1980.
149. **Willadsen, S. M.**, A method for culture of micro-manipulated sheep embryos and its use to produce monozygotic twins, *Nature (London)*, 277, 298, 1979.
150. **Spemann, H.**, cited by Raven, C. P., *An Outline of Developmental Physiology*, 3rd ed., Pergamon Press, Oxford, 1966.
151. **McKinnell, R. G.**, *Cloning — A Biologist Report*, University of Minnesota Press, Minneapolis, 1979.
152. **Bromhall, J. D.**, Nuclear transplantation in the rabbit egg, *Nature (London)*, 258, 719, 1975.
153. **Hoffner, N. J. and Di Berardino, M. A.**, Developmental potential of somatic nuclei transplanted into meiotic oocytes of Rana pipiens, *Science*, 209, 517, 1980.
154. **Cline, M. J., Stang, H., Mercola, K., Morse, L., Kuprecht, R., Browne, J., and Salser, W.**, Gene transfer in intact animals, *Nature (London)*, 284, 422, 1980.
155. **Seidel, G. E.**, Management of reproduction of cattle in the 1990's, in Special Report 77 of Carl B. and Florence E. King Visiting Scholar Lectures, Agricultural Experiment Station, University of Arkansas, Fayetteville, 1980, 51.

# INDEX

## A



## B

## C

## D



## E

## F



## G

## H



## I

## K



## L



## M

## N

## O



## P

## Q



## R

S

## T



## U



## V

W



X



Y



Z